E-Agriculture and E-Government for Global Policy Development: Implications and Future Directions

Blessing M. Maumbe
Eastern Kentucky University, USA

INFORMATION SCIENCE REFERENCE

Hershey · New York

Director of Editorial Content: Kristin Klinger
Senior Managing Editor: Jamie Snavely
Assistant Managing Editor: Michael Brehm
Publishing Assistant: Sean Woznicki
Typesetter: Sean Woznicki
Cover Design: Lisa Tosheff
Printed at: Yurchak Printing Inc.

Published in the United States of America by
Information Science Reference (an imprint of IGI Global)
701 E. Chocolate Avenue
Hershey PA 17033
Tel: 717-533-8845
Fax: 717-533-8661
E-mail: cust@igi-global.com
Web site: http://www.igi-global.com/reference

Library of Congress Cataloging-in-Publication Data

E-agriculture and e-government for global policy development : implications and future directions / Blessing M. Maumbe, editor.

p. cm.

Summary: "This book provides critical research and knowledge on electronic cultivation and political development experiences from around the world"-- Provided by publisher.

ISBN 978-1-60566-820-8 (hardcover) -- ISBN 978-1-60566-821-5 (ebook) 1. Electronic commerce. 2. Internet in agriculture. 3. Internet in public administration. 4. Globalization. I. Maumbe, Blessing M., 1966-
HF5548.32.E17364 2009
630.68'8--dc22

2009020542

British Cataloguing in Publication Data
A Cataloguing in Publication record for this book is available from the British Library.

All work contributed to this book is new, previously-unpublished material. The views expressed in this book are those of the authors, but not necessarily of the publisher.

Table of Contents

Section 4
Selected Readings

Detailed Table of Contents

improve rural livelihoods in developing countries. Over the past five years, South Africa has witnessed a swift ICT-led transformation of its public service delivery with major innovations in key development sectors. The growth of e-agriculture is seen as an engine to accelerate agriculture and rural development, promote food security, and reduce rural poverty. This chapter examines e-agriculture initiatives in South Africa. It describes ICT applications in improving the quality of on-farm management decisions, agricultural market information system, e-packaging, product traceability, and online marketing to access lucrative global wine markets. The chapter also highlights key constraints, and identifies considerations to enhance the future prospects for e-agriculture. Given the strategic importance of agriculture in supporting the livelihoods of the majority rural population in South Africa, the successful deployment and effective utilization of ICT is pivotal for sustainable agriculture development and raising the standards of living of marginalized communities. The results of the paper demonstrate that South Africa has made significant strides in e-agriculture and tangible benefits have accrued to the agricultural communities.

This study explores Nigeria's e-agriculture policies and those of other African governments. It also proposes what e-agricultural strategies these governments could adopt to enhance their agricultural output by examining the history of agriculture in Nigeria; the current status of Nigerian information and communication technologies and e-government policies with emphasis on agriculture. The analysis addresses the role of communications as instrument of national development. In view of the economic status of African countries, the study calls on Nigeria and other African governments to adopt a cautious approach as they embark on e-agriculture policies and acquisition of information and communication technologies to promote national development. The study calls on African governments to liberalize their agricultural policies, establish agricultural cooperatives, educate rural famers and offer telecommunication services in the rural areas if they hope to raise their agricultural productivity.

Establishment of the new institutions that could improve Polish agricultural market was one of the main goals of the Polish government agricultural policy in the period of economy transformation. The project of creating agricultural markets was successful. Several regional wholesale markets and commodity exchanges were established and most of them still function with good performance. These markets are the key important marketing channels for market-oriented farmers in Poland. They can also be seen as a source of electronic commerce innovations on the Polish agricultural market. The aim of the chapter is to present the process of establishment and first experiences of electronic market of agricultural products as one of the new e-commerce initiatives on the Polish agricultural market. The chapter also discusses conditions of the electronic exchange development and its impact on the Polish agricultural market.

Section 2
E-Agriculture Development: Conceptual Frameworks

The rapid growth of Information and Communication Technologies (ICT) has increased opportunities to improve agricultural production, distribution, and marketing activities in Sub-Saharan Africa (SSA). Such initiatives are expected to provide vast social and economic benefits to the agricultural community and help uplift standards of living of society in general. The process of how ICT should be applied in agriculture to raise living standards of millions of poor Africans is not yet well understood. Therefore, there is a need to deepen our understanding of the socio-economic benefits expected from ICT use in agriculture, and most importantly, how these benefits will be realized in SSA. Some green revolution technologies failed in Africa and parts of Asia because of inadequate attention to context specific issues, irrelevance, and prohibitive costs. In that regard, this chapter describes a framework for sustainable e-agriculture development in SSA. The proposed framework is divided into three main parts; (i) e-service delivery (ii) ICT development and diffusion pathway, and (iii) e-information flow and e-content development landscape. In order to facilitate the effective diffusion and adoption of e-agriculture, a set of preconditions and "e-value creation" opportunities are assessed. The preconditions filter out "irrelevant" ICT, and e-value creation facilitates context specific and demand driven e-innovations in agriculture. The chapter identifies and discusses ICT illiteracy, lack of ICT policies, infrastructural deficiencies, and poverty as key challenges affecting the future success of e-agriculture in SSA. The chapter recommends the development of e-policies and e-strategies on e-content, e-trust, e-security, e-value addition to promote sustainable e-agriculture development on the African continent.

Almost sixty five percent of Indian population is engaged in agriculture that contributes to food security of the world's second largest populated country. Though agriculture sector shares 26 percent of GDP, this sector is very crucial for the sustainable growth and development of India. The emerging agricultural challenges demand information intensive agriculture work and applications of state of the art knowledge to enhance agricultural productivity, but non-accessibility of information and subsequently awareness and knowledge gaps that exist in this sector, enormously affect agricultural productivity. Efforts are being made for e-communication of information in rural India. This chapter portrays such efforts of public and private sectors, pinpoints the problem areas for accessibility of latest agricultural knowledge and suggests an e-communication model suitable for transfer of agricultural knowledge in the rural areas of India.

E-government can benefit developing countries by enhancing the economy, increasing access to health care, improving bureaucracy, and consolidating democracy. Sub-Saharan countries have lagged behind the world in adopting this system of communications. A variety of reasons explain the lag, namely lack of national resources and an illiterate population. Zambia serves an example of democracy on a continent where freedom and peace are lacking, but also as a country where e-government is only beginning. This evaluation is the first to examine e-government there, and is carried out at five distinct levels: Current communication systems; Zambia's ICT policy; key central e-government websites; e-government at the provincial/municipal level; and at the individual level. As a result, this case study will evaluate how a developing country is struggling to provide government access and enhance the economy and suggests improvements needed if Zambia's e-government will become adequate and sustainable.

The skewed global workforce interactions during the agricultural and industrial revolutions still bother the antagonists of globalization but could be straightened by progressive workforce development policies that mutually benefit high and low income countries. In addition, the e- literacy and information technology boom have further narrowed spatial perception of geographic distance thus providing low-income countries insights on policy dynamics of high income countries and its impact on the rest of the world. Thus in order to attain equity and balanced global workforce development, this chapter explores the rational and different paradigms for capacity building on e-literacy in low income African countries so that their workforce would contribute to the globalized economy and civic responsibility. The chapter contends that e-literacy empowerment should be regarded as a human right issue and that through other ethical globalization efforts every person on earth should form part of the workforce for sustaining the global village.

This chapter presents the essence of critical success factors with focus on building capacity for electronic governance (eGovernance) in a developing country jurisdiction. The results are borne of the authors' years of experience with regard to national eGovernance implementations in developing member countries of the Commonwealth. Critical success factors (CSFs) denote those aspects of, or associated with,

extensively explore these pervading technologies as they impact on the education, commerce, social, cultural, and economic life of the poor Tanzanian people. The chapter looks at how Tanzania is coping with the issue of poverty eradication as one of the eight UN Millennium Development Goals (MDGs). It addresses the issue of digital divide and the role that ICTs can play in poverty reduction. Tanzania's efforts in embracing ICTs and the challenges facing the country in its efforts are also addressed. Overall, the chapter demonstrates that ICTs are a set of tools for knowledge sharing, which is a powerful means for poverty reduction. Furthermore, it is advisable to focus on information literacy rather than just focusing on computer literacy.

ICTs in general and e-governance in particular offer tremendous opportunities for improving demand-driven transparent and accountable service delivery targeting the underprivileged. The objective of this chapter is to examine the effects of E-government implementation in the context of widespread poverty in India through an extensive secondary data analysis on selected pro-poor initiatives in reducing poverty and improving rural livelihoods. Analysis also includes various contexts in which these ICT based interventions operate. Specific recommendations are made to involve the socially excluded groups in the design, implementation and access to e-government services. Governments to design appropriate public policies in implementing socially inclusive e-government strategies in the emerging information society draw the conclusion.

This chapter describes the problem state of organic farming development and procedures for modeling by the means of system dynamics, with emphasis on the organic products market. The modeling principles are described in the following steps: problem state formulation, development of causal loop diagrams, model development, scenario analysis and formulation of acceptable strategies. Basic structures developed by the system dynamics principle are presented. The concept of archetypes in the field of organic agriculture modeling is described. The simulation scenarios are formulated as a case study for the Slovenian organic agriculture.

Preface

The rapid diffusion and adoption of information and communication technologies (ICT) has provided opportunities to improve agricultural productivity, enhance access to markets, value-chain integration and coordination, poverty reduction, and raising standards of living of the majority poor. As the ICT revolution gains momentum, numerous governments around the world have established e-agriculture and e-government programs and projects. Over the past decade, the e-agriculture and e-government implementation pace has gained greater momentum especially in developing countries. The use of ICT to stimulate food and agricultural production and enhance the integration of domestic, regional, and global markets has provided widespread opportunities for socio-economic development through, agricultural employment creation, generation of new wealth, improved service delivery, and poverty reduction in most countries. Although fewer developing nations than developed countries have made great strides in modernizing their public service delivery to date, not much is known about the innovative programs, projects, and the policies and strategies that are guiding the implementation of ICT-based services especially in developing countries.

Isolated, undocumented, and oftentimes uncoordinated ICT initiatives have emerged in government service provision in many developing countries recently. The race to use ICT to facilitate socio-economic development and move from traditional face to face to online public service delivery has intensified around the world. There is need therefore, to capture the "infant phases" and systematically document the new knowledge in order to improve our understanding of the socio-economic circumstances surrounding the diffusion and adoption of these new tools that now form the basis for a new global development paradigm.

A number of technological revolutions have come and gone and there are ample lessons about missed opportunities, trade-offs, risks and challenges. As ICT applications in agriculture and rural development increase, and governments adopt the e-service agenda, it is imperative that we diagnose this relatively new phenomenon. It is also important from the onset to address some key issues about the motivation and expectations for writing this book. First, the editor recognizes fully that ICT are not a panacea for international global development problems. Second, the use of ICT will not make poverty and debilitating diseases that confront mankind disappear overnight. Third, the implementation of e-agriculture and e-government is associated with numerous risks and uncertainties. Fourth, there is clear competition for national and global resources for ICT investments with immediate societal basic needs such as water, food, and shelter. Nonetheless, the use of ICT will facilitate socio-economic development by providing opportunities to address poverty and the basic needs of society. This book aims to enlighten the world about the potential benefits arising from the application of these modern technologies, using practical examples of successes achieved elsewhere. In addition, the practical solutions to some of the intractable ICT problems and difficulties encountered provide clear lessons for those policy makers and other industry experts that are striving to develop and execute sustainable ICT policies and strategies in agriculture and rural development.

In order to accomplish that goal, this book documents early initiatives in e-agriculture development and e-government efforts around the world. The book synthesizes seminal e-agriculture development and e-government country case studies and provides insights into theoretical frameworks and policies that can be used guide the development of sustainable e-services in developing countries in particular. This book is a collection of chapters providing new knowledge on e-agriculture and e-government efforts around the world. The book brings together expert contributions from four continents, United States of America, Africa, Asia and Europe. The individual chapter contributions were drawn from case studies from ten countries around the world. The country case studies were drawn from Kenya, Namibia, Nigeria, South Africa, Zambia, and Zimbabwe (i.e., Africa), Poland (i.e., Europe) and India (i.e., Asia). The book chapters provide detailed insights, ideas, concepts, frameworks, policies and strategies, and balanced perspectives on the different e-agriculture and e-government initiatives, projects and programs that have been implemented globally. Collectively, the book provides a description of the key issues affecting the e-agriculture and e-government landscape that is unfolding worldwide, and it provides a sound basis to initiate a global policy dialogue on how to effectively execute the implementation of such programs and projects around the world. Shared experiences about success, failures, and challenges are critical for shaping both the current and future e-agriculture and e-government efforts in the twenty-first century. Understanding what works and what does not work, or what is inappropriate under certain conditions regarding the diffusion and adoption of modern ICT will be useful in guiding policy makers, industry experts, businesses, and society in general in making their technology choices.

The book is divided into three major parts. Section 1 describes selected e-agriculture country case studies. Section 2 examines proposed e-agriculture development frameworks, policies and strategies. Finally, Section 3 describes mainly e-government programs and projects and highlights implementation challenges. The effort to put this book together could not have been successful without the unique cooperation, hard work and resilience of many scientists, academics, researchers, and scholars around the world. The themes and topics that were selected for inclusion in this book highlight not only their importance, but also the focused attention that is being paid to these issues by e-agriculture and e-government professionals and experts around the world. This book highlights cutting edge initiatives in e-agriculture and e-government, best practices from the country case studies including social, economic, cultural, legal, technological factors driving success, e-challenges, and global policy development issues. The following section provides a brief narrative description of each specific chapter that has been included in this book.

Chapter 1 examines the awareness and use of mobile phones to enhance smallholder famer linkages to agricultural markets in Kenya. The chapter noted that in many developing countries smallholder farmer's participation in agricultural input and output markets is constrained by lack of market information. Actors in most developing country markets operate under conditions of information asymmetry which increases transaction costs and locks out smallholder farmers. The use of ICT to provide market information offers an opportunity to address the problem of information asymmetry and link farmers to product and factor markets. The chapter identifies a high level of awareness and widespread use of mobile phones for social purposes. Low educational levels, the cost of airtime, and the lack of electricity to recharge phone batteries were reported as major obstacles to mobile phone usage among Kenyan smallholders.

Chapter 2 describes South Africa's experiences with e-agriculture development using the Robertson Wine Valley located in Cape Town as the case study. In South Africa, the growth of e-agriculture is seen as an engine to accelerate agricultural industry development, promote e-commerce, and reduce rural poverty. This chapter examines e-agriculture initiatives such as computerized irrigation systems, e-record keeping, market information systems, e-packaging, product traceability, and online marketing to link wine farmers to both local and global markets. The chapter presents additional evidence which demonstrate South Africa's significant advances in e-agriculture and the tangible benefits that have accrued to the agricultural communities.

Chapter 3 explores Nigeria's e-agriculture policies and those of other African governments. It examines the history of agriculture in Nigeria; the current status of Nigerian information and communication technologies and e-government policies with emphasis on agriculture. It also proposes e-agricultural strategies that governments could adopt to enhance their agricultural productivity. The chapter highlights the importance of communication as an instrument of national development while recommending the need for caution in the adoption of e-agriculture policies and acquisition of ICT to promote national development.

Chapter 4 describes the early experience with electronic marketing of agricultural products- one of the new e-commerce initiatives in the Polish agricultural marketing system. The chapter also discusses conditions of the electronic exchange development and its impact on the Polish agricultural market. The Polish government's agricultural policy in the period of economic transformation was driven by the establishment of new marketing institutions such as regional wholesale markets and commodity exchanges. These markets serve as key important marketing channels for market-oriented farmers in Poland and are seen as a source of electronic commerce innovations in the Polish agricultural markets.

The three chapters presented in Section 2 of this book describe the theoretical frameworks, and the policy development processes and procedures for e-agriculture and e-government. The diffusion and adoption of e-agriculture and e-government in a policy vacuum will be a fad, if not a dismal failure and countries should avoid this risk. The three chapters included in the second part of this book highlight critical issues for consideration and the systematic steps that need to be followed by policy makers in various countries as they cautiously embark on their individual e-agriculture and e-government projects and programs.

Chapter 5 proposed a theoretical framework for e-agriculture development that is divided into three main parts; (i) e-service delivery (i) ICT investment, e-channel development and benefit articulation, and (iii) e-information flow and e-content development landscape. The chapter highlights basic preconditions and "e-value creation" considerations that must be integrated in the development of sustainable e-agriculture policies and strategies. The chapter identifies ICT illiteracy, lack of ICT policies, poor infrastructure, and poverty as key challenges affecting e-agriculture in Sub-Saharan Africa (SSA). The chapter recommends the need to develop systematic e-strategies and policies on e-content, e-trust, e-security, e-value addition to promote sustainable e-agriculture on the African continent.

Chapter 6 focuses on the development of an e-communication model for Indian agriculture. Almost sixty five percent of Indian population is engaged in agriculture that contributes to food security of the world's second largest populated country. Although the agriculture sector contributes 26 percent of GDP, this sector is very crucial for the sustainable growth and development of India. The emerging agricultural challenges demand information intensive agriculture work and applications of state of the art knowledge to enhance agricultural productivity, but non-accessibility of information and subsequently awareness and knowledge gaps that exist in this sector, enormously affect agricultural productivity. The chapter highlights public and private sector efforts being made in e-communication of information in rural India. Given that accessibility of latest agricultural knowledge is a major problem, there is need for an e-communication model suitable for the transfer of agricultural knowledge in the rural areas of India.

Chapter 7 is based on the premise that rural communities are being heavily influenced by the ongoing modernization process and there is a real fear that in the African context, the ensuing modernization will result in a paradox where modernization may lead to an increase in economic well-being, but have the unintended consequence of increasing alienation and reducing a sense of community that exists in rural villages. The chapter theoretically explores the possibility of using ICT to develop a sense of community in rural villages and thus offset and mitigate the more negative aspects of the modernization process. It also proposes a way to conceptualize this potential paradox by integrating the well established sociological concepts of Gemeinschaft (community) and Gesellschaft (individualism) with current paradox models of diversity and similarity curves. Such an approach has pedagogical utility in helping to describe and explain the modern paradox that African countries face. The chapter highlights a rare concern that modernization and economic well being of the citizenry might come about at the expense of increased alienation and a reduction in the sense of community or loss of social relationships normally found in rural villages.

In Section 3 of the book, e-government country case studies are presented. The common theme of these case studies is the effort by various governments to embark on innovative and more efficient ways to deliver services to their government departments, private businesses, and ordinary citizens. Collectively, the chapters present country specific e-government developments, constraints and challenges, proposes relevant best practices, and identifies critical success factors. The chapters also make an attempt to provide some of the policy implications for promoting sound e-government development in the respective countries.

Chapter 8 describes the challenges of developing e-government in a developing country context. To assist public administrators to think beyond traditional e-government, the chapter describes the concept of connected government, whose philosophy rests on the integration of back-end processes that facilitate collaboration among government agencies. Using Zimbabwe as a case study, the chapter describes five government-owned organizations where even the basic e-government services are barely available. The chapter highlights the challenges of taking the initial steps to establish e-government integration within and across government agencies in a developing country setting. The chapter exposes shortcomings to inter-departmental integration not only of the organizations under investigation, but also of other similar enterprises in developing countries within the same context. The chapter makes recommendations toward diffusing connected government applications for inter-organizational collaboration.

Chapter 9 is based on Zambia's experiences with e-government development. The chapter argues that e-government can benefit developing countries by enhancing the economy, increasing access to health care, improving bureaucracy, and consolidating democracy. Sub-Saharan African countries have lagged behind the world in adopting e-government. A variety of reasons explain the lag, namely lack of national resources and an illiterate population. Zambia is unique in that it serves as an example of democracy on a continent where freedom and peace are lacking, but also as a country where e-government is only beginning. The chapter is the first to examine e-government at five distinct levels: current communication systems; Zambia's ICT policy; key central e-government websites; e-government at the provincial/municipal level; and at the individual level. The case study results demonstrate how a developing country is struggling to provide government access and enhance the economy and suggests improvements needed if Zambia's e-government will become adequate and sustainable.

Chapter 10 describes the skewed global workforce interactions during the agricultural and industrial revolutions that still bother the antagonists of globalization but could be straightened by progressive workforce development policies that mutually benefit high and low income countries. The chapter argues that the e- literacy and information technology boom have further narrowed spatial perception of geographic distance thus providing low-income countries insights on policy dynamics of high income countries and its impact on the rest of the world. Thus in order to attain equity and balanced global workforce development, this chapter explores the rational and different paradigms for capacity building on e-literacy in low income African countries so that their workforce would contribute to the globalized economy and civic responsibility. The chapter contends that e-literacy empowerment should be regarded as a human right issue and that through other ethical globalization efforts every person on earth should form part of the workforce for sustaining the global village.

Chapter 11 presents the essence of critical success factors with a focus on building capacity for electronic governance (*e*-governance) in a developing country jurisdiction. The results are borne of the authors' years of experience with regard to national *e-g*overnance implementations in developing member countries of the Commonwealth. Critical success factors (CSFs) denote those aspects of, or associated with, the new ICT, which may be perceived as comprising core, critical factors against which the level of capability of National Capacity for ICT (*e*-governance) may be assessed, measured and interpreted. CSFs, perceived to be critical for the success of any *e*-governance initiative are best modelled as a three-tier minimalist framework, comprising CSFs at levels described as macro-, meso-, and *micro-* levels. The nature of any given ICT initiative which is appropriate nationally in central government, locally in local government, or in the public service, in the civil service, or in some selected sector or jurisdiction of the national economy, whether *existing or planned,* and whether *implicit or explicit*, must take cognisance of the need for the identification of CSFs at the inception stage of the initiative.

Chapter 12 explains how ICT have transformed health service delivery (HSD) in developing countries using the Namibian context. The chapter describes the Namibian Health Service Delivery Landscape (NHSDL) to provide an overview of the potential of ICT use by various actors in the health sector. The chapter describes ICT use patterns that are based on a primary survey of 134 patients and key informant interviews held with 27 health service providers (HSPs) in Khomas and Oshana regions of Namibia. The chapter highlights how Namibian patients are using diverse range of ICT to access health services including the traditional television and radio, and the more modern mobile phones and even computers. The chapter also indicates how HSPs have responded to ICT which is now deployed in various functional areas such as admission, clinical support, family planning, maternity, and emergency services. The chapter identifies key challenges and discusses policy implications to enhance the uptake of ICT-based health services in Namibia. The findings in this chapter will benefit HSPs and patients decide on affordable ICT choices; and policy makers as they design interventions to stimulate the use of ICT in HSD in Namibia. The results also provide key insights for other Sub-Saharan African countries contemplating ICT integration in HSD.

The afore-mentioned book chapters indicates the scope and scale of e-agriculture and e-government projects, programs, policies and strategies affecting different countries around the world. The development of e-agriculture and e-government is no longer an issue of whether or not, but a matter how best to proceed with the implementation of the new initiatives. This subject is now a reality and it is attracting major interest among academics, government policy makers, international development analysts, information, communication and technology specialists, and other scholars. As nations transform into knowledge societies, it is expected that the insights gained from reading the book chapters will help academics, policy makers, consultants, and other practitioners to craft more viable ICT polices, make adjustments to poorly conceived policies, and perhaps most importantly, introduce the philosophy of being proactive in constructing a sustainable global e-agriculture and e-government development policy in the 21st century.

This work is just the beginning of more work that still needs to be done to advance this huge knowledge frontier which has slowly evolved into a specialist discipline or field of research and practice in some Universities around the world. I trust that the readers of this book will find it extremely valuable as I find the narrative on various country case studies, theoretical frameworks, and discussions on ICT policy issues both stimulating and exciting. The combined effort of all the authors who have contributed chapters to make this publication a success has greatly expanded our understanding of e-agriculture and e-government and the associated global policy development issues. The successful development of e-agriculture and e-government around the world offers numerous opportunities to reduce global poverty and raise standards of living of millions of people through improvements in food and agricultural production, better linkages to input and product markets, value chain integration and coordination, and efficient public service delivery in the 21st century and beyond.

Blessing Mukabeta Maumbe, PhD
Editor
Eastern Kentucky University, USA, 2009

Section 1
E-Agriculture Development Country:
Case Studies

Chapter 1
Awareness and the Use of Mobile Phones for Market Linkage by Smallholder Farmers in Kenya

Julius Juma Okello
University of Nairobi, Kenya

Ruth M. Okello
Michigan State University, USA

Edith Ofwona-Adera
International Development Research Center, Nairobi, Kenya

ABSTRACT

In many developing countries smallholder farmer participation in agricultural input and output markets continues to be constrained by lack of market information. Actors in most developing country markets operate under conditions of information asymmetry which increases the costs of doing business and locks out smallholder farmers. Attempts to address this problem are currently focusing on the use of ICT technologies to provide market information and link farmers to markets. This study examines the awareness and use of one such technology – mobile phones. It finds for male and female smallholder farmers in Kenya a high level of awareness and widespread use of mobile phones, mainly for social purposes. This study further finds that a low level of education, the cost of mobile phone airtime recharge vouchers and the lack of electricity for recharging phone batteries are the major impediments to the ownership and use of mobile phones, with female farmers more constrained than males. A high awareness of mobile phones among smallholder farmers presents an opportunity to strengthen smallholder farmers' market linkage. However constraints to the usage of mobile phones will need to be addressed. The study findings indicate priorities for policymakers dealing with the specifics of ICT adoption as a tool to promote rural viability via rationalization of Kenyan agricultural markets.

DOI: 10.4018/978-1-60566-820-8.ch001

INTRODUCTION

Smallholder farmers account for more than 70 percent of the farming community in rural Africa. Yet, they continue to face significant problems in accessing production inputs and high value markets for their products. Some of these problems emanate from the bad past agricultural policies that entrenched the patronage of the state over marketing of agricultural produce in most African countries. The dominance of the state over agricultural marketing caused the private sector to retreat leading to lack of supportive services and inefficiency in the distribution of inputs (Poulton et al., 2006; Kirsten, Forthcoming). At the same time, the pricing and produce movement policies pursued by African governments made it difficult for private sector to profitably participate in the market.

Market liberalization policies of the 1980s and 1990s in many of the African countries removed some of the barriers to farmer participation in markets and their access to inputs. However, most smallholder farmers have largely missed the window of opportunity opened up by market liberalization (Jayne and Jones, 1997). Participation of many such farmers in better-paying markets and their access to inputs (information, credit, fertilizer, seeds, and insurance) has been dismal (Key and Runsten, 1999; Poulton, et al, 2006; Kydd and Doward, 2005).

One of the major contributors to poor smallholder farmers' access to markets is the high transaction costs in both the input and output markets (Omamo, 1998a & 1998b). High transaction costs in turn arise from poor access to market information or absence of the type of information the farmer needs, lack of needed skills, and the lack of access to risk insurance (Key and Runsten, 1999; Tollens, 2006). Lack of requisite infrastructure and assets further hinder smallholder farmers' access to markets (Barrett, 2008).

Market failure for smallholder farmers is especially more pronounced in cases where buyer(s) have quality and/or safety specifications that must met by farmers. Under such cases smallholder farmers face three distinct problems namely: i) how to meet buyer safety/quality specifications, ii) how to be recognized as producing safe food of required quality, and iii) how to maintain sustainable link with the buyer (Rich and Narrod, 2005). The challenge of maintaining a sustainable linkage with the buyer arises from the difficulties smallholder farmers face in producing consistent quality and volumes of the product (Poulton et al., 2006).

To enable farmers to overcome some of the constraints posed by market information inadequacies and induced input and output market failures a number of ICT-based projects are being implemented in several African countries. Examples include the Kenya Agricultural Commodity Exchange in Kenya, TradeNet in Ghana, Malawi Agricultural Commodity Exchnage in Malawi, Songhai Center in Benin, and Women of Uganda Network (WOUGNET) in Uganda (Ferris, et al, 2006). These initiatives include the use of mobile phones, internet, fax and CD Roms. Others use the more traditional radio and television options to provide market information. The shift to these technologies, especially mobile phones, has been driven by the rapid growth in the use of phones in rural Africa (Aker, 2007). A scoping study conducted in Africa finds widespread promotion of these ICT-based technologies in several eastern and southern African countries including Kenya, Uganda and Malawi (Munyua, 2009). Despite the mushrooming of ICT-based market information service projects that use ICT technologies, little is still known about the awareness and usage of such technologies by smallholder farmers in Africa. This chapter examines the awareness and usefulness of mobile phones among smallholder farmers in Kenya. Its broader purpose is to assess the level of awareness and use of ICT technologies by smallholder farmers and identify the opportunities (constraints) that should be exploited (addressed) by policymakers to promote greater

Table 1. Household characteristics and poverty levels in Kakamega district

Total no. of households	125,901
Average household size	4.8
No. of women groups	2,000
Absolute poverty	57.47 percent
Contribution to national poverty	5.89 percent
Share of smallholder farmers	82.5 percent
Proportion of women	51.90 percent

Source: Kakamega District Development Plan, 2002-2008.

use of the technology to link more farmers to markets. It specifically addresses the following research questions:

i) What is the extent of awareness of ICT-based technologies for provision of agricultural market information among rural smallholder farmers?
ii) What is the extent of usage of ICT-based market information technologies among such farmers?
iii) What is the effect of gender on awareness and usage of ICT-based technologies among rural smallholder farmers?

This paper is focused on the use of mobile phones by smallholder farmers in Lurambi North, Lurambi South and Navakholo divisions which fall under Kakamega district of Western province, Kenya. Kakamega district was selected for survey because a project involving the use of mobile phones and internet-based applications for linking farmers to input and output markets was earmarked for implementation in the district by an international NGO funded by the International Development Research Center (IDRC) Kenya. This study provides a baseline (i.e.,pre-project) conditions on the level of awareness and usage of mobile phones in the district. The rest of this paper is organized as follows: Section 2 gives the background information of the survey area; Section 3 presents the conceptual and empirical methods and also characterizes the respondents; Section 4 reports the results of the study. Section 5 summarizes the findings and concludes.

BACKGROUND INFORMATION

Kakamega district falls in the medium to highly populated zone of western Kenya. The total number of households in the district was 125,901 with an average household size of 4.8 persons (see Table 1). Approximately 57 percent of the population are classified as poor hence the targeting of the district for a development project.

Figure 1 gives the shares of household incomes earned from various employment opportunities in the district. As expected, agriculture is the major employer underscoring the importance of agricultural sector in current and future rural development and poverty eradication efforts. The contribution of rural non-farm employment opportunities to the overall household income is unknown. However, as Figure 1 indicates, less than 10 percent of household incomes are derived from rural self-employment opportunities and approximately 20% from wage employment.

As indicated by Table 1 above, agricultural sector in Kakamega district is dominated by smallholder farmers. The average farm size in the district is about 1.4 acres (GOK, 2002; Dose, 2007). Crop production is highly diversified and includes food crops (e.g., maize beans, finger

Figure 1. Employment opportunities in Kakamega district (percent of total population), 2006. Source: Kakamega District Development Plan, 2002-2008.

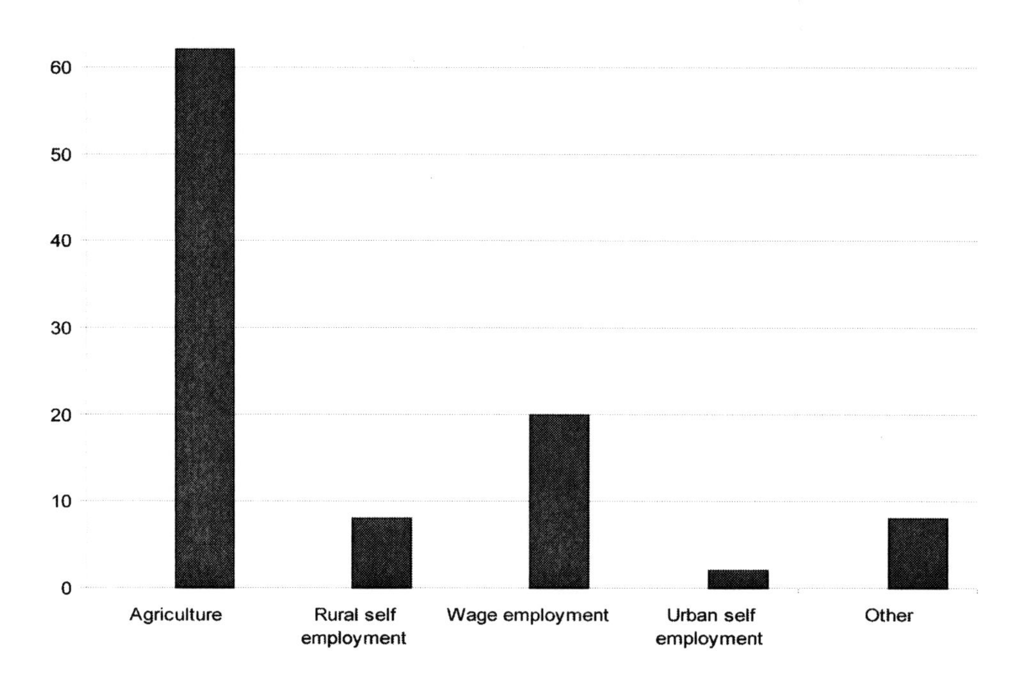

millets, cassava, sweet potatoes and sorghum, local vegetables, tomatoes, onions and kales) and cash crops (tea, sugarcane, coffee, bananas and sunflower). The high crop diversification is indicates that there is high degree of risk in farming in the district. Indeed, agricultural production in the district is predominantly rainfed. Nonetheless, it is the main source of income for 62 percent of the population (Dose, 2007).

Motivation for the DrumNet Project

This study is part of a wider study whose aim was to establish the baseline conditions prior to the implementation of a project that was to utilize mobile phones to link farmers in Kakamega district to input and output markets. The project, later implemented by an international NGO called Pride Africa, was funded through a grant from the International Development Research Center (IDRC). The district was targeted with the project due to poor access by smallholder farmers to markets and reliance on rainfed agriculture. The project (known as DrumNet project) aimed at linking sunflower farmers to sunflower production input markets (in particular seed, fertilizer, and credit) and to the leading sunflower buyer in Kenya, the BIDCO oil refineries limited. The project was to use a mobile phone and computer based platform to transmit information on inputs and output to the farmers and BIDCO, respectively. It therefore was to mediate the linkage between farmers to businesses that provides inputs (i.e., agro-input dealers), the bank, and to final buyer of sunflowers for a fee (see Figure 2). Under the DrumNet project farmers would get seed and fertilizer at designated agro-input dealers through a cash-less in-kind credit scheme. The loan was to be underwritten by a commercial bank namely, Equity Bank Limited. The DrumNet project would then provide other services including mobilizing farmers into groups while BIDCO provides the agronomic and other technical information farmers would need to produce sunflowers for oil

Figure 2. Schematic representation of the DrumNet model

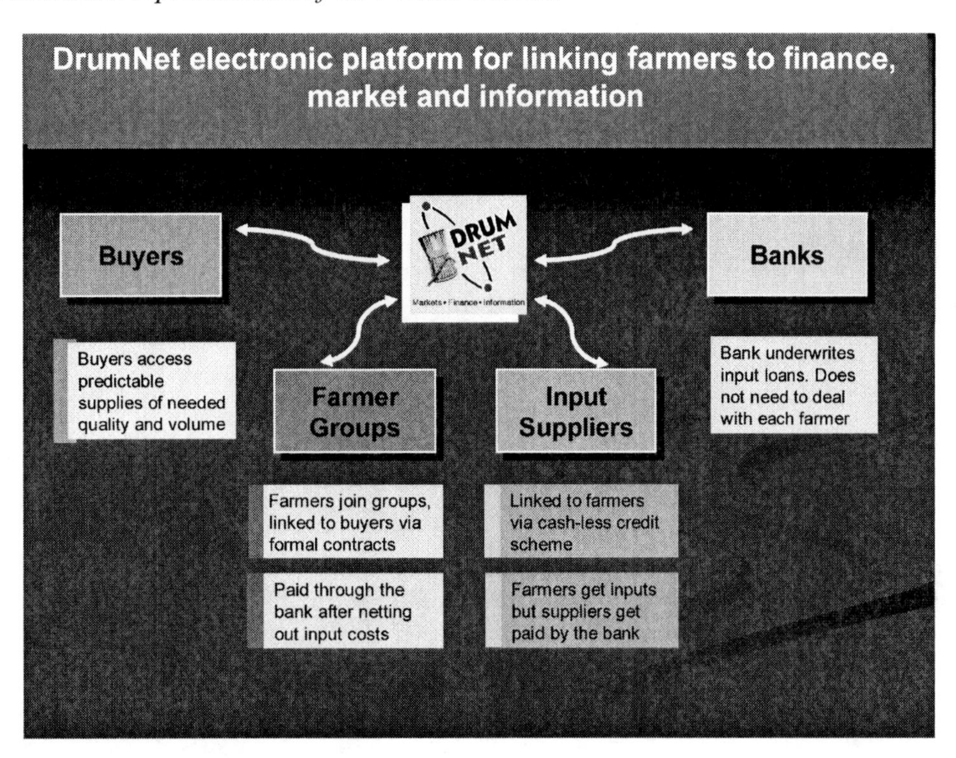

extraction. The DrumNet project would therefore essentially perform the role of an intermediary.

CONCEPTUAL AND EMPIRICAL METHODS OF ANALYZING SMALLHOLDER FARMERS' USE OF MOBILE PHONES FOR MARKET LINKAGE IN DEVELOPING COUNTRY CONTEXT

Conceptual Framework for Understanding Why Markets Fail for Smallholder Farmers

Rural input and output markets in most developing countries operate imperfectly due, among other factors, to information asymmetry that result in high transaction costs (Poulton et al, 2006). Input market failure in rural areas is attributed to information asymmetry (Besley, 1998). Buyers and sellers in rural markets operate under conditions of limited and/or no information (Barrett, 2008). This results in market failure in the input markets, especially in credit, and insurance markets (Key and Runsted, 1999).

Lack of well-defined universal grades for traded products that is endemic in most rural markets also limits the trade and hence participation of farmers in such markets (Fafchamps and Gabre Madhin, 2006). In many rural agricultural markets the rules/conventions (especially relating to product quality) that govern exchange in the output markets are either non-existent or poorly defined and hence only known by few traders (Fafchamps, 1996; Gebremedhin, 2005). Exchange in such market is therefore largely based on trust which can only build up with repeated transactions over many years (Fafchamps, 1996 & 2004; Fafchamps and Hill, 2005). Where loyalty and trust has not been built, transaction parties must take time to physically inspect the goods under exchange (Gebremedhin, 2005). The need to inspect every consignment to be traded increases the cost of transaction. Consequently exchange in rural markets tends to be characterized by high transaction costs that arise from the high costs of searching for and screening the exchange partners, the cost of negotiating the terms of exchange often with partners trading in small volumes, and the costs of monitoring the terms of exchange (i.e., by inspection) (Hueth, 1999).

Traders along the supply chain and processors are further constrained by poor and inconsistent quality, low volumes and generally unreliable supply. Smallholder farmers often produce small quantities that increase costs of assembling the produce. At the same time, where buyers have specific quality requirements, smallholder farmers face difficulties in assuring regular supply of consistent quality due to both environmental and idiosyncratic factors (Rich and Narrod, 2005; Poulton et al, 2006). This increases the costs to the buyers of monitoring and assuring quality. Where farmers are widely scattered, costs of monitoring and enforcing specific quality can build up rapidly and result in buyers' decision to abandon sourcing from smallholder farmers (Key and Runsten, 1999).

The effect of information asymmetry at the micro and meso levels (i.e., farm and village levels, respectively) can be understood by looking at simple stylized models relating the farmer income to the market prices and price differentials between two markets. Following Minot (1999), Larson (2006) and Barrett (2008) we argue that transaction costs at the farm/household (i.e., micro) level causes a wedge between the exogenous market price and the price farmers receive for their crop. The transaction costs depend, among others, on the range of public and private market information services available to farmers/traders (via e.g., agricultural extension services, broadcasts, etc). We also assume that some of these information services can be transmitted by mobile phones. Since high transaction costs reduce the net income to the farm household, lack of information can prohibit trade, for example, by limiting inter-household (micro) level trade.

At the meso level, the efficiency in inter-village level trade requires that the border price less the costs of moving the produce to the destination market (i.e., transfer cost) be equal to the price at the source market (Adbulahi, 2007). However, just as in the case of micro-level trade, the costs of transferring produce from one village market to another is affected by the availability of public and private market information services. Lack of market information increases the costs of exchange leading to a situation where price in the destination differs from that in the source market by more than the transfer costs. The high costs of moving the produce between markets dampens incentives for trade and can even eliminate trade (and/or participation in the market) altogether (Fafchamps and Hill, 2005; Barrett, 2008).

The importance of farmer access to information is in reducing the transaction costs of exchange caused by information asymmetry between actors (Bagetoft and Olesen, 2004). Provision of ICT-based market information services can improve farmers' access to market information and hence facilitate trade (de Silva, 2008). However, for rural farmers to benefit from such market information services, they need to be aware of their presence and to use them. Undoubtedly, farmers will use ICT technologies that provide agricultural information if they find it profitable to do so. In the case of mobile phones, the cost may include the expense of mobile phone calls seeking information, the cost of buying a mobile phone handset and the cost of recharging the phone battery.

The micro-level (farmer/household) benefits of using ICT technologies that provide information, on the other hand, includes a reduced cost for finding and selecting a trading/exchange partner (i.e., search and screening costs), the costs of negotiating and monitoring the terms of the transaction and the costs of adjusting the terms of exchange. It also reduces the price spread in the output market (Aker, 2008). The reduction in these costs increases the margins earned by farmers and hence the overall revenues/income from trade thus fostering greater participation in the market (Tollens, 2006). The use of ICT technologies can also reduce the costs of acquiring credit and other inputs (e.g., seed, fertilizer, and technical advice) by lowering search, screening, negotiation and monitoring costs thus increasing the margins and revenues assuming constant output price. Reduction in input costs and hence increased margins can, on the other hand, spur commercialization thus foster agricultural development and improvement in household welfare. It can also retard the exclusion of the economically less endowed households by improving the earnings of such households (Chigona et al., 2009).

The use of ICT technologies for market linkage has developmental benefits at the meso (i.e., village/community) levels well. It can reduce transaction costs by lowering the costs of agricultural exchange/trade (i.e., search, negotiation costs, and commodity inspection costs) between two villages/communities. Such market-information-access-driven reduction in transactions costs has been reported by studies in Sri Lanka and Colombo by De Silva (2008) and Financial Times (2008). Theoretically, for the commodity-source market, the reduction in costs of doing business benefits traders by increasing the net price earned and hence margins. Assuming efficient transmission of price to the farmers, ICT-mediated access to information can hence raise the price earned by farmers and result in household asset/capital accumulation. In the medium to long-run, household capital accumulation can in turn stimulate investment in agriculture, commercialization and improved household welfare. For instance De Silva (2008) finds that access to market information through mobile phones improves the welfare of small export vegetable growers by increasing their linkage to better paying export market and also reducing the losses they incur as rejects and uncollected produce in Columbia. Anderson et al (1998) and Aker (2008), on the other hand, suggest that increased availability of information improves the process of price discovery (by reducing search, negotiation and policing costs) and thus improves marketing efficiency and hence farmers incomes.

Sampling Procedure and Data of the Study

This study used data collected from smallholder farmers located in Lurambi South, Lurambi North and Navakholo divisions of Kakamega district. Since the DrumNet project was to work with farmer groups formed by the Kakamega Farmer Field School (FFS), the Food and Agriculture Organization (FAO) funded program, all the farmers that showed interest in participating in DrumNet project were members of the FFS groups. The respondents were therefore drawn from the FFS groups that intended to participate in DrumNet project and also from farmers (both in and outside the FFS groups) that did not intend to participate in the DrumNet project. The sampling from the two groups of farmers was doe as follows.

First, a list of all FFS groups that registered to participate in the DrumNet project was obtained. A list of farmers that that intended to participate in the DrumNet project was then obtained. Farmers were listed as having intention to participate in the DrumNet project if they had paid (in part or full) the 20 percent of the cost of the sunflower seed and fertilizer that the group they belonged to was to receive from DrumNet project under interlinked credit scheme once the project started. Consequently the sample of farmers with interest in participating in the DrumNet project was biased towards those who hard paid at least a part of the 20 percent fee. A second list comprising farmers that did not intend to participate in the DrumNet project was then compiled with the help of village leaders, FFS group leaders and project area agricultural extension officers. This gave rise to two lists of farmers – farmers interested in

Table 2. Characteristics of smallholder households in Lurambi and Navakholo, 2007

Characteristics	Mean	Std Dev	Min	Max
Age of female farmers	47.2	12.8	18	81
Age of male famers	51.9	14.7	22	85
Education of female famers (yrs)	6.5	4.2	0	17
Education of male family (yrs)	8.6	3.9	0	18
Household size	8.3	4.4	0	15
Non agricultural income (males)	32,106.9	47,435. 6	150	600,000
Total income (females)	22,357.8	47,387.4	150	336,000

participating in the DrumNet project and those that were not.

Second, the respondents were sampled from the two lists using probability proportionate to size sampling method. Thus more farmers were sampled from the list with more names, and vice versa. This procedure identified 94 farmers interested in participating in DrumNet project and 168 that are not. A total of 262 farmers were therefore interviewed in this study. The data was collected through personal interviews using pre-tested questionnaire and data entered and analyzed using SPSS data analysis package. The data collected included household characteristics, socio-economic indicators, household assets, information sources, ownership and use of mobile phones, problems related to the use of phones, sources and uses of income, among others. The household survey was conducted during April and May of 2007 but data was collected for the period January to December 2006. Three questionnaires were not adequately completed to warrant inclusion in the results and were therefore dropped. Consequently the results presented are based on 259 usable responses.

Household Characteristics

Table 2 presents the summary statistics on the social characteristics (including age, education and household size) of the 259 households interviewed.

The average age of female and male farmers in the sample was 47.2 and 51.9 respectively. The youngest farmer was 18 years while the oldest being 85 years indicating a wide range in the household farming experience. The average level of education of the farmers interviewed also differed between the two genders. Female farmers had, on average 6.5 years of education while the males had 8.6 years. About 10 percent of the respondents had no formal education of which 84 percent were women. The low level of education suggests that farmers in the study districts are likely to find some uses of the mobile phone that require literacy difficult. An example of such skill-intensive use of the mobile phone is the short messaging system (SMS) which require some level of literacy.

The results also show that respondents earned significant amount of income from non-agricultural sources. The mean income earned by female and male farmers from these sources was Ksh 22,357 and Ksh 32,357, respectively. The major non-agricultural sources of income for female farmers interviewed (in order of significance) were local small enterprise trading, remittances from family members who are away and off-farm employment. On the other hand, the major sources of non-agricultural income for male farmers were non-farm employment, business, and remittances for family members. Overall, female farmers earn, on average, lower income than their male counterparts, indicating that they have less capacity to invest in ICT technologies (e.g., the purchase of mobile phones) and hence may not be able use such technology to obtain agricultural information. Results also show that male farmers also earned a mean income of Kshs 36,666.8 and Ksh 19,927.9 from the sale and renting out of land, respectively, while female farmers only earned nothing from sale of land and only Kshs 10,375.0 from renting out family land. These results suggest that male farmers have greater control over the key household durable asset (land) and especially its disposal than their female counterparts. These findings corroborate those of Dose (2007) and indicate that female farmers do not have the flexibility to use the option of selling or renting out land to acquire the money they might need to invest in a new technology (such as mobile phones).

One of the social characteristics of rural households is the tendency to resolve their problems collectively (Rehber, 1998; Okello and Swinton, 2007). Such collective action takes the form of farmer participation in community based organizations, farmer marketing organizations/groups, rural savings schemes (e.g., merry-go-round[1] (MGR)) and/or religious organizations. Collective action in form of farmer groups can act as a source of market information thus reducing transaction costs and facilitating market participation by farmers (Okello, 2005). In this study, we asked (the 259) farmers about their participation in social and economic groups. Majority of the respondents participated in vegetable farming groups (42 percent), MGR, and funeral/welfare groups (18 percent) (see Figure 3). Figure 3 also shows that female respondents formed the majority in the farmer groups. Since female farmers have lower capacity to buy mobile phone handsets their participation in collective activities may help them access agricultural information from members of their organizations or even use such organizations (especially the MGR groups) to buy their own phones.

Sources Used by Respondents for Agricultural Market Information

Access to information is important in planning both production and marketing of agricultural products. Farmers need accurate information on the quality specifications of a buyer, timing

Figure 3. Farmer participation (counts) in major group activities by gender, 2007

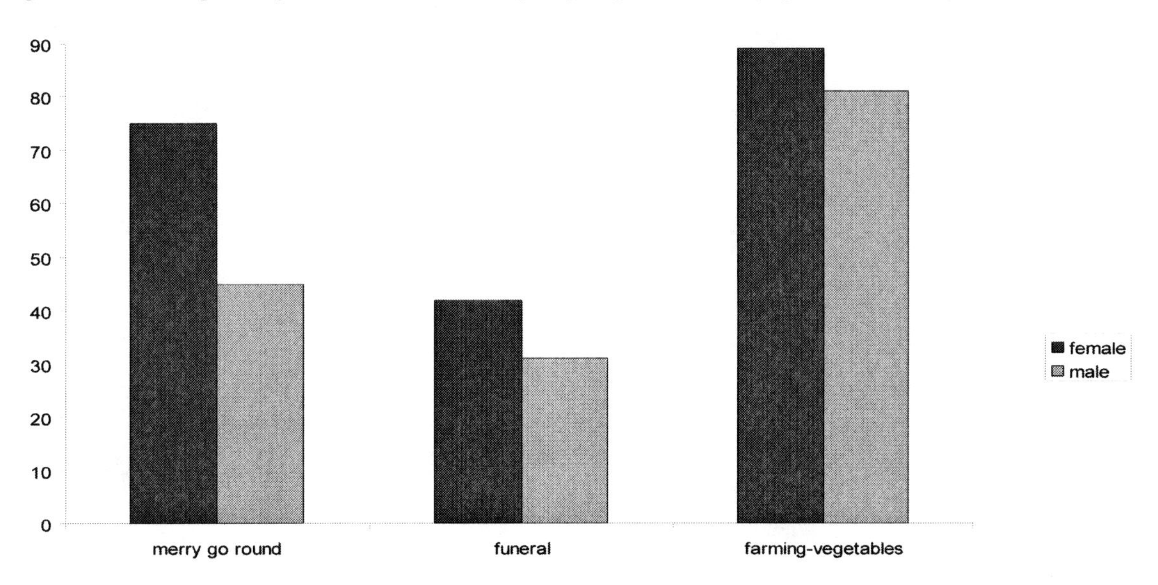

of sale, point of sale, and the volumes needed, among others. Some of the traditional sources of agricultural information used by rural farmers are radio and TV, agricultural extension agents, farmer field days, and farmer groups. Others include the use of mobile phones, internet, and fax. In some communities, NGOs and Church organizations provide information through bulletin boards posted at specific places (usually market place) in rural areas. Table 3 presents the sources of information used by smallholder farmers in Kakamega district particularly those interviewed in this study.

Most farmers obtain information relating to crop variety and input/output prices from traders and the Farmer Field School (FFS) network. The FFS network transmits information to farmers through its network of farmer groups on a weekly basis during farmer training sessions. This study however did not investigate the sources used by FFS network for information it transmits to farmers. Only a handful of farmers use mobile phones and KACE information service as sources of agricultural information.

AWARENESS AND USAGE OF MOBILE PHONES

In order to asses the level of awareness of mobile phones among rural farmers, we asked the 259 respondents to indicate if they have seen or heard of a mobile phone. The results indicate that 99 percent have seen and/or heard[2] of mobile phones. This high level of awareness is the same for both male and female farmers in the sample. The finding that there is high level of awareness of mobile phones in rural areas corroborates past studies (Spence, 2003). It also points to the existence of an opportunity that can be exploited to link more farmers to better paying markets (de Silva, 2008). Nonetheless, the use of mobile phones to transmit agricultural information to farmers and hence foster such market linkage is likely to be affected by the ownership of phones among the farmers. We thus investigated the ownership of mobile phones among the respondents. Overall, 84 percent of the farmers interviewed have used mobile phones from various sources (their own, leased, borrowed, etc) in the past. However, only 94 (36.3 percent) of all the 259 respondents owned mobile phones. Of the 94 respondents that owned mobile phones, 33 (35 percent) were female farmers. These findings imply that the level of awareness of cell phones among the 259 surveyed farmers is very high, however, that the high level of awareness has not translated into high ownership.

Given the low level of ownership of mobile phones by rural farmers, we investigated the various means used by mobile phone owners to acquire them. The results indicate that, of the 94 farmers that owned mobile phones, 30 farmers (12 percent) received the phones as gifts while 64 (88 percent) purchased them. Among the 64

Table 3. Sources of market information used by farmers in Kakamega district (count), 2007

Type of information	Daily Newspaper	Trader	Mobile phone	Neighbors	Government extension agent	KACE*	Farmer group	Farmer Field School
Agricultural credit	3	2	0	12	23	0	17	26
Input price	14	62	0	15	16	0	16	25
Output price	11	44	1	19	18	1	15	30
Weather conditions	12	5	0	6	16	1	12	15

*KACE stands for Kenya Agricultural Commodity Exchange. (Total N = 259

Figure 4. Sources of money used for purchasing mobile phones, 2007

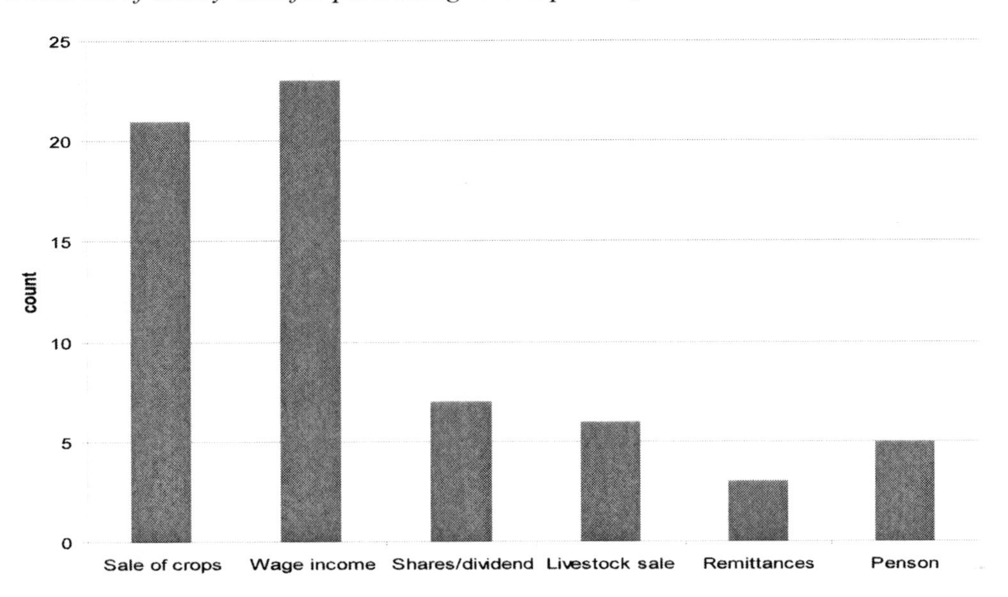

farmers that purchased the phones only 13 (20 percent) were females. Figure 4 shows the major sources of funds used in purchasing mobile phones by those who bought them. Majority of farmers that purchased their phones depended on wage income (from farm and off-farm sources) and the sale of crops.

The finding that there is low ownership of mobile phone is largely due to the inability of smallholder farmers to afford the handsets. Indeed, a recent study (Chigona et al, 2009) finds that the use of mobile phone-mediated internet among individuals with low incomes in developing countries remains low. Other studies in developing countries (Barrett, 2008; Poulton et al., 2006) also find that lack of assets (also known as asset poverty) constrains smallholder farmers' ability to adopt new technologies. This suggests that the usage of mobile phones among smallholder farmers, who usually tend to have low income, might be low due to their lack of capacity to buy the phones. Indeed, the results indicate that the mean price paid for mobile phone handsets by the respondents in this study at the time of the survey was Kshs 3,769.

Among the 94 of farmers that owned mobile phones, only 17 (18 percent) used such phones to enquire about agricultural information (see Table 4). Female farmers accounted for 29 percent of these respondents that used their phones to seek agricultural information. Some additional 9 farmers used phones owned by other people to seek agricultural price information. In particular, they turned especially to family members (spouse, daughter and/or son) for the mobile phones they used in seeking agricultural information. One respondent used a pay phone popularly known as *simu ya jamii*[3] to obtain agricultural information. In total, only 26 farmers sought agricultural information using mobile phones. The finding that 35 percent of the farmers that used mobile phones to obtain agricultural information depended on phones owned by other people further suggests that the high cost of mobile phone handsets is constraining its ownership and hence use for market linkage.

Given the low use of mobile phones for business, we investigated the general usage of mobile phones by the farmers interviewed. Approximately 75 percent of the 259 farmers used mobile phone for social purposes. These included calling friends and family for non business/economic related reasons. Only approximately 17 percent of the

Table 4. Ownership and use of mobile phones by smallholder farmers to get agricultural information

Price information sought on	Who the phone used by the farmer belonged to				Total
	Respondent	Spouse	Son/daughter	*Simu ya Jamii*	
Maize	3 (50)	2(33)	1(16)	0(0)	6(100)
Millet	1(100)	0(0)	0(0)	0(0)	1(100)
Beans	2(68)	1(33)	0(0)	0(0)	3(100)
Tomatoes	1(100)	0(0)	0(0)	0(0)	1(100)
Sugarcane	1(50)	0(0)	1(50)	0(0)	2(100)
Sunflower	1(50)	0(0)	1(50)	0(0)	2(100)
Livestock	2(68)	0(0)	0(0)	1(0)	3(100)
Milk	2(100)	0(0)	0(0)	0(0)	2(100)
Other	3(100)	0(0)	0(0)	0(0)	3(100)
Total	17(65)	5(19)	3(12)	1(4)	26(100)

Numbers in parentheses are percentages.

259 interviewed farmers indicated that they have never used mobile phones before. More than half (i.e., 60 percent) of all the respondents that used mobile phones (for the various non-business purposes) depended on phones owned by friends and family. Of the 60 percent that used phones owned by other people, 38 percent were female farmers. Together, these findings indicate that majority of the smallholder farmers that got to use mobile phones in 2006 used it largely for social purposes. Thus, although there is widespread awareness of mobile phones and relatively high usage, majority of rural farmers use mobile phones for strengthening social ties with friends and family. These findings also suggest that mobile phones play much less role of as an agricultural market information source. The findings further indicate that smallholder farmers depend to a large extent on others for the mobile phones used to contact friend and family.

The main approach that the DrumNet project planned to use to transmit market information to smallholder farmers was the short messaging system popularly known as SMS. We thus investigated the effectiveness of such information medium in reaching smallholder farmers with market information. In particular, we asked the respondents who used mobile phones in 2006 what function (voice or SMS) they employed. Results showed that majority (81 percent) of those who used mobile phones in 2006, depended on the voice system to communicate messages. Approximately 51 percent of these farmers that relied on the voice option were female farmers, indicating that the preference for voice among the smallholder farmers is the same across both genders. The preference for voice over SMS may be attributed to low level of education among the farmers and the difficulty of writing and sending a text message.

Part of the difficulty of writing and sending the SMS springs from the nature of the phone instructions system in the mobile phone handsets used in Kenya. The instructions for navigation of the phone message menu to the message pad are all in the English language. This makes it hard for farmers who can not speak or read the English language to write the text message. The simplest instruction runs thus: MESSAGES > WRITE MESSAGE. At this point the user can write the message. After writing the message, the user then has to navigate through a second set of instructions which are again in English language namely, SEND > PHONE NUMBER. This pro-

Figure 5. Gender differences (count) in the mobile phone function used by smallholder farmers to obtain agricultural information in western Kenya, 2007

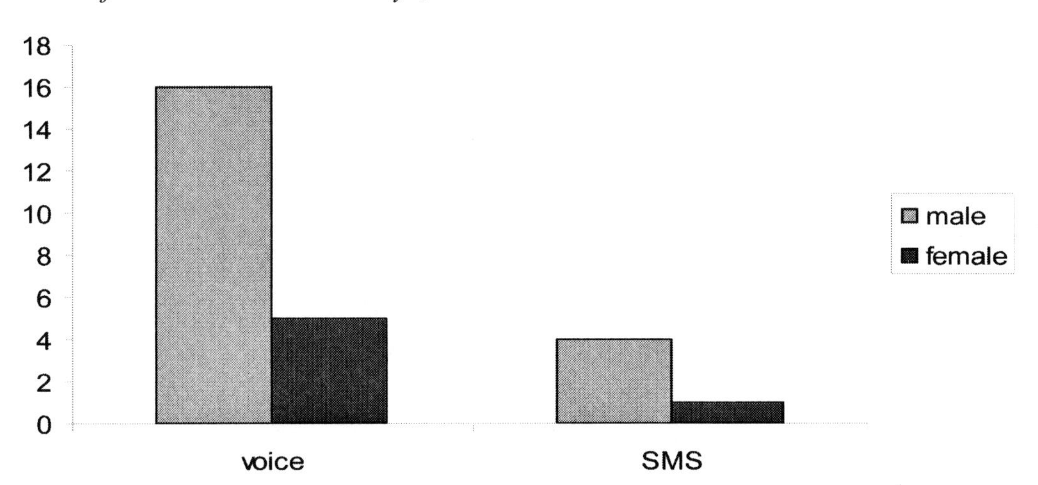

cess is, in most cases, too complex for the little or non-educated smallholder farmers. Indeed, the mean number of years of formal schooling in the whole sample of interviewed farmers was 7.5 (8.1 for males and 6.4 for females). The average education was even smallholder when only the respondents that used mobile phones to obtain agricultural information are considered. Among this latter group of farmers, the average education was 5.5 years These findings suggest that sending an SMS (i.e., a text message) using a mobile phone requires both writing and reading skills above what is possessed by most of the smallholder farmers interviewed in this study.

There were, however, differences between male and female smallholder farmers in the usage of mobile phones to obtain agricultural information with regard to the function used (i.e., voice and SMS). More male farmers used both the voice and SMS options in seeking information on prices of agricultural commodities (Figure 5). At the same time male farmers were more likely to borrow phones from other people to seek agricultural information than females (Figure 6). These differences may be due to the fact that male farmers tend to dominate decision-making regarding the sale of crops, especially those grown for cash, in most households in the area (Dose, 2007). Indeed the results of this study show that there are gender differences in the kind of commodity whose price is sought using mobile phones.

Male farmers that used mobile phones for agricultural price information used it on maize and sugarcane which are cash crops whose marketing is normally controlled by male household heads. Women on the other hand used the phones to seek prices on beans and maize[4] which are food crops and are therefore under the control of women. These findings suggest that there are gender differences in the usage of mobile phones for market linkage with female farmers being more conservative especially with the source of phone used for seeking agricultural price information. The finding that fewer female farmers use SMS to obtain agricultural price information than male farmers and that the average education of female farmers is lower support our earlier argument that low level of education is the reason why smallholder farmers prefer to use the more expensive voice option to the SMS in communicating with their contacts.

We also investigated the other factors that affect the usage of mobile phones by the respondents. Results indicate that one of the most important factors constraining the use of mobile phones by smallholder farmers was the high cost of airtime (i.e., recharge vouchers). At the time

Figure 6. Gender differences (count) in the ownership of mobile phone used for obtain agricultural information, 2007

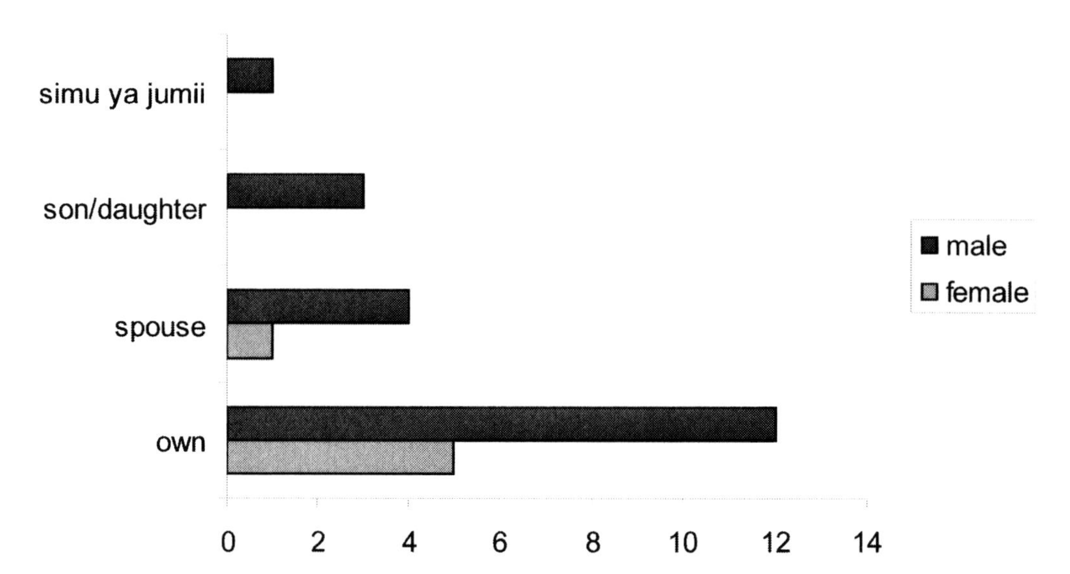

of this study, the the two mobile phone network providers (Safaricom and Celtel) charged a rate between Ksh. 16 and 25 per minutes. This rate escalated for calls across the two networks and to the landline. Among the respondents that used mobile phones for various purposes in 2006, 75 (59.5 percent) listed the high cost airtime recharge vouchers as the first most important impediment to their usage of mobile phones. Twenty five (one-third) of these farmers that found cost of airtime a major impediment to use of mobile phones were females while the rest (two-thirds) were males. Recently, the mobile phone service providers have substantially revised their tariffs downwards making it slightly cheaper to call or send a text message. Smallholder farmers could therefore substantially reduce the costs of using mobile phones if they used SMS (i.e., text messages) to communicate with others. However, as discussed earlier, this option is constrained by low level of education.

The second most important impediment to the use of mobile phones cited by the farmers interviewed in this study was the difficulty of recharging the phone battery. Approximately 62 percent of the farmers who owned mobile phones indicated that they encountered difficulties keeping their phones active due to lack of electricity/power for recharging their batteries (34 percent) and also the high cost of recharging their phones in market places (28 percent). Most farmers do not have electricity in their homes. Therefore they have to walk, in some cases, many kilometers to reach electricity hook ups. Among the farmers that listed difficulty of recharging phone batteries as the second most significant impediment to phone use, 29.4 percent were females while the rest were males.

A significant proportion (39 percent) of farmers also listed frequent borrowing of mobile phones by friends as a disadvantage of owing and keeping the phone functional. Of these farmers, 28.5 percent were males while the rest were females further indicating, as discussed earlier, that male farmers were more likely to borrow and use mobile phones from friends and family. The close ties that exist among community members in the study area (Dose, 2007) usually mean that a farmer can not deny a neighbor with a pressing need the use of his/her phone. However, such frequent borrowing can escalate the costs of maintaining the mobile phone.

SUMMARY, CONCLUSION AND POLICY IMPLICATIONS

This study assessed the level of awareness and usage of mobile phones among smallholder farmers in Kenya. It finds that awareness of mobile phone is quite high with 99 percent of the farmers indicating that they have seen or heard of a mobile phone. However, the use and ownership of mobile phones for economic activities (in particular agricultural information) in the study area is quite low and differs by gender. Fewer female farmers owned phones than male farmers. Consequently, more male farmers used mobile phones to obtain agricultural information than their male counterparts. However, there were no gender differences in use of mobile phones for social purposes such as contacting friends and family.

Overall, only 18 percent of the respondents who owned mobile phones used them to obtain agricultural information, with even fewer female than male respondents using them to enquire about agricultural market price information. The low usage of mobile phones for market linkage is due the high cost of airtime and difficulty of recharging the phone batteries (both recharging cost and access to electricity hook up). The study also finds that the use of mobile phones by both male and female farmers for market linkage is constrained by the low level of education which forces farmers to use the more expensive voice option in communicating with their contacts rather than text messaging.

These findings imply that strategies aimed at improving ownership and use of mobile phones by farmers for market linkage should focus on identifying cheaper mobile phone handsets and less expensive and accessible alternatives of recharging batteries, such as use of solar chargers, to keep the phones operable. The low ownership of mobile phones by the majority of interviewed smallholder farmers further implies that programs that target the use of mobile phones to link farmers to markets through provision of agricultural information may be limited in the number of farmers they reach out to. Such a program should consider loaning mobile phones to farmer groups through interlinked[5] credit scheme in cases where the program involves working with groups of farmers. Farmer groups would then act as an information hub from where members can obtain market information they need. The disproportionately low use of mobile phones by female smallholder farmers for market linkage can then be resolved by targeting women farmer organizations or organizations with greater participation of women. The study findings indicate priorities for policy makers dealing with the specifics of ICT adoption as a tool to promote rural viability via rationalization of the agricultural markets in Kenya.

REFERENCES

Abdulahi, A. (2007, March). Spatial and vertical price transmission in food stapples market chains in eastern and southern Africa: What is the evidence? A paper presented at the FAO *Trade and Markets Division Workshop on Staple Food Trade and Market Policy Options for Promoting Development in Eastern and Southern Africa*, Rome.

Aker, J. C. (2008). *Does digital divide or provide? The impact of cell phones on grain markets in Niger*. Job Market Paper, University of California, Berkeley, CA.

Anderson, J. D., Ward, C. E., Koontz, S. R., Peel, D. S., & Trapp, J. N. (1998). Experimental simulation of public information impacts on price discovery and marketing efficiency in the fed cattle market. *Journal of Agricultural and Resource Economics*, *23*, 62–278.

Ashraf, N., Gine, X., & Karlan, D. (2005). *Growing export oriented crops in Kenya: Evaluation of the DrumNet services*. Unpublished.

Bagetoft, P., & Olesen, H. B. (2004). *The Design of Production Contracts*. Denmark: Copenhagen Business School Press.

Barrett, C. (2008). Smallholder market participation: Concepts and evidence from eastern and southern Africa. *Food Policy, 34*, 299–317. doi:10.1016/j.foodpol.2007.10.005

Besley, T. (1998). How do financial failures justify interventions in rural credit markets? In C.K. Eicher & J.M. Staatz (Eds). *International Agricultural Development,* (3rd Ed.). (pp. 370-389).

Chigona, W., Beukes, J., Vally, D., & Tanner, M. (2009). Can Mobile Internet Help Alleviate Social Exclusion in Developing Countries? *The Electronic Journal of Information Systems in Developing Countries, 36*, 1–16.

Coulter, J., Millns, J., Tallontire, A. & Stringfellow, R. (1999). Marrying farmer co-operation and contract farming for agricultural service provision in liberalizing economies in Sub-Saharan Africa. *ODI Natural Resources Perspectives, 48.*

de Janvry, A., Fafchamps, M., & Sadoulet, E. (1991). Peasant household behavior with missing markets: some paradoxes explained. *The Economic Journal, 101*, 1400–1417. doi:10.2307/2234892

de Silva, H. (2008). Scoping Study: ICT and rural livelihoods in South Asia. Draft report to IDRC.

Dorward, A., Poole, N., Morrison, J., Kydd, J., & Uray, I. (2003). Markets, institutions and technology: missing links in livelihoods analysis. *Development Policy Review, 21*(3), 319–332. doi:10.1111/1467-7679.00213

Dose, H. (2007). Securing household income among smallscale in Kakamega district: possibilities and limitations of diversification. *GIGA Working Paper No. 41*. Hamburg. Germany.

Fafchamps, M. (1996). The Enforcement of Commercial Contracts in Ghana. *World Development, 24*, 427–448. doi:10.1016/0305-750X(95)00143-Z

Fafchamps, M. (2004). *Market institutions in sub-Saharan Africa*. Cambridge, MA: MIT Press.

Fafchamps, M., & Gabre Madhin, E. (2006). Agricultural markets in Benin and Malawi. *African Journal of Agricultural and Resource Economics, 1*(1), 67–94.

Fafchamps, M., & Hill, R. V. (2005). Selling at the farm gate or traveling to the market. *American Journal of Agricultural Economics, 87*(3), 717–734. doi:10.1111/j.1467-8276.2005.00758.x

Ferris, S., & Engoru, P. &. Kanganzi, K. (2006, October). *Making market information services work better for the poor in Uganda.* A paper presented at Research Workshop on Collective Action and Market Access for Smallholder Farmers, Cali, Columbia. *ICT to substantially reduce agricultural costs to farmers* (2008). *Financial Times, 2*(40), March 2.

Government of Kenya. (2002). *Kakamega District Development Plan 2002-2008*. Nairobi, Kenya.

Hueth, B. (1999). Incentive instruments in fruits and vegetables: Input control monitoring, measuring and price risk. *Review of Agricultural Economics, 2*, 374–398. doi:10.2307/1349886

Jayne, T. S., & Jones, S. (1997). Food marketing and pricing policy in Eastern and Southern Africa: A survey. *World Development, 25*(9), 1505–1527. doi:10.1016/S0305-750X(97)00049-1

Key, N., & Runsten, D. (1999). Contract farming, smallholders, and rural development in Latin America: The organization of agroprocessing firms and scale of outgrower production. *World Development, 27*, 381–401. doi:10.1016/S0305-750X(98)00144-2

Kindness, H., & Gordon, A. (2001). *Agricultural marketing in developing countries: the role of NGOs and CBOs*. Policy Series No. 13. Social and Economic Development Department, Natural Resources Institute, University of Greenwich.

Kirsten, J. (in press). The case for interdisciplinary research and training in agricultural economics in Southern Africa. *Nature Sciences Sociétés*.

Kydd, J., & Dorward, A. (2004). Implications of market and coordination failures for rural development in least developed countries. *Journal of International Development, 16*, 951–970. doi:10.1002/jid.1157

Larson, D. F. (2006). Measuring the efficiency of food markets: The case of Nicaragua. In A. Sarris & D. Hallam (Eds), *Agricultural Commodity Markets and Trade: New Approaches to Analyzing Market Structure and Stability*. Rome: FAO.

Minot, N. (1999). *Effects of transaction costs on supply response and marketed surplus: Simulations using non-separable household models*. IFPRI Discussion Paper No. 36. Washington DC.

Moser, C., Barrett, C., & Minten, B. (2005, July). *Missed opportunities and missing markets: Spatio-temporal arbitrage of rice in Madagascar*. Selected Paper presented at the American Agricultural Economics Association Annual Meeting, Providence, Rhode Island, July 24-27, 2005.

Munyua, H. (2007). *ICTs and smallscale agriculture in Africa: a scoping study*. Unpublished Report 1, Submitted to International Development Research Center.

North, D. C. (1990). *Institutions, Institutional Change and Economic Performance*. Cambridge, MA: Cambridge University Press.

Okello, J. J. (2005). *Compliance with international food safety standards: The case of green bean production in Kenyan family farms*. Unpublished Dissertation, Michigan State University, East Lansing, MI.

Omamo, S. W. (1998a). Transport costs and smallholder cropping choices: an application to Siaya district, Kenya. *American Journal of Agricultural Economics, 80*, 116–123. doi:10.2307/3180274

Omamo, S. W. (1998b). Farm-to-market transaction costs and specialization in smallscale agriculture: explorations with non-separable household model. *The Journal of Development Studies, 35*, 152–163. doi:10.1080/00220389808422568

Poulton, C., Dorward, A. & Kydd, J. (1998). The revival of smallholder cash crop in Africa: public and private roles in the provision of finance. *Journal of international International Development, 10*(1), 85-103.

Poulton, C., Kydd, J., & Doward, A. (2006). Overcoming market constraints on pro-poor agricultural growth in sub-saharan Africa. *Development Policy Review, 24*(3), 243–277. doi:10.1111/j.1467-7679.2006.00324.x

Rehber, E. (1998). *Vertical integration in agriculture and contract farming*. Working Paper No.46. Department of Agricultural Economics, Uludag University, Turkey.

Rich, K. M., & Narrod, C. (2005). *Perspectives on supply chain management of high value agriculture: The role of public-private partnerships in promoting smallholder access*. Unpublished paper.

Shepherd, A. W. (1997). *Market information services: Theory and practice*. FAO Agricultural Services Bulleting No. 125. Rome.

Spence, R. (2003, July). *Information and communication technologies (ICTs) for poverty reduction: When, where and how?* Unpublished IDRC Background paper, Ottawa, Canada.

Tollens, E. F. (2006). Market information systems in sub-Sahara Africa challenges and opportunities. *Poster* paper prepared for presentation at the International Association of Agricultural Economists Conference, Gold Coast, Australia August 12-18, 2006.

World Bank. (2002). *World Development Report 2002: Building Institutions for Markets*. New York: Oxford University Press.

ENDNOTES

[1] A merry-go-round is social grouping in which members contribute uniform amount of money at regular intervals but the money is given to members in turns to use in resolving any pressing need rather than being banked.

[2] In this study farmers that have seen or heard of mobile phones are taken to be aware of mobile phones. While this is not a perfect measure, it enables us to differentiate awareness from usage which encompasses knowledge of how to operate the phone (e.g., send a text/short message).

[3] *Simu ya Jamii* is a mobile pay phone usually found in trading centers

[4] Maize acts as both cash and food crop in most rural communities in Kenya.

[5] That is, the project buys a group a mobile phone and recovers its costs from group sale of produce.

Chapter 2
E-Agriculture Development in South Africa: Opportunities, Challenges and Prospects

Rachael Tembo
Cape Peninsula University of Technology, South Africa

Blessing Mukabeta Maumbe
Eastern Kentucky University, USA

ABSTRACT

The global development of information and communication technologies (ICT) has created a new agricultural development paradigm that promises to transform the performance of the agricultural sector and improve rural livelihoods in developing countries. Over the past five years, South Africa has witnessed a swift ICT-led transformation of its public service delivery with major innovations in key development sectors. The growth of e-agriculture is seen as an engine to accelerate agriculture and rural development, promote food security, and reduce rural poverty. This chapter examines e-agriculture initiatives in South Africa. It describes ICT applications in improving the quality of on-farm management decisions, agricultural market information system, e-packaging, product traceability, and online marketing to access lucrative global wine markets. The chapter also highlights key constraints, and identifies considerations to enhance the future prospects for e-agriculture. Given the strategic importance of agriculture in supporting the livelihoods of the majority rural population in South Africa, the successful deployment and effective utilization of ICT is pivotal for sustainable agriculture development and raising the standards of living of marginalized communities. The results of the paper demonstrate that South Africa has made significant strides in e-agriculture and tangible benefits have accrued to the agricultural communities.

INTRODUCTION

The global development of information and communication technology (ICT) has created a new agricultural development paradigm that promises to transform the performance of the agricultural sector and improve rural livelihoods in most developing countries. The diffusion and adoption of the modern technologies which started in industrialized countries has now spread to developing countries. Today, technological advancement in agriculture is evidenced in both developed and developing

DOI: 10.4018/978-1-60566-820-8.ch002

countries (Economic and Social Commission for Asia and the Pacific, 2008). As the use of these ICT becomes widespread, pressure has mounted on identifying not only new users but also newer uses for these modern technologies. Over the past five years, South Africa has witnessed a swift ICT-led transformation of its public service delivery with major innovations in key developments sectors such education, health and agriculture. The growth of e-agriculture is now perceived as both an instrument and a viable option to accelerate agricultural development, promote food security, and help alleviate rural alleviate poverty in the New Millennium. Whether these promises will be turned into tangible or perceptible benefits especially for the majority poor agricultural communities in Africa and other developing countries remains to be seen. Nonetheless, understanding early initiatives in the practical use of ICT will help both current and future generations to devise new and better ways to transform agricultural development using these modern technologies.

South African agriculture provides an interesting case study to assess the current use and future potential for ICT utilization to advance socio-economic development. The agricultural sector in South Africa is characterized by extreme dualism and inequality. The dualistic agricultural sector is characterized by a well-developed commercial sub-sector that co-exists with a predominantly large subsistence or communal farming sub-sector. The later sub-sector is located in the remote and historically disadvantaged rural areas (South Africa, 2005). The former homelands established during apartheid comprise the backbone of what forms the current largely backward rural sub-sector in a relatively advanced South African economy.

Given the great potential and significance of agriculture to most African economies including South Africa on one hand, and the rising population with approximately 70% poor people on the continent, the need to effectively deploy ICT in the agricultural sector in order to increase food production, reduce poverty, improve the people's livelihoods and attain agricultural development cannot be under-estimated. By fully utilizing ICT and taking into consideration people's different needs, ICT can be a powerful tool for economic and agricultural development aimed at eradicating poverty (World Bank, 2003). It cannot be overemphasized that ICT plays a crucial role in facilitating communication and access to information for agriculture and rural development (Kapange, 2006). This makes access to, and exchange of information to be of utmost importance in a country such as South Africa with over 47 million people who need to be fed through agriculture. ICT offers new opportunities that enhance the quality (e.g. timeliness, availability, relevance) of critical information. This has led to the emergence of global agricultural production chains that are interlinked by digital networks, and therefore has important implications for the livelihoods of farmers. ICT are driving agricultural production and supply chain innovations globally, but their full potential is yet to be realized in developing countries' agriculture communities.

More studies on ICT deployment have focused on other sectors such as education, health, banking, e-government and tourism. Fewer studies have focused on ICT deployment in South African agriculture. Yet the use of ICT is now viewed as a basis for a new development paradigm with immense opportunities to provide escape routes from hunger and poverty to marginalized communities. How ICT will alleviate poverty, improve on-farm productivity in South Africa, and or enhance domestic and global supply chain coordination remains largely unknown. The objectives of this chapter are therefore to (i) describe the key developments in e-agriculture in South Africa (ii) describe the socio-economic benefits and key constraints arising from ICT utilization in the agricultural sector and (iii) identify major considerations for future development of e-agriculture and suggest practical solutions including future pathways for successful deployment and growth of e-agriculture in South Africa.

It is important to note that this chapter provides a very important contribution to the e-agriculture revolution unfolding on the African continent. First, the early lessons from e-agriculture development in South Africa will provide critical lessons for other African countries that may want to leap-frog their e-agriculture development efforts. Second, as an economic giant on the continent, South Africa's experiences with e-agriculture development will provide vital lessons for other nations that are still contemplating embarking on a major e-agriculture uptake. Thirdly, e-agriculture development pathways in Africa are new and unique and for South Africa, the knowledge gained from this chapter will provide "learning by doing" approaches that will be beneficial in reshaping its own e-agriculture development pathways, identify potential risks, detours, and areas where success needs to be optimized. Overall, the fact that e-agriculture is a relatively new phenomenon in most developing countries suggests that the experiences highlighted in this chapter will provide a rare opportunity for those who are still uncertain about what this new development paradigm entails even though it is still early days to describe its full potential and development impact.

Background Information

Agriculture is an important sector in most developing countries and the majority of the rural population depends on it (Stienen, Bruinsma & Neuman, 2007). However, the contribution of agriculture to rural development is highly dependent on the generation and delivery of new agricultural technologies and most of these new technologies can be described as information-intensive (Tripp, 2001). In addition to land, labor, capital and entrepreneurship, information has become a critical factor of production in agriculture (Rao, 2006). Like in any other fields such as mining, engineering and commerce, computers and electronic-based mechanisms are now used to collect, manipulate and process information automatically to control and manage agriculture production and marketing. Therefore, agricultural development can be accelerated and livelihoods can be supported by ICT in so many ways.

In South Africa, mobile technologies have the highest penetration rates than any other ICT. Mobile technologies have enabled the previously marginalized millions of people to have access to personal communication (Esselaar, Gillwald & Stork, 2006). In a related study, Zama and Weir-Smith (2006) investigated ICT accessibility in all South African provinces. The results on Table 1 show the percentages of households who have access to different types of ICT in each Province.

On a provincial level, the Western Cape Province (WCP) has the highest penetration rates for landline phones, personal computers and Internet. Gauteng Province leads in mobile technology accessibility.

According to a comparative analysis of trends in fixed telephones line, Internet and mobile phones reported by the International Telecommunication Union (ITU), Africa still has a great potential to improve its telecommunications usage. On a continental level, South Africa is among the top 15 African countries that are actively involved in facilitating and promoting the use of ICT (ITU, 2007). Figure 1 shows the 15 largest African markets in terms of total (fixed and mobile) telephone subscribers per 100 inhabitants. Among the top 15 African countries with the highest penetration rates, South Africa is ranked number 4.

In terms of Internet penetration, South Africa is ranked number 6 among the top 15 African countries with highest penetration rates. Figure 2 shows the Internet users per 100 inhabitants for top 15 African countries.

The rapid penetration of ICT in South Africa offers a clear opportunity to use ICT. Firstly, ICT are essential for the procurement, distribution, and marketing of farm inputs (e.g. seed, fertilizers, and chemicals) and farm equipment, tools, and spare parts from downstream industries. Secondly, ICT can be used to monitor and evaluate on-farm

Table 1. Percentages of households with access to ICT in South African provinces

Province	Landline at home %	Access to PC %	Access to internet %	Access to cell phone %
Western Cape	40.4	23.8	14.1	38.0
Eastern Cape	10.6	4.5	3.1	23.0
Northern Cape	17.6	9.3	3.6	18.9
Free State	18.5	10.2	4.5	26.3
Kwazulu-Natal	12.5	6.6	2.8	25.3
North West	11.1	3.7	1.9	27.9
Gauteng	17.3	15.5	11.3	43.8
Mpumalanga	14.6	6.3	3.8	25.5
Limpopo	9.5	4.7	2.7	29.1

Source: Adapted from Zama and Weir-Smith, 2006

production inventories. Thirdly, ICT can be to support the production, processing, distribution and marketing services for agricultural goods until they reach the final consumer. With this increasing penetration, it is logical to investigate how ICT penetration is changing lives of agricultural communities and benefiting the agribusiness industry in general. If ICT have managed to find

Figure 1. 15 African countries with highest penetration. Source: Adapted from International Telecommunication Union, 2007

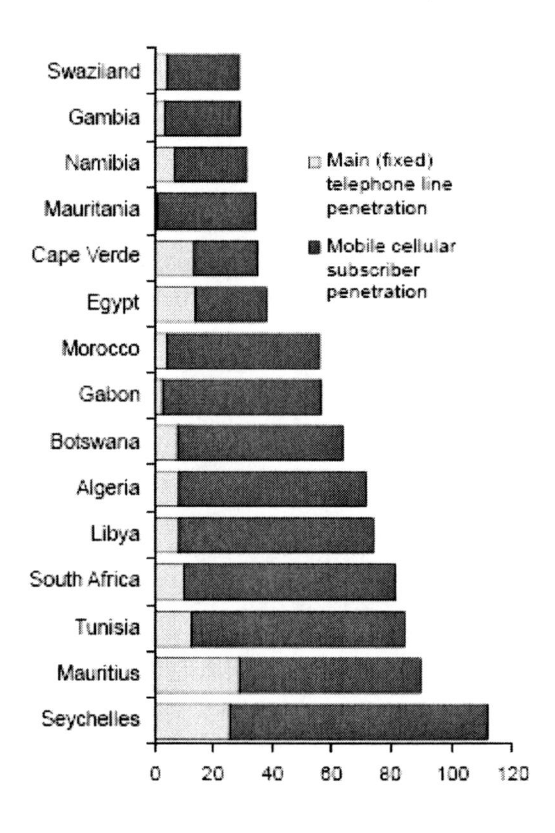

Figure 2. Top 15 African countries-Internet users per 100 inhabitants. Source: Adapted from International Telecommunication Union, 2007

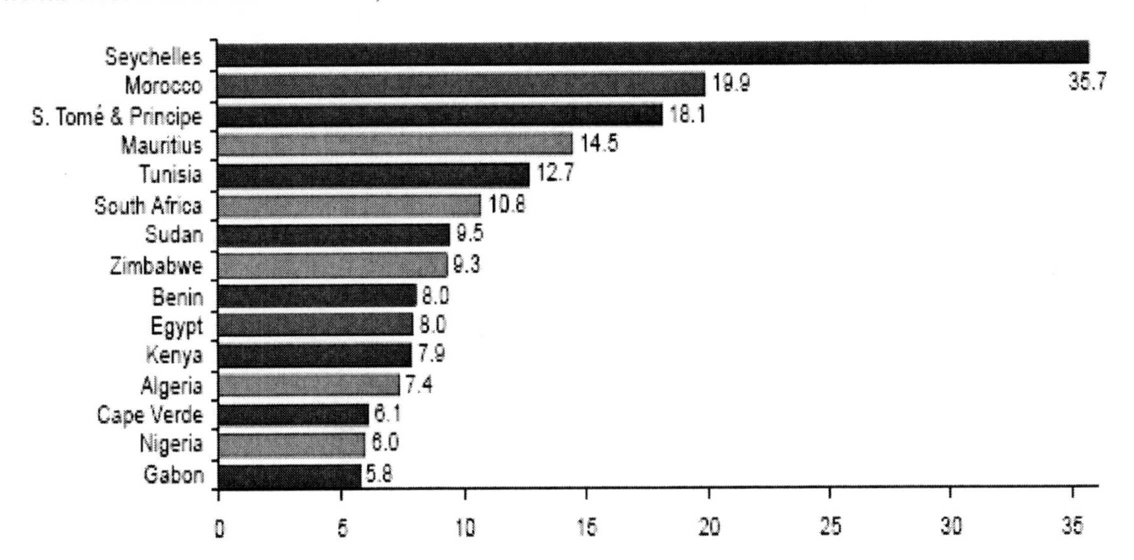

their way into different agricultural communities, then it is likely that they are being used in these communities and are impacting livelihoods of the members of these communities. Given the high ICT penetration rates presented in this section, ICT use is also expected to be high in the respective communities.

An Overview of the ICT Policy Environment in South Africa

ICT policies are put in place for a number of reasons. One of the main purposes is to provide direction and guidance for sustainable development and applications of ICT in a country (Zimbabwe, 2005). Galloway and Mochrie (2005) also state that ICT policies are formulated among other things, to raise ICT awareness, and to provide advice and support for ICT use and infrastructure development. An ICT policy can be defined as an integrated infrastructural, social, economic, environmental, legal and institutional framework that provides strategic direction and guidance for sustainable national development through the development and application of ICT (Zimbabwe, 2005). According to James (2001), ICT policies are in three levels namely, infrastructural policy (deals with development of national ICT infrastructure), vertical information policy (addresses sector needs) and horizontal information policy (impacts broader aspects of the society). All these levels should be integrated in the development of sustainable ICT policies that ensure the achievement of twin goals (i) to empower communities and (ii) improve people's quality of life. In South Africa, the government initiated a number of policies to support and empower local industries, communities and individuals. Table 2 shows a chronology of ICT-related policy initiatives developed in South Africa from 1995 to 2005.

Overview of Agricultural Importance of Western Cape Province and ICT Potential for Rural Development in South Africa

Western Cape agriculture contributes 14.5% of the total GDP and generates about 23% of the total gross income of the country's agricultural sector. The province accounts for more than half of South African agricultural sector's exports (Elsenburg, 2007). Commercial agriculture is the leading export sector in the province. Western Cape agriculture is based mainly on the production of fruits (fresh, dried, processed and juices), wine, vegetables, livestock and winter grains. Fruits and nuts dominate the first position in the top 20 export products for the province and contribute 19.4% of the total agricultural produce from the province. The majority of the export potential for South African agriculture is concentrated in the wine grape producing areas of Worcester, Robertson

Table 2. Chronology of ICT Policy Initiatives in South Africa, 2009

Year	Policy Initiated	Brief Policy Description
1995	Green paper on Telecommunication policy	Outline different policy options to stimulate public debate on aspects of telecommunications.
1996	Telecommunications Act (No. 103 of 1996)	Make provision for an independent regulator of telecommunication activities, create universal service agency and universal service fund to extend network to unprofitable areas.
	White paper on Telecommunications	Recommends setting up of a universal service agency that will monitor universal service obligations of network operators.
2000	Independent Communications Authority of South Africa Act (No.13 of 2000)	Merge telecommunication and broadcasting regulatory functions for convergence in the ICT sector, and coherent regulation.
2001	Telecommunications Amendment Act (No. 64 of 2001)	Amend Telecommunications Act, 1996 (Act No. 103 of 1996), make provision for radio frequency access, make provision for new kinds of licenses, make further provision for license applications, provide public switched telecommunication services & networks, regulate mobile cellular telecommunication services.
1996-2001	Consolidated telecommunications Act	Promotes universal and affordable provision of telecommunication services, provision of a wide range of telecommunication services in the interest of the economic growth and development, make progress towards the universal provision of telecommunication services.
2002	Electronic communications and transactions Act (No. 25 of 2002)	Provides for the facilitation & regulation of electronic communications & transactions, provides for the development of a national e-strategy, promotes universal access to electronic communications and transactions and the use of electronic transactions by SMMEs, provide for human resource development in electronic transactions, prevents abuse of information systems; to encourage the use of e-government services.
	Regulation of interception of communications and provision of communication-related information Act (No. 70 of 2002)	Regulates the interception of communications, monitors signals and radio frequency spectrums, provides communication-related information, authorizes interception of communications, and prohibits the provision of telecommunication services which do not have the capability to be intercepted.
2005	Independent Communication Authority of South Africa Amendment Bill	Amendment of the Independent Communications Authority of South Africa Act, 2000, provides for substitution of the Postal Regulator for the Authority, determines the functions of the Authority; consolidates certain powers and duties of the Authority, provides for inquiries by the Authority, amends the procedure for the appointment and removal of councilors, and it regulates the financing of the Authority.
	Electronic Communication Act (No. 36 of 2005)	Promotes convergence in the broadcasting, broadcasting signal distribution & telecommunications sectors, provides the legal framework for convergence of these sectors, makes new provision for the regulation of electronic communications services, electronic communications network services & broadcasting services, granting of new licenses and new social obligations, provide for continued existence of the Universal Service Agency and the Universal Service Fund

Source: Adapted from ICT Portal, 2009; James, 2001

and Stellenbosch, the deciduous fruit areas such as Caledon, Paarl and Ceres and the citrus producing areas such as Citrusdal (McDonald & Punt, 2001). The Western Cape is the country's largest producer of citrus fruit contributing about 90% of the country's total exports. Many fruit farms produce deciduous fruits that have been of major importance. Examples include grapes, apples, cherries, pears, peaches, and other fruits, which are being exported in great quantities mainly to European markets. According to Cape Gateway (2003), South African citrus contributes up to 2% of world citrus production, but accounts for more than 8.5% of total world exports. An estimated 60% of the annual crop is exported, 25% locally consumed and 5% processed to juice. This demonstrates the comparative advantages of the WCP in agricultural production and hence it's relative high potential for ICT applications.

The South African wine industry dates back to more than three centuries ago when the first vines were grown in the Cape in 1655. In 1659, wine was processed from grapes for the first time in the Cape (Flockemann, 1972). Now South Africa is a major global player in wine production and marketing. In wine production, South Africa is ranked the 8th largest wine producer in the world (Bruwer, 2003). The wine industry is one of the major contributors to South Africa's economic growth, especially in the Western and Northern Cape provinces (Esterhuizen & Van Rooyen, 2006). The wine industry has prospered the most among all the commercial farming sectors in the WCP (Moseley, 2007). This is because of the favourable climatic and soil conditions in the province. Exports from this industry have increased ten-fold in the past decade.

One of the major constraints facing emerging farmers in WCP is the lack of market information (Moloi, 2007). Without market information, these farmers will remain dependent excessively on "exploitative" middlemen, making it difficult for them to adapt to ever-changing market and environmental conditions and to get better profits from their yields. Given the WCP sophisticated, wine, dairy, wheat and livestock operations, tremendous opportunities exist for the application of ICT in the agricultural sector. The wine industry is vertically integrated with both regional and global markets. Perishable products such as meat and milk products which are marketed domestically offer new opportunities for using ICT to quicken deliveries and reduce spoilage. The collection, processing, and exchange of key decision information is critical to the improvement of the production frontier of the WCP in South Africa. Figure 3 shows a summary of major agricultural production output in South Africa for the past ten years.

ICT Applications at Input and Farm Production Levels

Farmers use farm automation that involves the use of control systems, such as computers to derive higher yields, through farming processes that are more efficient, less labour intensive and less-time consuming. Precision farming, biotechnology, computerised irrigation systems, global positioning systems and weather forecasting and monitoring all depend on sophisticated ICT. In the area of agricultural production, prices of inputs are the most frequently telecommunicated information (Economic and Social Commission for Asia and the Pacific, 2008). Some farmers in the wine industry use "ezywine" and "winems" software to consolidate all the winery procedures into one comprehensive package that can be easily utilized on the farm.

ICT Applications in Processing, Distribution and Marketing

ICT has improved the effectiveness and speed of packaging and labelling of agricultural products. In addition, ICT has improved the ability for information dissemination to agriculture stakeholders to enhance market participation. A number of commercial farmers now use websites for advertising their agricultural products and e-commerce for carrying out transactions online e.g. input procurement systems. The transportation and logistics systems of agricultural input and output benefit from ICT at various levels, from communication to locate destinations to automatic vehicle tracking (Economic and Social Commission for Asia and the Pacific, 2008). Computer and microprocessor technologies have led to drastic changes in processing, quality control and safety, standardization of weights, volumes, sizes, automated handling and stock control, and monitoring of storage conditions of agricultural products (Selamat & Shamsudin, 2005).

ICT Applications in Food Retailing and Global Supply Chains

Due to European Union regulations pertaining to wine exported to the EU, wine farmers are obliged to use tracking systems. The most popular ICT used for traceability are bar codes and tracking software. Using these ICT, the farmers can trace the fruits from the orchard to the final consumer (Matthee, 2004). The efficiency of various contacts with the potential and existing customers has also been enhanced by the use of ICT. Globally, consumers are increasingly concerned about who produces their food, and how their food is processed. Identity preservation and food tracking systems are utilizing ICT in response to growing consumer demands for information about the origin of their food.

RESEARCH METHOD

This chapter is based on research findings from a survey that was conducted in the fall of 2007 in WCP in South Africa. A structured questionnaire was developed and later administered on commercial farmers in the wine industry and their farm employees in the Robertson Wine Valley. The questionnaire covered the following major sections (i) the type of ICT they use for agriculture activities, (ii) how the commercial farmers and their workers use ICT in wine production and marketing (iii) constraints associated with use of ICT (iv) recommendation for improving ICT in wine production systems. The sample size for the commercial farmers interviewed was fifteen and that for employees was sixty. This case study was selected because of (i) evidence of widespread use of ICT by both commercial farmers and workers (ii) proximity of the farm location and cost implications (iii) willingness by the farmers to assist the researcher after having faced difficulties with similar study with citrus growers association and deciduous fruit growers. The results from the survey and other related literature were used to report the key findings in this chapter.

LITERATURE REVIEW

Role of ICT in Poverty Alleviation and Promoting Food Security

The development of ICT provides new opportunities for poverty alleviation in developing countries. Poverty has different definitions and different dimensions. According to Obayelu and Ogunlade (2006) poverty is the lack of what is considered as a minimum requirement to sustain a livelihood in a given society by a household or an individual. Poverty is not only about lack of food and income. It is multidimensional, complex and extends beyond what is mostly interpreted as poverty. It encompasses several dimensions such as lack of access to goods and services, income, nutrition, opportunities, political freedom and equal rights. These are the basic necessities that the largest percentage of people living in most developing countries lack. These dimensions of poverty should be addressed in order to reduce the extent and severity of poverty (Smith, 2004).

There is growing evidence that ICT plays an important role in eradicating poverty. If ICT are

Figure 3. Summary of major agricultural production output in South Africa for the past ten years, 2009. Source: Statistics South Africa, 2009

Year	1997/98	1998/99	1999/00	2000/01	2001/02	2002/03	2003/04	2004/05	2005/06	2006/07	2007/08[1]
						R1 000					
Maize	4 454 363	5 397 112	6 289 664	6 928 826	13 614 457	9 134 479	8 217 126	7 490 874	6 870 617	10 772 965	20 798 560
Wheat	1 986 183	1 529 183	1 664 750	2 829 568	3 559 642	3 832 257	2 209 104	1 841 746	1 978 498	3 222 667	4 794 331
Oats	9 842	10 756	12 361	24 002	42 895	55 306	36 164	38 784	31 675	39 371	70 422
Barley	145 600	152 866	70 068	92 961	131 400	215 863	343 920	248 326	261 642	372 036	430 772
Rye/Canola[2]	1 920	1 614	2 219	24 522	42 179	96 368	71 568	56 852	73 951	97 090	98 803
Grain sorghum	197 156	163 177	245 837	156 691	452 985	360 238	404 226	140 863	131 593	300 246	507 738
Hay	1 519 642	1 607 911	1 528 601	1 826 497	1 608 979	1 969 412	2 532 059	2 402 649	2 365 232	2 174 874	2 483 645
Lucerne seed	13 271	3 541	510	2 629	6 112	9 851	8 096	8 256	4 339	3 461	2 243
Dry beans[3]	146 227	283 554	265 579	287 419	281 064	286 017	297 908	244 412	324 729	261 354	428 872
Dry peas	4 625	2 376	1 179	1 754	1 936	4 186	1 630	1 406	995	1 020	1 263
Lentils	11	7	15	23	80	26	30	33	26	27	34
Sugar cane	2 638 855	2 865 781	2 575 636	3 115 839	3 359 987	3 963 173	3 452 433	3 046 569	3 654 463	4 030 981	4 116 551
Chicory root	19 576	22 563	27 567	22 215	39 703	31 576	35 365	31 010	36 200	26 873	14 416
Tobacco	590 113	472 749	437 720	621 580	609 173	749 026	456 445	413 001	262 787	230 672	194 742
Cotton	249 436	331 169	150 365	232 671	164 913	151 759	228 208	119 041	90 251	87 979	113 077
Groundnuts	143 603	212 281	334 937	454 720	324 410	338 009	367 305	178 456	238 569	385 121	675 156
Sunflower seed	797 340	1 525 064	518 609	893 156	2 163 162	1 349 131	1 231 010	1 018 516	1 045 966	794 616	3 646 068
Soya-beans	220 086	226 029	197 294	280 934	448 442	339 547	469 643	347 293	622 195	480 379	1 134 293
Wattle bark	66 626	49 119	53 880	57 433	81 757	54 286	47 866	49 934	75 665	97 408	93 250
Sisal	2 710	2 949	2 482	1 732	2 311	2 157	1 248	953	950	81	24
Other field crops	30 342	36 260	49 366	68 266	92 742	119 762	107 716	90 704	81 191	91 369	119 146
TOTAL FIELD CROPS	**13 237 730**	**14 916 241**	**14 428 858**	**17 923 440**	**27 258 329**	**23 072 445**	**20 518 850**	**17 768 698**	**18 151 936**	**23 470 590**	**39 922 426**
Viticulture	1 412 591	1 435 912	1 451 024	1 561 712	1 921 946	2 415 893	2 721 920	2 678 432	2 615 611	2 686 885	2 973 822
Citrus fruit	1 238 247	1 892 803	2 333 706	1 674 909	2 761 070	2 968 208	3 793 309	3 964 584	2 773 400	3 366 492	5 013 128
Subtropical fruit	719 459	937 954	975 172	1 008 261	1 152 126	1 379 704	1 334 842	1 439 726	1 480 895	1 634 191	1 617 820
Deciduous and other fruit	3 911 435	3 053 638	2 965 721	3 634 863	4 051 384	4 775 553	5 810 171	5 016 434	4 766 187	6 101 516	6 425 416
Dried fruit	167 590	287 534	237 594	193 678	177 787	245 402	241 207	243 180	286 496	337 996	402 702
Nuts	28 676	32 836	37 601	42 225	47 292	47 600	53 606	56 737	58 731	66 529	78 505
Vegetables	2 007 964	2 041 841	2 338 630	2 462 457	3 031 049	3 880 487	4 047 862	3 845 707	4 169 526	4 892 108	5 851 695
Potatoes	1 431 257	1 588 050	1 631 553	1 832 351	2 018 442	3 012 209	2 522 415	2 507 395	2 969 268	2 941 100	3 546 523
Flowers and bulbs	416 040	463 573	492 283	490 094	536 343	617 523	663 808	694 285	760 375	756 866	720 076
Rooibos tea	38 250	25 000	25 670	38 059	72 302	101 657	121 811	126 659	117 961	141 467	150 000
Tea	192 757	169 708	158 537	168 663	173 483	145 396	67 917	23 577	33 166	46 493	72 426
Other horticultural products	17 892	19 498	20 412	22 692	27 414	32 506	35 389	33 318	35 704	36 631	44 606
TOTAL HORTICULTURE	**10 582 178**	**11 948 347**	**12 667 903**	**13 140 184**	**15 970 637**	**19 622 140**	**21 614 257**	**20 630 134**	**20 087 220**	**23 012 474**	**27 096 919**
Wool	595 616	482 321	523 253	546 721	677 043	1 241 936	947 821	737 611	690 064	1 131 931	1 499 505
Mohair	151 653	160 420	226 722	194 532	186 523	217 557	167 548	177 216	236 372	252 948	242 942
Karakul pelts	1 766	2 386	3 162	2 890	3 674	4 969	2 900	4 541	4 686	7 896	8 168
Ostrich feathers and products	226 236	175 827	247 270	320 678	327 581	3 62 267	299 578	337 648	416 502	343 964	370 270
Fowls slaughtered	7 380 437	5 566 788	6 887 688	7 690 014	9 214 912	10 408 100	10 580 828	10 999 428	12 204 286	13 965 725	16 666 289
Eggs[4]	1 908 708	1 829 451	1 887 402	2 157 850	2 402 656	2 991 957	3 061 541	3 321 472	3 804 532	4 714 904	5 429 214
Cattle and calves slaughtered	3 169 070	3 151 246	4 187 646	3 445 050	4 632 129	5 753 004	6 411 735	7 329 051	9 530 481	12 514 286	12 982 639
Sheep and goats slaughtered	935 487	1 020 252	1 150 732	1 266 639	1 316 449	1 614 523	1 715 955	1 780 906	1 990 076	2 381 827	2 369 439
Pigs slaughtered	793 452	646 683	784 079	803 050	918 280	1 346 371	1 385 670	1 490 305	1 466 134	2 066 507	2 581 506
Fresh milk	3 644 082	3 268 080	3 105 690	3 734 537	4 257 484	4 880 873	5 198 597	5 324 423	5 311 544	6 027 899	9 006 891
Other livestock products	538 904	612 186	702 876	823 922	999 622	1 198 146	1 241 854	1 363 026	1 572 176	1 518 066	1 979 220
TOTAL ANIMAL PRODUCTS	**19 345 821**	**17 915 846**	**19 706 522**	**21 086 293**	**25 138 353**	**30 019 705**	**31 014 226**	**32 865 626**	**37 246 758**	**44 925 953**	**53 136 103**
GRAND TOTAL	**43 165 729**	**44 780 434**	**46 803 283**	**52 149 917**	**68 367 319**	**72 714 290**	**73 147 373**	**71 264 458**	**75 485 914**	**91 409 017**	**120 155 448**

appropriately deployed to take into consideration people's different needs, they can become a powerful economic, social and political tool for the poor and those who work to eradicate poverty (World Bank, 2003). Some authors argue that the poor do not need ICT but that they need improved nutrition and other basic necessities that assure them of a sustainable livelihood instead. Some even argue that the poor do not eat computers and telephones; they don't need them, as they are a luxury (Daly, 2003). There may well be some validity to these arguments, but McNamara (2003) however, argues that ICT may not be the immediate solution or goal but are a tool to help achieve what the poor people actually need, combat poverty and achieve economic growth. Above all, the poor have information, communication and knowledge needs like any other people in society and those needs deserve to be addressed (Marker, McNamara & Wallace, 2002).

The World Development Report 2000/01 identified three priority poverty–reducing potential of ICT which include: (i) increasing opportunities for the poor to access markets and expand their assets, (ii) enhancing empowerment through better coordination in the provision of basic needs of the poor and removal of social barriers by state institutions and agencies, and (iii) improving security of poor people through better risk management and improved access to micro-finance and global markets (Cecchini & Scott, 2003). Richardson (2005) provides a key insight about the potential of ICT interventions in the context of developing countries by stating that; " Any ICT intervention that improves the livelihoods of poor rural families will likely have significant direct and indirect impacts on enhancing agricultural production, marketing and post-harvest activities – which in turn can further contribute to poverty reduction". Richardson's observation implies that any ICT intervention will result in multiplier effects, which will ultimately lead to poverty reduction. For example, a simple rural payphone can play a significant role in enhancing the ability of poor rural families to enhance their contribution to national agricultural production and post harvest activities though better input and output market access coordination.

The expected outcome of these ICT interventions is not only improved agricultural production or post harvest activities, but also poverty reduction via increased incomes and improved livelihoods. Batchelor, Evangelista, Hearn, Peirce, Sugden & Webb, (2003), argue that the provision of transparent and timely market information to the role players (buyers and sellers) in the agricultural supply chain will enhance efficiencies in agricultural markets. Therefore, ICT intervention will promote better coordination and communication between buyers and sellers, thus leading to competitive prices in the inputs and products markets. In the short to medium term, as farmers' household incomes increase, the net result will be a reduction in poverty and an improvement in the living standards of the rural people.

Marker, *et. al*., (2002) argue that ICT play a vital role in poverty reduction by improving the flow of both information and communications. The Food Agricultural Organization's ICT agenda is "fighting poverty with information" and it stresses the development of appropriate databases that facilitate agricultural information services to reduce poverty and its effects. So far, it has provided technical assistance to member states to develop and apply a variety of media in support of agriculture and food security (Winrock, 2003). Khaketla (2004) argues that ICT can revolutionize the management of the agricultural sector and improve food security by significantly improving information flows between agricultural role players, leading to higher crop yields, production that is more responsive to consumer needs and greater market access. Dissemination of information regarding crop production, cost-effective means of production and methods of preserving food is made easy by using ICT.

Harris (2004) proposed a framework that illustrates how ICT can be used for poverty alleviation by governments and institutions, which have ICT policies and are developing their ICT infrastructure. Figure 4 shows the framework. The ICT policy framework shown in Figure 4 specifically targets poverty alleviation and aims at addressing the cause of poverty. Development strategies are there to propose how poverty can be reduced. There is need for well-managed and organized local ICT access facilities. Government is a major ICT user, policy implementer and supplier of public services. Information and physical infrastructure should provide for the dissemination of information in the society. Methods of implementation of ICT use should be demand-driven and use bottom-up approach for sustainable results (Gebremichael & Jackson, 2006; Moodley, 2005). Respective institutions are there to use ICT approaches and deliver services to the public. ICT should also be directed to specific services that will be delivered to the target population for the purpose of poverty alleviation (Harris, 2004).

Moodley (2005) however argues that in order to reduce poverty, ICT projects should be driven by development objectives and information needs not by technological concerns. The concerns should be poverty reduction not the spread of technology or reducing the digital divide. To make ICT-based system effective for poverty reduction programs, there is need for relevant data to be available to the poor; relevant skills, technology and money to access the data; motivation, knowledge and confidence to access and apply the data. Without these, ICT investments will not accomplish their purpose. As a result the poor will remain poor.

Figure 4. Framework for poverty alleviation using ICT. Source: Adapted from Harris, 2004

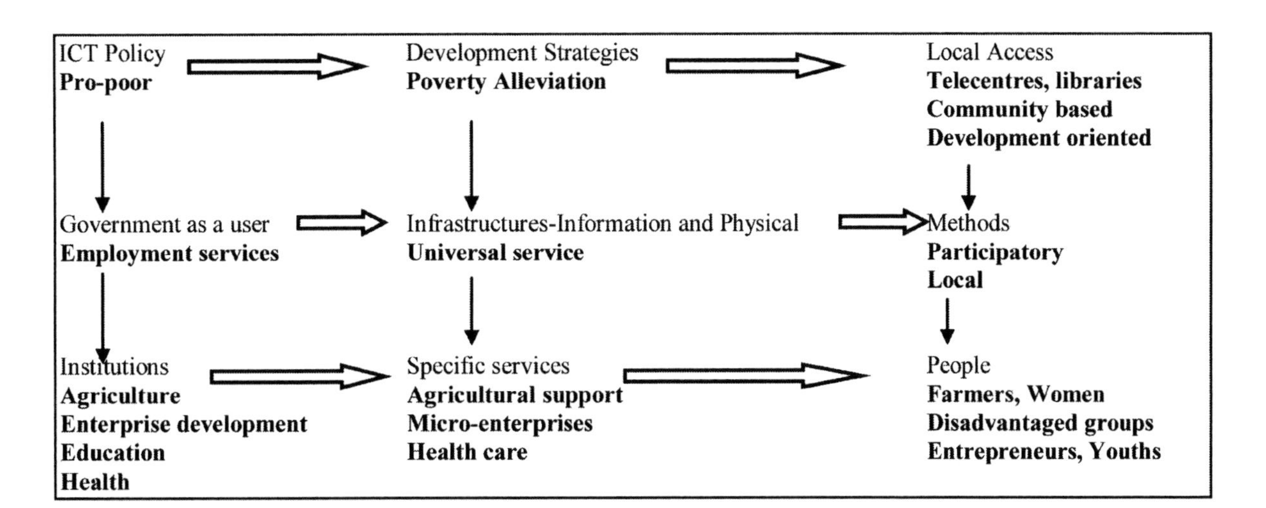

One of the key benefits of ICT applications in agriculture lies in meeting the farmer's needs for market information (Meeira, Jhamtani & Rao, 2004). However, if modern ICT are not adequately built into the agricultural supply chains, there is a probability of stagnation in the dissemination, utilization and application of new scientific agricultural information for development purposes. Kalusopa (2006) study on Zambia stresses that agricultural producers with limited access to ICT are at the mercy of the global market giants due to their competitive disadvantage. The potential use of ICT by agricultural communities should not be viewed in isolation to other economic agents as ICT plays an important role in facilitating communication and transactions with other agricultural stakeholders, their suppliers and customers. For instance, poor information flows within and between government institutions lead to inefficient institutions, poor policies, poor service delivery and the inability to meet the needs of vulnerable and poor communities (McNamara, 2003). Information communicated by ICT can support livelihoods through the dissemination of information on better use of resources, markets, commodity prices, income generation projects and support services (Nikam, Ganesh & Tamizhchelvan, 2004).

Empirical evidence has shown that there is a direct relationship between effective flow and transfer of information and agricultural development (Kalusopa, 2006). Agricultural development is severely constrained when markets perform poorly due to lack of appropriate information and communication flows. Without proper communication of useful information, consumers may not be in a position to fully realize the advantages of technologies (Lee, Lee & Schumann, 2002). ICT can accelerate agricultural development by providing more accessible, complete, timely or accurate information to those making key decisions. Enhanced information flows help farmers to make more informed timely decisions. In addition, ICT enable larger amounts of information to circulate and be stored at a much higher speed, diminishing the effect of distance. Among other uses, it enhances the provision of essential information on market prices, weather conditions and credit facilities. Accessibility of information on prices do not only give farmers more bargaining power but also allows them to explore alternative market opportunities beyond the farm gate (ICT Update, 2002). The question that arises is how many agricultural communities are fully participating in this information age and enjoying the benefits of access to information in developing countries and specifically in South Africa?

E-Commerce Applications in Agriculture

Recently, e-commerce has been used in agriculture by a number of companies in developing countries. It has been used successfully in the fishing and agricultural sectors in the Philippines where a trading portal b2bpricenow.com has been launched to provide price updates and market information for agriculturalists, consumers and manufacturers (Batchelor, et. al, 2003). Processing a purchase order include paperwork, data entry, phone calls, faxes and approval requests, and this can be quite expensive when done manually. On the other hand, performing these transactions online reduces costs substantially.

ICT Applications in Farm Record Keeping and Analysis

Besides accessing the Internet, computers still play an important role for farmers in their management of information needs. Some farmers have realized that computer operations are useful for collecting, entering and interpreting data, thereby very instrumental in making decisions (Batte, 2005). Besides, computers can be used for crop, livestock, field and farm record keeping. Field activity records include information on field identification, tillage practices, crop progress, weather, fertilizer and lime applications, general planting information, chemical application records and general harvest information. Computers can also be used for keeping financial records, budgets, accounts and business plans.

RESULTS AND DISCUSSION

This section presents the empirical results of the main findings from the analysis of data that was collected from surveys conducted in Robertson Wine Valley. The scope of this study is limited to descriptive statistics. Table 4 describes the demographic profile of farmers and farm employees surveyed in the Robertson wine valley.

In contrast, the majority of the farm employees surveyed were females (52%) while the majority respondents from the farmers were males (57%). However, both the majority of the farmers and farm workers surveyed were aged between 31-40 years old and married. The proportion of farm employees was equally divided between blacks, coloreds (i.e., mixed race), and whites. These farm employees worked in the vineyards seasonally (17%) or on a permanent basis (28%). Other farm responsibilities were divided into administrative (28%) and winery operations (27%). Most of the farm employees had worked for 10 years or less while majority of the famers had more than 10 years experience in the wine business. The farm workers had matriculation (42%) and college (40%) level of education compared to the farm owners who had mostly college diploma (53%) and University degree (27%).

ICT Use Patterns in Robertson Wine Valley

The results of the survey show a high level of ICT use among farmers and their workers according to Table 5. The findings indicate 100% penetration rate for mobile phone use among both the farm owners and their workers. It is also important to observe that all the farmers interviewed reported using a computer, Internet, email, fax and printer. Despite their relatively low incomes, farm workers also reported using a computer (58%) and Internet (53%). Such a high use can be attributed to access via local Internet cafes. Other ICT that are being used by workers include phone fax, printer, email and photocopier. Table 6 provides a description of the various ICT applications that are being made along the supply chain. Clearly, the results show that computers, Internet, email, mobile phones,

Table 4. Demographics of study area for both commercial farmers and farm workers, 2009

	Category of respondents	Variable Categories	Number of respondents	Percentage of respondents
Gender	Farm employees	Male Female	29 31	48 52
	Farmers	Male Female	8 7	57 43
Age	Farm employees	30 and less 31- 40 41- 50	12 29 19	20 48 32
	Farmers	30 and less 31- 40 41- 50	1 12 2	7 80 13
Marital status	Farm employees	Married Single	43 17	72 28
	Farmers	Married Single	12 3	80 20
Race	Farm employees	Black Coloured White	22 20 18	37 33 30
	Farmers	Black Coloured White	0 0 15	0 0 100
Occupation on farm	Farm employees	Seasonal/vineyard Perma- nent /vineyard Administrative Winery	10 17 17 16	17 28 28 27
Experience on farm	Farm employees	5 and below 5- 10 years 10- 20 years 20 and above	29 24 6 1	48 40 10 2
	Farmers	5 and below 5- 10 years 10- 20 years 20 and above	2 3 8 2	13 20 53 14
Average salary per month	Farm employees	Less than R4 000 R4 000- R8 000 R8 000 and above	28 24 8	47 40 13
	Farmers	Less than R4 000 R4 000- R8 000 R8 000 and above	11 4 0	73 27 0
Permanent off-farm income	Farm employees	Yes No	9 51	15 85
	Farmers	Yes No	5 10	33 67
Education	Farm employees	Primary Matric College University	7 25 24 4	12 42 40 7
	Farmers	Primary Matric College University	0 3 8 4	0 20 53 27

continued on following page

Table 4. continued

	Category of respondents	Variable Categories	Number of respondents	Percentage of respondents
Type of agricultural qualification	Farmers	Diploma Degree	6 4	40 27
Household size	Farm employees	3 and below 4- 6 6 and above	24 26 10	40 43 17
	Farmers	3 and below 4- 6 6 and above	7 8 0	47 53 0

Source: Survey Data, 2007

and fax machines have been widely adopted along the wine production and marketing supply chain in the Robertson Wine Valley. It is also important to note that computerized irrigation equipment is widely being used in wine production in Robertson Wine Valley. Bar coding is being deployed in wine production especially at the packing stage of the supply chain. The wine sales utilize e-commerce applications in terms of website marketing and advertising of different wine brands that are available to both domestic and global customers. The ICT use patterns illustrate that both farmers and their employees in the Robertson Wine Valley have made significant strides in using ICT at all the different stages of the wine supply chain. It is also important to observe that the farmers and their workers are using multiple and specialized ICT to improve their daily farm decision making and to enhance the profitability of their enterprises.

Table 5. Different types of ICT being used by commercial farmers and their farm employees, 2009

Type of ICT	% of farm employees using ICT	% of farmers using ICT
Computer	58	100
Internet	53	100
Photocopier	48	87
Landline Phone	62	93
Cell phone	100	100
Two-way radio	13	33
PDA	5	7
Computerized irrigation	7	27
Bar code system	33	87
Email	47	100
Fax	50	100
Printer	52	100

Source: Survey Data, 2007

Table 6. Different areas of ICT Use, 2009

Area of ICT use	ICT used in each area
Input procurement	Computer, cell phone, internet, fax, email, printer, land line phone
Production	Bar code system, computer, cell phone, fax, printer, land line phone, computerized irrigation
Packing	Bar code system, computer, printer
Marketing	Bar code system, computer, cell phone, internet, e-commerce, fax, email, printer, land line phone
Sales	Bar code system, computer, cell phone, internet, e-commerce, fax, email, printer, land line phone

Source: Survey Data, 2007

Summary of Key Constraints of ICT Use in Robertson Wine Valley

The farmers and workers in Robertson Wine Valleys encountered various problems when using ICT in the wine production and marketing activities. Most of the respondents indicated that the cost of using ICT is very high, beyond the means of many farm employees. The cost of airtime and Internet cafe charges was identified as the main obstacles and cannot be afforded every time. Poor network and reception (i.e. low transmission signals) have been identified as the second major obstacle to ICT use in the region. Some farms are located in valleys and uneven landscapes where telephone and mobile reception is so poor. It causes communication problems. This situation worsened with the Eskom load shedding which started in 2006 and worsened in 2007. Due to the fact that most of the farm employees are not highly educated, some find it difficult to use English as a medium of communication. Therefore, the language itself is a barrier to the use of ICT. Most prefer their mother languages like Afrikaans and Xhosa to English. Some of the highlighted problems of ICT use are shown in Figure 5. The response percentages of the different problems encountered when using ICT are indicated accordingly.

Besides problems encountered when using ICT, there are other constraints that inhibit the acquisition and adoption of ICT by the farm employees (see Figure 6). Some of the reasons

Figure 5. Problems of ICT use in Robertson Wine Valley, 2009. Source: Survey data, 2007

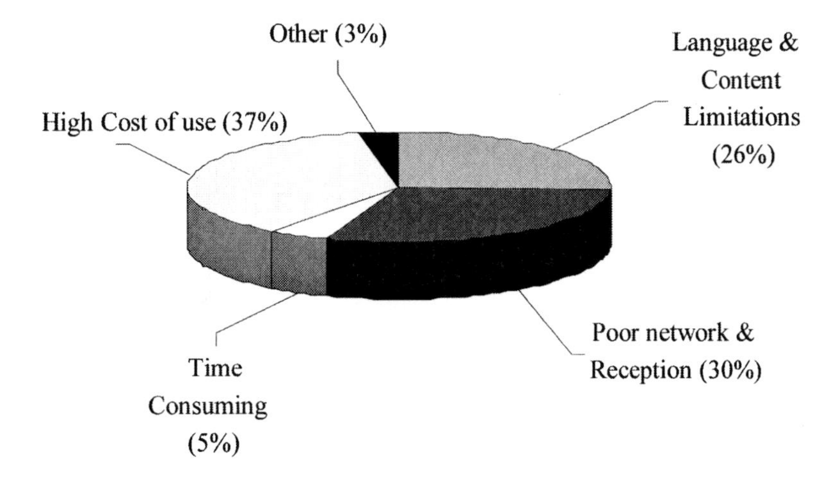

Figure 6. Key constraints inhibiting the acquisition and adoption of ICT. Source: Survey data, 2007

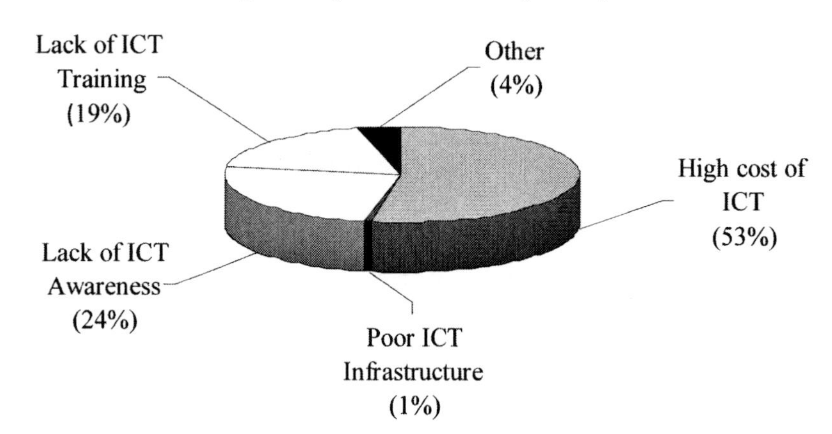

that were highlighted by the farmers and workers include high cost of ICT, lack of ICT awareness and lack of ICT training.

Perceptions on Strengths and Weaknesses of Using ICT in Wine Supply Chain

The farmers and workers expressed different perceptions and views on a number of strengths and weaknesses of using ICT. Table 7 highlights the identified strengths and weaknesses. Some of the key strengths that were identified can be summarized as follows:

- ICT assist in farm decision making on input procurement by increasing timeliness and cost efficiency.
- ICT assist in farm production decisions by improving timeliness of operations and saving labor costs.
- ICT help in improving packaging efficiency and it helps with product traceability.
- ICT lowers transaction costs, improves access to global markets and facilitates market information updates.
- ICT makes communication with suppliers, customers and other stakeholders faster and easier.

The farmers also pointed out some of the weakness or limitations with ICT use, and these included:

- Poor infrastructure to support the smooth transmission of telecommunication signals in the valley.
- High cost of some of the ICT such as ones used to support irrigation systems.
- Unreliable energy supply following the introduction in 2006 of load shedding by Eskom South Africa's electricity supply company.
- The loss of face to face contact with key marketing agents.
- The failure by key market participants and business partners to invest in ICT can dilute the potential benefits.

Additional Experiences of ICT use in South African Agriculture

Applications of ICT in Agriculture

ICT can be used in almost every step in the input procurement, production, distribution and marketing of agricultural produce. Information and the technologies that facilitate its use, exchange and reliability have been important aspects of agriculture and agriculture-related activities. Knowledge,

Table 7. Perceptions on strengths and weaknesses of ICT use along the wine supply chain, 2009

Stage	Strengths	Weaknesses
Input procurement	-Assist in decision making on how much quantities of inputs to buy -Orders arrive on time -Timing efficiency -Fast and saves time	-If supplier does not have same communication technology, it is impossible to use ICT - Affected by poor networks
Production	-Assist in decision making on how much inputs to apply -Using machinery reduces labour costs by more than R30 000 -Reduce production costs -Fast and saves time	-High costs of use -Expensive to purchase -Affected by power load shedding
Packing	-Makes packing faster and saves time -Ensures standards consistence -Enables traceability efficiency control	-Affected by power load shedding
Marketing	-Cheap and easy to use -Facilitates easy and fast communication with International customers and agents -Enables information update and dissemination -Reduces paperwork and transactions costs -Saves lots of paper	-E-commerce is still poor in South Africa -Affected by power problems -High costs of use -No face to face contact with customers and marketing agents
Sales	-Facilitates traceability of products -Timing efficiency -Makes things fast and cheap	
All stages	-All work is made easier and faster -Communication is made easier	-Power problems may affect ICT use

Source: Survey Data, 2007

communication and information exchange have influenced decisions on what to plant, when to plant it, how to cultivate and harvest and where to store and sell and at what price. Recent ICT have provided new opportunities to increase the timeliness and availability of critical information and to improve its quality and relevance. As compared to traditional methods, ICT add value when information is time sensitive and is customized to satisfy a client's needs (Winrock, 2003). ICT offer more cost-effective feedback methods required to empower previously marginalized agricultural communities. There is an enormous potential in the use of ICT to provide information, access resources, enhance learning and research sources. The use of ICT is a rapidly growing method of finding relevant, reliable, timely and accurate agricultural information, as well as well quality products and services.

According to Ortmann (2000), the use of ICT is becoming popular as more and more farmers are no longer interested in the traditional ways of keeping farm data and records. This is because ICT are relatively simple to use and satisfy the numerous recording and information requirements for many farmers. ICT are very useful for keeping financial and physical (crops, livestock) records for business planning purposes and for payroll preparation. They are also highly rated for providing up-to-date, more usable, easily accessible information and for saving time compared to traditional manual records. In projects, the integration of ICT can help to improve the (pace and) performance of agricultural development (Woodburn *et al*, 1994).

Major benefits have emanated from using ICT in diverse areas of the food and agricultural industry including precision agriculture, product traceability and identity preservation (e-supply chains), animal husbandry, computerized irrigation systems, digital financing for farmers (i.e., Internet and mobile banking support), and computerized farm record keeping systems among others. The ICT applications and their benefits are described briefly in the following section.

Product Traceability

In South Africa, the fruit farmers in the Western Cape who market their produce to Tesco Supermarkets in the United Kingdom are expected to employ ethical labor standards and traceability systems for their fruits to be accepted in Europe. Since 2005, all South African wine exported into the EU was required to comply with their regulation which states that all key players in the industry must have traceability systems in place for each product to facilitate its traceability. In addition, each supply chain participant was expected to identify any person or business dealt with one step forward and one step back along the supply chain (Matthee, 2004). In Fort Beaufort in the Eastern Cape Province, citrus farmers use computerized bar code system to track each farmer's produce in the supply chain back to the orchard where the fruit was picked.

Computerized Irrigation

A computerized system of irrigation in South Africa is currently in use in the mango and tomato farming region of Blydepoort in Mpumalanga Province. The farmers installed a computerized private irrigation network, which also uses Short Message System (SMS) to notify operators of water pressure if it is too high or low (Information for Development, 2005). A study by SAITIS (2002) found that the South African agriculture industry has been a major user of ICT more than any other sectors. Among the key ICT users in the agricultural sector is the deciduous fruit industry that deploys more specialized and sophisticated software to support its business. The industry uses ICT for computerized irrigation, global position sensing, remote sensing, farm support operations, field data collection, agricultural mapping and communication with the export and local market.

Agricultural Information Dissemination

A Computer-Mediated Communication Services (CMS) enabled by the Internet is currently being used to disseminate key market intelligence to agricultural stakeholders. E-mail system enables farmer service providers and intermediary agencies working with farmers to exchange ideas, transmit data, access and disseminate information. However, e-mail technology requires efficient power supply and a reliable and affordable telecommunications system (Munyua, 2000). The printing medium also plays a critical role for the farmers who can read, evidenced by the agricultural newspapers, posters, pamphlets and booklets that have been used successfully in the past.

Online Banking

Online banking in South Africa is the fastest growing service that banks are offering in order to gain and retain new customers (Singh, 2004). It provides increased convenience and transaction cost savings to many farmers. Farmers can pay accounts, check account balances, transfer funds, create orders and carry out transactions without leaving the farm premises, thereby saving time and money. With electronic banking, consumers can carry out fast and convenient financial transactions thus enjoying the benefits of convenience and cost savings (Lee, Lee & Schumann, 2002).

Mobile Banking for Disadvantaged and Under-Privileged Communities

In South Africa, banking service providers have targeted the "unbanked" people by using mobile or m-banking. In South Africa, a mobile banking firm called Wizzit has a target market estimated at 16 million "unbanked" clients. These clients comprise the disadvantaged and under-privileged communities which include groups such as farm workers, township residents, and emerging rural entrepreneurs (ICT Update, 2007). Wizzit's clients use mobile phones to make person-to-person payments, transfer money to and from a savings account, pay utility bills and buy prepaid airtime. Debit cards are also issued to help clients pay for purchases at any shop with a point-of-sale terminal, or withdraw cash from any automated teller machine in South Africa. A key competitor to Wizzit is MTN whose mobile banking services are directed at the same lucrative market comprising poor communities in South Africa.

Agri-Hubs

ICT use in South Africa has given rise to development of agri-hubs. One agri-hub was launched in Mokopane, a town in the Limpopo Province of South Africa. The agri-hub offers diversified agricultural and pastoral activities. The European Union (EU) along with the Limpopo government launched the agri-hub which is now benefiting more than 300 farmers. Same initiatives are being replicated in other agricultural communities for the benefit of emerging farmers (e-agriculture, 2008).

Role of ICT in Promoting Food and Agricultural Marketing Linkages

Market transactions are heavily dependent on information. Most remote rural areas suffer from paucity of market information due to poor infrastructure. The poor integration of remote villages to domestic and global markets negatively affects economic development in such areas (Lio & Liu, 2006). Therefore, ICT promotes greater economic benefits from easier networking, and faster and more effective information transfer. Rural agribusinesses in developing countries stand to benefit from ICT use through better access to global markets. Through direct marketing, remote rural agribusinesses are likely to derive tangible benefits from lower transaction costs and enhanced profits from elimination of the middlemen (World Bank, 2002).

One interesting example of the use of e-commerce in South African agriculture is that of women in Pietersburg who sell chickens over the Internet to a local market and wealthy communities in the area. They are now operating as wholesale suppliers to hotels, hospitals and the nearby town (CTA, 2002). ICT reduces transaction and marketing costs and combat risk before, during, and after carrying out the business transaction (Lucking-Reiley & Spulber, 2001). Before the transaction, ICT lowers the cost of searching for suppliers or buyers, making price and product comparisons relatively easier, as most of the information will be available on the Internet. During the transaction, e-commerce reduces the cost of communication with counterparts in other companies regarding transaction details. The computer-based transactions eliminate market search costs, need for physical space for product displays, meetings and processing paper work. After the transaction, ICT allows companies to lower costs of communication, monitor contractual performance and confirm delivery. By lowering the costs of transactions, ICT also lowers the costs of production and marketing, thus improving profitability and overall performance of agribusinesses. Both established and emerging agribusinesses have a lot to gain from ICT in terms of improved market access, reduction in transaction and marketing costs, and promotion of competitiveness in the global market. To remain in business in this globalised world, ICT have to be utilized to develop sustainable competitive advantage against numerous competitors.

CONCLUSION AND FUTURE PROSPECTS

The chapter highlights the fact that ICT has played a major role in South African agriculture, especially in the commercial agricultural sector. However, there is an inherent risk that some farmers or farm workers might be left out by this technological revolution. Currently, opportunities of the digital age are not equally accessible in South Africa, and the gap is even greater in some developing countries. In South Africa, there is a digital divide between the commercial and the small scale subsistence farmers. The wine farmers exhibited very high ICT penetration rates but the same cannot be said for their farm workers who tend to adopt less sophisticated and cheaper ICT such as mobile phones. Their use of the Internet is mainly supported by Internet cafes not ownership of computers at home.

In order for South Africa to expand its current use of ICT in agriculture and to increase the number of intended beneficiaries, it is imperative that a number of key challenges are addressed. These include but are not limited to the problem of poor ICT infrastructure in most rural areas, development of relevant e-content to support the country's 9 different official languages, improvements in ICT functional literacy among disadvantaged communities, and the development of a sound e-agriculture policy. Equitable access to information is one of the most vital principles in the emerging global information economy but this is lacking in the agriculture sector of South Africa. While it is probably impossible to achieve one-hundred percent equitable access in ICT use, realizing the need to address the current imbalance is quite critical especially if e-agricultural development is to be achieved at a national scale. The South African government could play a vital role by creating conditions and incentives that allow small-scale subsistence farmers, and the farm workers to be incorporated into this ever-growing information society. The government could help facilitate ICT training of small scale farmers, improvement of telecommunication infrastructure in poorly developed agricultural communities and provide tax-benefits to small scale farmers who purchase ICT for farm use. Such interventions must be coordinated and should be guided by a systematic e-agriculture development policy process.

In South Africa, the farm workforce is recruited from the poorest black and colored communities, where the majority of the people are illiterate, poor and uneducated. As a result, using ICT for specialized farm activities is a great challenge to most of the employees. Although few farmers are providing training and teaching farm workers to be literate, there is still need for more farmers to take part. From the results of the study, most farm employees need ICT training and awareness in order to develop the necessary skills to use more sophisticated technologies on the farms.

The problems of electricity supply that are currently affecting South Africa are likely to slacken the country's pace in developing new ICT applications in agriculture. However, the broad-based application of ICT in the wine supply chain in South Africa clearly demonstrated that the country stands to benefit if this e-agriculture momentum is not disturbed. Opportunities for ICT use have been reported among emerging rural entrepreneurs in South Africa as well. In summary, access to ICT is not only a question of connectivity, but also of capability to use the ICT and provision of relevant content in accessible and useful forms. In future, this entails the need for government to provide the necessary ICT facilities, technical support, and relevant content. In addition, the issue of optimum use of ICT requires critical consideration as network disruptions and poor telecommunication signals reported in the Robertson Wine Valley and similar locations elsewhere tend to create situations where farmers underutilize the agricultural potential of their ICT. The chapter demonstrates that numerous activities along the wine supply chain involve the use of ICT in some ways. In addition to the use of traditional ICT such as radio, landline phone, fax, and printers, more modern ICT have been developed and deployed in agriculture and these include lap-top computers, Internet, mobile phones, blackberries, and personal digital assistants among others.

Given South Africa's comparative advantage in wine production, it is important to note that some farmers are demanding an ICT-based world stock exchange system specifically for bulk wine, where daily trading activities are made available to the rest of the world. This will be a giant step in the dissemination of transparent information about wine production, grading, pricing, packaging, and marketing trends. This will also allow farmers to have more options in terms of better pricing for their bulk wine, and also additional options in terms of reliable markets to sell their wine. Finally, the evidence presented in this chapter underscores the importance of ICT applications in agriculture and its potential to transform the livelihoods of the marginalized communities, and boost the performance of agribusinesses in South Africa and the rest of the world.

REFERENCES

Batchelor, S., & Evangelista, S. Hearn, S., Peirce, M., Sugden, S. & Webb, M. (2003). *ICT for development contributing to the millennium development goals: Lessons learned from seventeen InfoDev Projects.* Washington, DC: World Bank.

Batte, M. T. (2005). Changing computer use in agriculture: Evidence from Ohio. *Computers and Electronics in Agriculture, 47*, 1–13. doi:10.1016/j.compag.2004.08.002

Bruwer, J. (2003). South African wine routes some perspectives on the wine tourism industry's structural dimensions and wine tourism product. *Tourism Management, 24*(4), 23–435. doi:10.1016/S0261-5177(02)00105-X

Cape Gateway. (2003). *Strategic plan for the Department of Agriculture, Western Cape*. Retrieved May 3, 2006, http://www.capegateway.gov.za/text/2003/strategic_plan_2003b.pdf

Cecchini, S., & Scott, C. (2003). Can information and communication technology applications contribute to poverty reduction? Lessons from rural India. *Information Technology for Development, 10*, 73–84. doi:10.1002/itdj.1590100203

CTA, (2002). *Gender and agriculture in the information society: A special report of a CTA meeting. Wageningen,* the Netherlands, September 11-13.

Daly, J. A. (2003). *Information and communications technology and the eradication of hunger.* Washington, DC: Development Gateway.

E-agriculture. (2008). *ICT-based Agri-hub in Mokopane to aid subsistence farmers*. Retreived February 25, 2009 from http://www.e-agriculture.org

Economic and Social Commission for Asia and the Pacific. (2008). Information and communications technology for food security and sustainable agriculture in the knowledge economy. In *Proceedings of Word summit on the information society*, November 19-21, 2008.

Elsenburg, (2007). Department of agriculture: Western Cape. Retreived June 28, 2007 from http://www.elsenburg.com/economics/ statistics/start.htm

Esselaar, S., Gillwald, A., & Stork, C. (2006). South African Telecommunications Sector Performance Review 2006. *LINK Centre Public Policy Research Paper No. 8.* Johannesburg: LINK.

Esterhuizen, D., & Van Rooyen, C. J. (2006). An enquiry into factors impacting on the competitiveness of the South African wine industry. *Agrekon, 45*(4), 467–485.

Flockemann, H. M. E. (1972). South Africa. In A. H. Gold (eds.), *Wines and spirits of the world.* Surrey, UK: Virtue & Company Limited.

Galloway, L., & Mochrie, R. (2005). The use of ICT in rural firms: A policy-orientated literature review. *Info*, 7(3), 33–46. doi:10.1108/14636690510596784

Gebremichael, M. D., & Jackson, J. W. (2006). Bridging the gap in Sub-Saharan Africa: A holistic look at information poverty and the region's digital divide. *Government Information Quarterly*, *23*, 267–280. doi:10.1016/j.giq.2006.02.011

Harris, R. W. (2004). *Information and communication technologies for poverty alleviation*. Kuala Lumpur, India: UNDP-APDIP.

ICTPORTAL. (2009). *Information and communication technology penetration in South Africa.* Retrieved March 27, 2009 from http://www.ictportal.org.za/news.php

Information for Development, (2005). Agriculture: Computerised system of irrigation in South Africa. *I4D, 3*(8), 23-26.

International Telecommunication Union (ITU). (2007). *Telecommunication/ ICT markets and trends in Africa*. Geneva: ITU.

James, T. (2001). *An information policy handbook for Southern Africa*. Ottawa, Canada: International Development Research Centre.

Kalusopa, T. (2006). The challenges of utilizing information communication technologies (ICTs) for the small scale farmers in Zambia. *Library Hi Tech*, *23*(3), 414–424. doi:10.1108/07378830510621810

Kapange, B. (2006). ICTs in agricultural development: The case of Tanzania. In *Proceedings of the IST-Africa 2006 Conference*, Pretoria, May 3-5, 2006.

Khaketla, M. (2004, October). *A national ICT policy for Lesotho: Discussion draft*. Retreived April 26, 2006 from http://www.lesotho.gov.ls/articles/ 2004/Lesotho_ICT_Policy_Draft.pdf

Lee, E., Lee, J., & Schumann, D. W. (2002). The influence of communication source and mode on consumer adoption of technological innovations. *The Journal of Consumer Affairs*, *36*(1), 1–27.

Lio, M., & Liu, M. C. (2006). ICT and agricultural productivity: evidence from cross-country data. *The Journal of the International Association of Agricultural Economics*, *34*(3), 221–228.

Lucking-Reiley, D., & Spulber, D. F. (2001). Business to Business Electronic Commerce. *The Journal of Economic Perspectives*, *15*(1), 55–68.

Marker, P., McNamara, K., & Wallace, L. (2002). *The significance of Information and Communication Technologies for Reducing Poverty*. London: DFID.

Matthee, A. (2004). *Traceability in the wine industry. Wynboer: A technical guide for wine producers*. Paarl, South Africa: Wineland.

McDonald, S., & Punt, C. (2001). *The Western Cape of South Africa: Export opportunities, productivity growth and agriculture*. A paper prepared for the 4th Global economic analysis conference, Global trade analysis project. Purdue University, West Lafayette, Indiana. June 2001.

McNamara, K. S. (2003). Information and communication technologies, poverty and development: Learning from experience. *A background paper for the InfoDev Annual Symposium*. Geneva, Switzerland. December 9-10, 2003.

Meeira, S. N., Jhamtani, A. & Rao, D. U. M. (2004). *Information and communication technology in agricultural development: A comparative analysis of 3 projects from India. AgREN*, Network paper 135.

Moloi, M. J. (2007). Commercialisation of emerging vegetable producers in the Western Cape province of South Africa. *Proceedings of the 17Th International Food and Agribusiness Management Association (IAMA) Conference*, Parma, Italy. 23-26 June 2007. Moodley, S. (20050. The promise of e-development? A critical assessment of the state ICT poverty reduction discourse in South Africa. *Perspectives on Global Development and Technology*, *4*(1), 1–26.

Munyua, H. (2000). Application of ICT in Africa's agricultural sector: A gender perspective. In Rathgeber, E. M., & Adera, E. O (eds). *Gender and the information Revolution in Africa*. Ottawa: IDRC.

Nikam, K., Ganesh, A. C., & Tamizhchelvan, M. (2004). The changing face of India. Part 1: bridging the digital divide. *Library Review*, *53*(4), 213–219. doi:10.1108/00242530410531839

Obayelu, A. E. & Ogunlade, I. (2006). Analysis of the uses of information and communication technology for gender empowerment and sustainable poverty alleviation in Nigeria. *International Journal of Education and Development using Information and Communication Technology*, *2*(3), 45-69.

Ortmann, G. F. (2000). Use of information technology in South African agriculture. *Agrekon*, *39*(1), 26–35.

Rao, N. H. (2006). A framework for implementing information and communication technologies in agricultural development in India. *Technological Forecasting and Social Change*, *74*, 491–518. doi:10.1016/j.techfore.2006.02.002

Rao, S. S. (2004). Role of ICTs in India's rural community information systems. *Info*, *6*(4), 261–269. doi:10.1108/14636690410555663

Richardson, D. (2005). How can agricultural extension best harness ICTs to improve rural livelihoods in developing countries? In E. Geld, & A. Offer, (eds), *ICT in agriculture: Perspectives of technological innovation*. Jerusalem: EFITA.

SAITIS. (2002). *International Scan for ICT Usage and Diffusion*. Pretoria, South Africa: Miller Esselaar.

Selamat, J., & Shamsudin, M. N. (2005). *Sustainability of the Malaysian agri-food sector: issues at hand and the way forward*. University of Putra, Malaysia.

Singh, A. M. (2004). Trends in South African Internet banking. *Aslib Proceedings*, *56*(3), 187–196. doi:10.1108/00012530410539368

Smith, L. E. D. (2004). Assessment of the contribution of irrigation to poverty reduction and sustainable livelihoods. *Water Resources Development*, *20*(2), 243–257. doi:10.1080/0790062042000206084

South Africa. (2005). *South Africa yearbook 2005/2006*. Pretoria, South Africa: GCIS.

Statistics South Africa. 2009. *Abstract of agricultural statistics*. Pretoria, South Africa: DoA.

Stienen, J., Bruinsma, W., & Neuman, F. (2007). *How ICT can make a difference in agricultural livelihoods. The commonwealth ministers book-2007*. Netherlands: International Institute for Communication and Development.

Tlabela, K., Roodt, J., Paterson, A., & Weir-Smith, G. (2007). *Mapping ICT access in South Africa*. Cape Town, South Africa: HSRC Press.

Tripp, R. (2001). Agricultural technology policies for rural development. *Development Policy Review*, *19*(4), 479–489. doi:10.1111/1467-7679.00146

Update, I. C. T. (2002). *Agricultural market information services, Issue 9*. Wageningen, The Netherlands: CTA.

Update, I. C. T. (2007). *Financial services, Issue 36*. Wageningen, The Netherlands: CTA.

Winrock, J. (2003). *Future directions in agriculture and Information and Communication Technologies (ICT) at USAID*. Arkansas: Winrock.

Woodburn, M. R., Ortmann, G. F., & Levin, J. B. (1994). Computer use and factors influencing computer adoption among commercial farmers in Natal Province, South Africa. *Computers and Electronics in Agriculture*, *11*, 183–194. doi:10.1016/0168-1699(94)90007-8

World Bank. (2002). *Information and communication technologies: A World Bank group strategy*. Washington, DC: World Bank.

World Bank. (2003). *ICT and MDGs: A World Bank group perspective*. Washington, DC: World Bank.

Zama, S., & Weir-Smith, G. (2006). *ICT Penetration in South Africa*. Pretoria, South Africa: Universal Service Agency.

Zimbabwe e-readiness survey report. (2005). *Information and communication technologies project Zim/03/003.* Retrieved August 21, 2007 http://www.ict.org.zw/Zim%20E-Readiness%20Survay%20Report.pdf

Chapter 3
Agriculture, Communication, and Socioeconomic Development:
A Proposal for e-Agriculture in Nigeria

Emmanuel C. Alozie
University of Southern Mississippi, USA

ABSTRACT

This study explores Nigeria's e-agriculture policies and those of other African governments. It also proposes what e-agricultural strategies these governments could adopt to enhance their agricultural output by examining the history of agriculture in Nigeria; the current status of Nigerian information and communication technologies and e-government policies with emphasis on agriculture. The analysis addresses the role of communications as instrument of national development. In view of the economic status of African countries, the study calls on Nigeria and other African governments to adopt a cautious approach as they embark on e-agriculture policies and acquisition of information and communication technologies to promote national development. The study calls on African governments to liberalize their agricultural policies, establish agricultural cooperatives, educate rural famers and offer telecommunication services in the rural areas if they hope to raise their agricultural productivity.

INTRODUCTION

With a population of about 140 million, Nigeria ranks as the tenth most populous nation in the world (*The Guardian Online*, 2001, Aug. 13). Barring wars, political disintegration, and social upheavals, Nigeria's population will surge to 204 million by 2025, thereby making it the world's seventh most populous nation. By 2050, the population will soar to 337 million to become the fourth most populated country in the world while India, China, and the United States will be the top three (Eribo, 2001). If so, a great challenge faces this diverse nation where national coherence remains fragile and where food scarcity and economic deprivation could lead to national upheaval. This situation possesses international implications because Nigeria is regarded as one of the two economic anchors in sub-Saharan Africa. But Nigeria may not reach

DOI: 10.4018/978-1-60566-820-8.ch003

her promise without returning to agriculture as its economic mainstay.

It will be a shameful thing and a very unfortunate scenario to see a prince of a well-known, highly influential, connected king with unquantifiable and stupendous wealth begging for food or crying of hunger despite his father's wealth and popularity. (Adeniyi, 2008a)

Foluso Okunmadewa, a Nigerian professor of agricultural economics and World Bank consultant, used this illustration to explain the precarious situation facing Nigerian citizens who reside in a nation blessed with abundant agricultural, natural, mineral and human resources, but has found it increasingly difficult to feed her people (Obi, 2008; BBC News, 2007, July 12). Okunmadewa offered this perspective while addressing the press after delivering a paper entitled "Food prices crisis and the challenge of sustainable development in Nigeria" at the 40th Interdisciplinary Research Discourse by the Postgraduate School of the University of Ibadan, Oyo State, Nigeria.

Critics contend that despite the nation's massive wealth—especially the wealth generated from oil over the past 50 years—about one in three of Nigeria's 140 million people goes to bed hungry every night. To promote awareness about widespread hunger in Africa's most populous nation, Action Aid, a Nigerian non-governmental pressure group, organized a march in Abuja, the national capital. During the march, Action Aid described the plight of Nigerians who suffer from hunger as "an indictment of those who have more than enough to eat in a country with more than enough resources and potential to banish the hunger it breeds," (BBC News, 2007, July 12). Like other marches to raise awareness on social issues, the police dispersed the march before the protesters reached their goal of presenting petitions to President Umaru Yar'Adua and parliamentarians to encourage policies to end hunger in the country, which ranks fifth in the world hunger index (Punch On the Web, 2008, Oct. 8) despite her rich natural resources (Obi, 2008; BBC News, 2007, July 12)

Nigeria has earned about $340 billion from oil revenue in forty years (IRIN, 2002, July 4), but she has been ranked as the 13th poorest country in world since the 1990s after a brief spur in economic vitality in the 1970s (Over, 2001). The average citizen survives on $1 a day, however the country boasts a large number of millionaires whose wealth it has been alleged was derived from corrupt practices (BBC News Online, 2002, Jan. 16; Alozie, 2005). The wealth derived from oil has been blamed for Nigeria's economic stagnation and the prevalence of corruption (BBC News Online, 2002, Jan. 16).

Federal, state, and local governments have neglected almost every other aspect of social economic development including industrial, human, telecommunications, and agricultural. Prior to the discovery of oil, the Nigerian economy was agrarian. At the time, about 70% of Nigerians were engaged in agriculture; the sector provided more than 75% of its income and the country could feed itself and export agricultural products (Alozie, 2005). Presently, oil generates more than 85% of the nation's earning (BBC News Online, 2002, Jan. 16; Alozie, 2005), and it has become increasingly dependent on food imports (Gumm, 2008) at an annual cost of $3 billion (*Business Day Nigeria*/Africa News Network, 2007, Nov. 22). While agriculture presently contributes 40% against oil's 13% to the Gross Domestic Product (GDP), it also employs about two-thirds of the nation's total labor force (Mohammed, 2008).

If Nigeria is to produce enough food in the future to feed her growing population and overcome her dependence on oil revenue, economic stagnation, high youth unemployment, food importation and other vexing socioeconomic problems, her economy must be diversified. In past 50 years since oil became its economic mainstay, successive Nigerian governments have been advised to use the revenues derived from oil to expand its economy in order to ameliorate the sufferings of her people and avoid a future political and socio-economic upheaval. A key area of diversification would be agricultural innovation. Heeding this advice calls for the adoption of e-agriculture in view of the growing role that information and telecommunication technologies are playing in the global economy and their contributions to agricultural innovations and output.

Thus, this chapter is aimed at exploring and proposing a guide regarding how Nigeria and other African countries could enhance their e-agriculture policies to assist with agricultural output. To examine this strategic proposal, the study will explore the following: (a) history of agriculture and agricultural research in Nigeria; (b) the role of communications as an instrument of national development by drawing from the theoretical frameworks of development communications and communication development; (c) Nigerian information and communication technologies (ICTs) policies and status; (d) Nigerian government e-policy with emphasis on agriculture; and (e) proposals for enhancing Nigerian e-agricultural and food production policies.

History of Agriculture and Agricultural Research in Nigeria

Like most African societies, evolution of Nigerian agriculture has three distinct eras: pre-colonial, colonial and post colonial/independence (Nigeria's Agriculture Sector, 2007, May 21). The pre-colonial era started and lasted for thousands of years when man settled in a place to engage small-scale sustenance farming and hunting instead of moving seasonally from place to place foraging for food. Agriculture remained the mainstay of most traditional communities for thousands of years until the arrival of Europeans and the colonization of Nigeria in the mid-1800s, which marked the beginning of the colonial era, 1860–1960. Organized agriculture, research and extension services began in the colonial era. The colonial administration's agricultural policies remained informal, while research concentrated on export crops—palm oil, cotton, cocoa, groundnut and rubber—which benefited the British economy at the expense of staple food crops—rice, maize, taro, yams, cassava, sorghum and millet (Idacha, 1980). During the colonial and early post-independence period, Nigeria produced enough staple food crops, livestock and fish to feed itself (Ogen, 2007; Babatunde & Oyatoye, 2005).

After 1960, the federal government initiated the First National Development Plan, 1962-1968, in which agriculture represented a key component of the plan. In addition to producing enough food to feed itself and cash crops for export, the plan's agricultural objectives included introduction of modernized agricultural methods through farm settlements, co-operative (nucleus) plantations, supply of improved farm implements (e.g. hydraulic hand presses for oil palm processing), and an expanded agricultural extension service. Due to the foreign exchange generated from export crops, the first, second, 1970–1974, and subsequent plans paid extensive attention to export instead of production and yield performance of staple food crops (Idacha, 1980; Nigeria's Agriculture Sector, 2007, May 21). This is often regarded as one of the biggest errors in Nigerian agricultural history. It marked the genesis of the lack of attention to problems that have plagued Nigerian agriculture.

At independence, the agricultural sector contributed over 60% of the GDP in the 1960s. Despite the reliance of Nigerian peasant farmers on traditional tools and indigenous farming methods, they produced 70% of Nigeria's exports and 95% of its food needs (Ogen, 2007). Agriculture suffered-intense neglect as the country moved from an agricultural-based to mono-cultural petro-based economy in the 1970s. As the shift occurred, the contribution of agriculture to GDP declined progressively. Estimates of the agricultural sector's contribution ranged from less than 5% to about 30% (Ogen, 2007; Babatunde & Oyatoye,

2005). At present, it is estimated that agriculture contributes more than 30% of the total annual GDP, employs about 70% of the labor force, accounts for over 70% of the non-oil exports and, perhaps most important, provided over 80% of the food needs of the country (Babatunde & Oyatoye, 2005).

It should be noted that Nigerian agricultural development and production have not kept pace with population growth. Although the country has embarked on a number of agricultural projects including Operation Feed the Nation, OFN (1976); River Basin Development Authorities, RBDA (1977); Directorate of Food Road and Rural Infrastructure, DFRRI (1986); National Agricultural Land Development Authority, NALDA (1992); National Fadama Development Project, NFDP (1992); Nigerian Agricultural Cooperatives and Rural Development Bank, NACRDB (2000); National Agricultural Development Fund, NADF (2002); Commodity Marketing and Development Companies, CMDC (2003), most of these and other agricultural development programs have failed woefully.

Okuneye (2002) attributed these failures to a variety of factors including, but not limited to, the following factors: (a) infrastructural facilities (poor road and networks, lack of appropriate on-farm and off-farm storage facilities, poor rural electrification, inadequate irrigation facilities); (b) manpower/skill development (high illiteracy among farmers, poor training extension works); (c) socio-cultural factors (land tenure system, discrimination against women); economic factors (unavailability and high cost of inputs, increased demand for higher wages in rural areas); government/regulatory policies (Land Use Act, importation tariff, unprotective policies, and lack of support for research, corruption in government); and environmental factors (high incidence of pests and diseases; erosion, drought, pollution, desert encroachment).

However, Okuneye (2002) pointed out that whenever trade liberalization is seriously implemented, it stimulates the growth of the export of Nigeria's agricultural commodities, but takes a downturn when it is neglected and relaxed. He reported that the advent of democracy leads to massive imports of luxury goods which began to dominate the economic scene. He noted that since 1986, when the Structural Adjustment Program (SAP) came into existence, Nigeria witnessed a gross neglect of the food production sector, thus bringing about a significant increase in food prices. The rise forced policy makers to resort to food imports to deal with shortages due to poor response and mechanism to enhance food production locally.

These failures have made Nigeria a net food importer, since she is unable to feed herself. She spends about $3 billion annually on basic food imports (*Business Day Nigeria*/Africa News Network, 2007, Nov. 22), while the average family spends about 73% of its budget on food (*The New York Times*, 2008, April 10) leaving about 65% of all Nigerians food insecure (*Nigerian Tribune*, 2008, April 29). Okuneye (2002) reported that Nigerian food imports continue to rise in value from N3.47 billion in 1990 to N113.63 billion in 2000.[1] Babatunde & Oyatoye (2005) reported that in 1989, Nigeria's food import bill was about N2.1 billion (about 0.8% of the total GDP) while it stood at about N226 billion in 2003 (about 3.1% of the total GDP. Since the 1990s Nigeria has spent about $60 million annually on rice importation only (Ogen, 2007). When a ban was placed on rice importation, food shortage ensued and groups rose in protest as its price and prices of other staple food products rose (BBC News, 2007, July 12).

Despite these challenges, it should be noted that Nigeria is blessed with vast agricultural space for cash and food crops and can achieve self-sufficiency and security in the areas of food production, storage and distribution (Mohammed, 2008; Nigeria's Agriculture Sector, 2007, May 21; Idacha, 1980). Nigeria has favorable climate conditions to engage in agriculture year round. While the rainy season (March through

November), offers abundant rainfall sufficient for planting and harvesting for farmers in southern Nigeria, the dry season (December to April) offers northern farmers the ability to farm by irrigation (Mohammed, 2008). Nigeria has a land area of 92.4 million hectares from the mangroves of the Niger Delta, through the rain forests to the savannah plains. Of this area, 91 million hectares is adjudged to be suitable for cultivation. Half of this cultivable land is effectively under permanent and arable crops, while the rest is covered by forest woodland, permanent pasture and built up areas (Nigeria's Agriculture Sector, 2007, May 21).

The food problem is not peculiar to Nigeria. It is estimated that some 850–967 million people around the world suffer from hunger and malnutrition (eJournalUSA, 2007, September); with 10,000–24,000 of them dying of starvation each week in Africa, Latin America and other parts of the world (*The New York Times*, 2008, April 10; Oxfam, n.d.). About 2.7 billion people live on less than $1–2 a day. They spend almost 80 per cent of their income on food. The rising cost of basic foods has grown as much as 300% pushing families in most of the Third World to the brink of starvation (Oxfam, n.d.). Inadequate food supplies have led to protests and riots have taken place in several developing countries (James, 2008). Governments in the developing world are spending billions and billions of their incomes on food importation—some up to a quarter of their annual budgets (*The Economist*, 2008, April 17; *The New York Times*, 2008, April 10; Oxfam, n.d.). The World Food Program (WFP) asked for $700 million in additional funds in 2008 to ameliorate the crisis. The United States has donated $120 million and other western nations are also offering assistance. Critics contend that donations from developed nations remain inadequate considering their policies, influence, and adoption of the conduct of international agricultural trade which they advocated contribute to the problem (*The Guardian*, 2008, Dec. 11; *The Economist*, 2008, April 17).

Decrying western reaction to the current situation as unacceptable, Jacques Diouf, head of the Food and Agriculture Organization (FAO), urged the developed nations to invest $30 billion annually in agriculture in developing countries. He said that "thirty billion dollars is nothing compared to subsidies and support in OECD (Organization for Economic Co-operation and Development) countries . . . nothing compared to the billions of dollars being spent in all developing countries to face the ongoing global financial crisis" (*The Guardian*, 2008, Dec. 11). He pointed out that there are about 963 million undernourished people in the world; with another 75 million pushed into chronic hunger since 2005, while the number of the hungry in Africa and Asia in 2008 is expected to reach 40 million due to rising food prices. Diouf contends that the Millennium Development Goal (MDG) of many countries to curb hunger by half by 2015 is getting increasingly unattainable (*The Guardian*, 2008, Dec. 11). The rise in food prices has been attributed to growing bio-fuel usage, lower food reserves, growing consumer demand in emerging economies, oil price rises, and changes to the global economy. Agricultural subsidies in developed nations, and decreasing interest rates as investors shift from financial to investing in food commodities which instigated an increase in demand are also blamed for the rise in food prices (Corcoran, 2008; Guardian.co.uk, 2008, May 10).

Theory and Role of Communications as an Instrument for National Development

Information and communication technologies (ICTs) have made the world a smaller sphere, thus creating and ushering in globalization and the information age (Robertson, 1992; Castells, 2004). Globalization refers to the transformation of the world's economy into an interconnected and information-and-technologically-based one (Ernst & Lundvall, 1997; Hamelink, 1997; Pendakur &

Harris, 2002; D'Costa, 2006; Ayers, 2005). The promises of a globalization include the transfer and diffusion of ideas, knowledge, technological know-how, instant financial transactions, worldwide availability of goods and services, reduced tariffs, increased foreign investments, increased productivity and managerial efficiency (Ernst & Lundvall, 1997; Hamelink, 1997; Pendakur & Harris, 2002). Reliance on technology to relay and store information has reduced paperwork, thus reducing costs for business and helping to save the environment (Minasyan, 2006). Despite these promises, the transition toward a global economy based on information and technologies innovation has produced some challenges. These challenges included dependence of developing nations on developed nations, unbridled world economic restructuring, outsourcing, spamming, employment displacement, growing influence of western capitalism, homogenization of world cultures, explosion of crooked business practices and financial transactions (D'Costa, 2006; Ayers, 2005; Pendakur & Harris, 2002).

With regard to e-agriculture, it has been argued that globalization offers promises and challenges for farmers in developing and developed nations. Globalization will enable farmers in developed and developing nations to access and exchange information about new farm techniques and other relevant facts that will help improve production. Information about the availability of equipment, machinery and crops needed for large and small-scale farming will be available. Prices for such materials will be reduced since they are being marketed worldwide. The income of farmers will rise because they are able to market their products globally. Despite these promises, it has been argued that farmers in the developing world will not be able to compete with their counterparts in developed nations because they lack their financial resources, infrastructure and government backing.

Following the end of World War II and emancipation of societies in Africa, Asia and South America from colonialism, governments in developing nations have viewed ICTs as vehicles for modernization because western governments and international supra-nation organizations instilled in them that they helped Western Europe to overcome the ravages of the war. As a result, most of the newly independent nations embarked on two complementary communication-oriented theories: communication development and development communication, with the aim of using them to implement the concept of development communication. Communication development refers to both the acquisition, and also installation of communication facilities such as radio, television, telecommunication, and print outlets (Alozie, 2005; Moemeka, 1994). Upon independence, many Third World countries embarked on expensive projects to acquire communication facilities. Having been left with little infrastructure, financial and human resources, the development of communication facilities were sporadic. Most developing nations are plagued with a lack of infrastructure due to the high cost of acquisition. However, the development of wireless communication such as satellites, cell-phones, and the information super-highway is helping to reduce the cost of acquisition. Since wireless communications do not need a great deal of land infrastructure, Third World residents are increasingly accessing and using them.

Development communication refers to the practice of employing the principles of strategic communication, propaganda techniques and social marketing through the mass media to promote socioeconomic development and modernization habits. Development communication uses mass media artifacts such as drama, news and advertising to educate, instill occupational skills and build national oneness in an effort to help a society evolve from a lower to a higher stage of development (Jallow, 2004; Singhal & Domatob, 2004). Third World governments welcomed these concepts enthusiastically, but half a century after it was introduced, development communication failed to fulfill expectation. The failure was blamed on wholesale importation of western-oriented messages at the expense of using local values and traditional media. The lack of success could be also be attributed to failure to realize that communication facilities in most developing nations are not as sophisticated as those in the developed nations also contributed.. In recent years, the concept has been evolving by employing traditional media, relying on locally produced media artifacts and encouraging those being targeted to get involved (Patterson, n.d.). The principles of these complementary concepts can be found in most e-agricultural models and policies. While it is essential to incorporate these models, it must be done with caution; earlier failures are to be avoided in this era of globalization.

Nigerian Information and Communication Technologies Policies and Status

Like most developing countries, Nigeria and other African countries have long recognized the need to employ ICTs in poverty eradication, provision of healthcare, good governance, better education, science, promotion of agriculture, tourism, and preservation of environment and culture if she is to promote the nation's socioeconomic development in order to raise the welfare of her masses (Tamukong, 2007; Dada, 2007). The ITU Telecom Africa (2004) described the approaches African countries have undertaken to develop their ICTs in the last fifteen years of pioneering. Those efforts have been described as innovative and have been bearing fruit.

Agriculture represents a sector where the Nigerian government is increasingly carrying out projects to integrate with ICTs in order to raise productivity, thus enhancing the economic welfare of the more than 70% of her population employed in the sector, who largely reside in the rural areas where telecommunication facilities are acutely lacking. Considering the failures of successive administrations to meet the goals of past socioeconomic projects, critics question whether Nigeria's current efforts to integrate agriculture and ICTs could be attained. This is in view of the huge challenges such programs face, including inadequate infrastructure, insufficient human resources, inadequate access, finance, and education as well as increased global competition.

Dada (2007) reported that Nigeria's telecom penetration has improved from 400,000 lines in 1996 to 4.7 million in March 2004—making her one of the fast growing markets in Africa. Her teledensity rose dramatically from 0.4% in 1996 to 3.92% in 2004. She places 31st with an index of 0.15 on Digital Opportunity Index. With regard to ICT diffusion ranking, Nigeria ranked 161st. Dada pointed out that mobile telephony holds a lot of promise for increasing access for marginalized sectors of the population. Cell phone growth has grown exponentially with all 36 states having some degree of coverage. In 2004, Nigeria had about three million subscribers, while recent reports indicate the number has risen to 54 million active subscribers (Vanguard, 2008, Sept. 2). Comparing the status of ICTs in Nigeria with countries of comparable population and economic status in developing countries worldwide, Dada opined that the country has a long way to go.

Nigeria hopes to overcome these aforementioned challenges and achieve her vision of harnessing the promises that ICTs offer through liberalization reform policies that would open up the telecommunication sector to competition from private concerns within and outside the countries and by providing adequate infrastructural development. Since independence, the government has dominated telecommunication. A commission set up by the federal government to study and develop telecommunication policies to address these challenges identified the following specific objectives which the federal government has adopted (National Telecommunication Policy, 2000). The short objectives include:

(i) To implement network development projects which shall ensure that the country meets and exceeds the ITU recommended minimum teledensity of 1 telephone to 100 inhabitants. This means that a minimum of 2 million fixed lines and 1,200,000 mobile lines should be provided within 2 years;
(ii) To promote widespread access to advanced communications technologies and services, in particular the Internet and related capabilities;
(iii) To develop and enhance indigenous capacity in telecommunications technology;
(iv) To participate effectively in international telecommunications activities in order to promote telecommunications development in Nigeria, meet the country's international obligations and derive maximum benefit from international cooperation in these areas;
(v) To establish a National Frequency Management Council (NFMC) within the Ministry of Communications with the responsibility for the planning, coordination and bulk allocation of the radio spectrum in the interest of efficiency, transparency and accountability;
(vi) To ensure that the government divests its interest in the state-owned telecommunications entities;
(vii) To promote competition to meet growing demand through the full liberalization of the telecommunications market;
(viii) To review and update telecommunications laws in order to bring all telecommunications operators under the regulatory control of NCC;
(ix) To resolve with dispatch all licensing problems that are existing in the most equitable and transparent manner. (National Telecommunication Policy, 2000).

The medium term objectives include:

(i) To provide a new regulatory environment that is sufficiently flexible to take into account new technological development and the international trend towards convergence;
(ii) To meet telecommunications service needs of the social, commercial and industrial sectors of the economy;
(iii) To ensure that public telecommunications facilities are accessible to all communities in the country;
(iv) To encourage domestic production of telecommunications equipment in Nigeria; and development of related software and services;
(v) To establish and meet aggressive targets for the installation of new fixed and mobile lines;
(vi) To protect the integrity, defence and the security of the state and its citizens.
(vii) To encourage Nigerian telecommunications operating companies to become global leaders in the industry;
(viii) To encourage the development of an information super-highway that will enable Nigerians to enjoy the benefits of globalization and convergence;
(ix) To create the enabling environment, including the provision of incentives, that will attract investors and resources to achieve the objectives earlier stated.

To attain these set objectives, the federal government has identified a number of structural components charged with the special responsibilities to implement them. These include:

(a) Government: Provide directions, enact and establish consistent policies for telecommunication development;

(b) Ministry of Communications: Implement government policies and ensure universal access;
(c) Nigerian Communications Commission (and other related agencies): To serve as independent regulator of telecommunication and issue license;
(d) Telecommunications service providers: Private companies involved in the provision of services to the public. (National Telecommunication Policy, 2000).

Since the adoption of these policies in 2000, Dada (2007) has identified several initiatives that were ongoing or initiated since then. He noted that most of these projects involve government-private partnership and they have enjoyed the support of pan-African bodies like the UN Economic Commission for Africa (UNECA), with its National Information and Communication Infrastructure (NICI) process, and the New Partnership for Africa's Development (NEPAD), with its eSchools Initiative. Supra national organizations like the World Bank, UNESCO, International Telecommunication Union (ITU) as well as international and national non-governmental organizations (NGO) have offered strategic assistance with regard to funding and training (International Finance Corporation/World Bank, 2008). However, Dada (2007) was troubled by what he called "the near absence of the voice of Nigerian civil society in the nation's policy development processes has deprived the nation of much-needed robust consultation and discussions." Some of the initiatives Dada (2007) identified include:

Universal Service Provision Fund (USPF): All licensed telecoms providers are required to contribute 2.5% of their annual financial turnover to the Fund to be used to link up all the country's states with fiber optic cable, and to provide wireless broadband services in Nigerian cities.

Broadband infrastructure: Aimed at developing broadband backbone infrastructure **to** overcome the major constraints to the growth of rural telephony and internet connectivity and to improve access to the 774 local governments in the country. However, there are as yet no installations in place.

Computers for All Nigerians Initiative (CANI): Aimed at raising Nigerians' access to computer hardware loans at low rate.

Universities Bandwidth Consortium: A pilot program that allows six of the nation's universities to bulk purchase bandwidth for academic purposes. If successful, the scheme may be expanded to the nation's over 600 higher education facilities.

National Rural Telephony Project (NRTP): Aimed at providing 500,000 connected lines to 343 local governments in Nigeria within one year. However, the project was only started in August 2004 and has been implemented in 108 of 218 targeted local government headquarters in Nigeria.

Internet Exchange Points (IXPs). Established internet exchange points to help keep local internet traffic within the country, which reduces the need to use international bandwidth and thus significantly lowers costs.

Tele-centre Network of Nigeria (TNN): Aimed at providing one tele-center in each of the country's 774 local government areas.

With regard to mass communications, when Chief Chukwuemeka Chikelu, Minister of Information and National Orientation, set up a committee consisting of government and private sector representatives to review Nigerian current mass media policy that was developed during the military era that lasted for more than two decades, he told the committee the review is needed "because of the dynamism in global information management and the need to evolve an enduring National Mass Communication Policy that befits the country's status as a virile international player and an emergent regional power" (Afrol News, 2008, June 7). The Minister stated that he hoped the committee's work would assist Nigeria's communication sector "towards becoming productive, citizen-friendly and technology-driven"

(Afrol News, 2008, June 7). Chikelu reminded the commission of the need to review and orient the nation's mass media policy when he stated that: Mass communication policy has a decisive influence on how the Nigerian media sector is organized. (Afrol News, 2008, June 7).

The minister added that national regulatory bodies, such as the National Broadcasting Commission (NBC), are required to implement the new policy which calls for: (a) operational boundary for the entire media industry and provides guidelines for the government's information workers and its own media; (b) for the promotion of national ideology and consolidation of the nation-building process through the development of an authentic Nigerian culture; (c) for the mass media to serve as a vehicle for national mobilization in support of policies and programs aimed at improving the standard of living of Nigerians as well as and champion and enhance the positive aspect of our national values, image, corporate unity and stability; (d) provides for regulation of media ownership, but opens the media industry for both government and private ownership by outlining the need for legislation hindering a monopoly situation. A key goal is to prevent the abuse of the print media by a few rich or powerful individuals or group of Nigerians (afrol News, 2008, June 7).

However, the failure of the Nigerian government to pass the Freedom of Information (FOI) bill, which has been lingering for years in the national assembly, calls into question the intentions of the government. Critics contend the government is trying to control the mass media instead of allowing them to play their watchdog role of monitoring the society in order to promote democracy and socioeconomic development (Dada, 2007).

Nigerian Government e-Policy with Emphasis on Agriculture

In 2004 Nigeria took a giant step toward the application of information and communication in socioeconomic development and governance when she launched e-government. Speaking at the launching, former President Olusegun Obasanjo, whose administration initiated the program, defined e-government as:

The use of information and communication technologies to improve efficiency, effectiveness, transparency and accountability of governance through implementing data warehousing and an integrated decision support system to manage modern economy for the benefit of the governed.

He stated that the aim of adopting e-government is to "to inject ICT to drive economic transformation in all the sectors, from primary education, or primary health care to agriculture, solid minerals development, tourism, virtual libraries, due process and contract monitoring" (Obasanjo, 2004). Describing the benefits of e-government, the former president explained that: Implementing e-government offers an opportunity for improving the way government businesses are carried out, how services are delivered and how avenues are created for the citizens to interact with government. It draws from the openness of the ICT world and impacts on government activities and operations breaking barriers of hierarchical traditions, secrecy, and bureaucracies that are naturally not compliant with 21st century aspirations and practice. It is an umbrella for institutionalizing due diligence, due process and hence transparency.

Apart from being the vehicle for developing and strengthening the National Information and Communication Infrastructure and its Architecture, e-government permeates seamlessly through activities involving Government-to-Government, Government-to-Business and Government-to-Citizens. This is ensured through e-Tourism, e-Judiciary, e-Agriculture, e-Taxation, e-Revenue, e-Pricing, e-Procurement, e-Learning etc; all of which together make government more compact, effective and efficient. I should not conclude this address without mentioning the revenue generation aspect of e-government programs. It is however, not a main product of e-government. It is a side-effect. (Obasanjo, 2004). Obasanjo explained that enacting the National Telecommunications Act, establishing an independent regulatory body, launching of a data and research satellite in 2003, as well as the communications satellite in 2007, represents his government's measures to integrate ICTs applications to agriculture, health, education, military, and to other economic sectors. Other measures taken in this respect consist of setting up the National Information Technology Development Agency, NITDA, an agency which is "to implement related national policy as well as making it the custodian of the nation's Top Level Domain (.ng); actively promoting cyber-specific laws to ensure security in the use of email and other operations originating from electronic or Internet related facilities, Cyber cafés, ISPs or personal mobile or fixed telephones etc" (Obasanjo, 2004).

It should be noted that the Nigerian communication satellite (NigComSat-1) launched in May 2007 malfunctioned in November 2008. It was shut down to prevent it from spinning out of control and damaging others in the orbit. The $340 million satellite was built to provide communications for government agencies and broadband internet (BBC News, 2008, Nov. 13).

Obasanjo hinted that all tiers of government were urged to join this campaign. States have encouraged the State Information Technology Development Agency (SITDA) by developing programs that bind the State assemblies, local government councils, judiciaries and the citizens. Local governments were being particularly enjoined to develop and implement on-line and off-line computer-assisted solutions to render basic and people-centered operations services. They have been encouraged to set up Geographical Information System (GIS) based applications to deliver primary education, healthcare delivery and two-way citizens' interaction with Government (Obasanjo, 2004).

Private sector investment and participation represents a key factor in the evolution and implementation of Nigeria's e-government program. Underscoring the importance of private sector involvement, Obasanjo stated:

This administration will actively encourage Public-Private-Partnership for the development of ICT and related applications, customization, commissioning and maintenance of software, hardware and networks as may be necessary. In this regard, we have set the pace by consigning development and implementation of e-government programs to a special purpose vehicle. This is a Joint Venture between Government, technical experts and financial partners wherein the Government has only nominal shares and the bulk of the equity, loans and other costs are generated through private sector sources put into practice through the National e-Government Strategies Limited which is registered under the Company's Act. (Obasanjo, 2004)

Obasanjo's successor, Yar'Adua and his administration, has largely adopted the e-government policies the former administration set up. It is important to note that the former and present administration identified e-agriculture as a vital component of the government's e-program. Adegite (2006) defined e-agriculture as the application of "agricultural services and information delivered or enhanced through the internet and related technologies. E-agriculture, therefore, involves the conceptualization, design, development, evaluation and application of innovative ways to utilize existing or emerging information and communications technologies (ICTs)." He identified mobile phones, digital personal assistants (PDAs), smart cards, CD-ROM, Geographic Information Systems (GPS), radio, Radio-Frequency Identification Devices (RFID), imaging and acoustic technologies, websites and weblogs, and email-based information services as forms of ICTs required for implementing e-agriculture.

According to Adegite (2006) areas where information and communication technologies can be applied in the implementation of e-agriculture include mapping natural resources, using participatory approaches to empower local communities to manage their own resources; creating business opportunities by providing agricultural market information for farmers and traders; speeding up application procedures in agricultural credit programs; protecting natural resources, such as fish stocks and forest resources from illegal poachers and loggers; forecasting weather conditions and pest outbreaks; making information more easily available; and enabling communication and knowledge exchange in online communities. Presently Nigeria has embarked on a number of e-agriculture programs. These programs include:

Agrovision: An Information and Communication Technology driven program that provides farmers with the agro-ecological mappings of the crops and soil types in Nigeria. In every farm, there will be an automatic weather station that captures the relevant meteorological data such as the intensity of rainfall, soil moisture, radiation, wind direction, atmospheric pressure, temperature and other agro-meteorological data. These data will be collated and then transmitted to a central collation facility in the National Weather Centre. There is a super data base and server that will relate the captured data from the farm and use it for computation at the central location of Agrovision at the National Weather Centre. All these are done online—'real time.' Agrovision was initiated by Teledom Group, a leading indigenous ICT solutions provider, along with NITDA and the National President of Association of Telecommunication Companies of Nigeria (ATCON). (i4d, 2008, February)

Nigeria Stored Products Research Institute (NSPRI): Stores food products such as cowpeas in the inert atmosphere silos by utilizing modern technologies. It was originally launched in 1981; the primary goal is guaranteed stored food safety and prevention of rampant food poisoning and to provide protection from pesticides-toxicity which many stored grains are exposed to, thus allowing year-round storage. (Adeniyi, 2008b)

Market Information Systems (MIS)/Regional Network of Information Systems of Market and Agricultural Trade in West Africa (MISTOWA): Aims to increase regional agricultural trade and food security by improving and linking the existing regional efforts to generate, disseminate, and make commercial use of market information. Assists to address other constraints, so that strong and dynamic commodity chains emerge that will use the information to enhance production, handling, credit, and trade; and value added services such as post-harvest, processing, packaging, and quality control. (Tadégnon, 2007)

Tradenet: An ECOWAS region-based program offers strong integration of data and mobile technology marketing information on agricultural products (Tadégnon, 2007; Tradenethome, n.d.).

Nigeria's Agricultural Extension Service [Information Services]: Relies on a myriad of approaches ranging from the pre-colonial export commodity approach to the present variants of the Training and Visit (T&V) System, to reach its diverse clients [farmers] with their various needs. The application of ICTs in the extension delivery in Nigeria has similarly developed with the growth of the extension service, with radio and television making the most important contributions even with their overload of provider-driven rather than demand-driven information that appropriately addresses the needs of the clients. Use of ICTs has the potential to enhance the performance of Nigeria's extension service as it will allow for pluralistic flows in an agricultural innovation system to ensure effective and efficient sharing and exchange of information, knowledge and skills among all stakeholders. Unfortunately however, ICTs use in Nigeria's extension service has been severely limited principally because of the problems of access, connectivity, low-level literacy and computer skills of the major stakeholders, costs, and the country's poor ICT infrastructural development (Aroyo, 2006).

Anambra Government to Train Youths in Agriculture: The Anambra government of Nigeria has earmarked N45 million for the training of 6,000 youths in agriculture-related skills. The program will train youth in aquaculture, horticulture, domestic science, poultry, ICT, and bakery, as a part of a pilot project. Once the pilot scheme succeeds, the young graduates would be encouraged to start their own ventures, and the training program would become an annual event to upgrade their skills (i4d, 2008, April).

Judging from this and other studies on Nigerian e-agriculture programs, it could be argued that Nigeria has a long way to go to fully implement a constructive and consistent e-agricultural program. Some of the areas where ICT can be engaged to practice e-agriculture are almost non-existent in Nigeria. Current programs seem to be haphazard and lack coordination within the different tiers of government. Relationships between government and related private and non-governmental agencies seem to be ad hoc. The implementation of e-agricultural programs seems not to have reached the rural areas where most of the farming is conducted and where it is needed to improve the economic welfare of the farmers. It could be argued that like other government socioeconomic programs, it could be described as paying "lip-service" to e-agriculture. Considering Nigeria's poor infrastructural facilities, poor financial resources and the failures of past socioeconomic programs, it is imperative for government to find ways to develop a workable program.

Proposals for Enhancing Nigerian e-Agricultural and Food Production Policies

To turn Nigeria into one of the world's top 20 economies by 2020, the country has adopted the goals of MDG, an UN-instigated program aimed at improving the lot of Third world masses by striving to eliminate poverty in order to raise their social welfare by promoting economic development. The program which was adopted in 2000 calls for eradicating extreme poverty and hunger, achieving universal primary education, promoting gender equality and empowering women. It calls for reducing child mortality; improving maternal health; combating HIV/AIDS, malaria and other diseases; ensuring environmental sustainability; and developing a global partnership for development (The World Bank, 2004). Nigeria launched Vision 20/20 as a mechanism of reaching her goal of becoming a developed economy by 2020 (People's Daily Online, 2008, Nov. 7). In this regard, President Yar'adua has initiated a seven-point agenda that serves as the basis for attaining Vision 20/20. Nigerian Embassy Vienna (n.d.) stated that the president's seven-point agenda consists of:

Power and Energy: The infrastructural reforms in this critical sector through the development of sufficient and adequate power supply will be to ensure Nigeria's ability to develop as a modern economy and an industrial nation by the year 2015.

Food Security: This reform is primarily agrarian based. The emphasis on the development of modern technology, research, financial injection into research, production and development of agricultural inputs will revolutionize the agricultural sector leading to a 5 to 10-fold increase in yield and production. This will result in massive domestic and commercial outputs and technological knowledge transfer to farmers.

Wealth Creation: By virtue of its reliance on revenue from non-renewable oil, Nigeria has yet to develop industrially. This reform is focused on wealth creation through diversified production especially in the agricultural and solid mineral sector. This requires Nigerians to choose to work, as hard work by all is required to achieve this reform.

Transport Sector: The transportation sector in Nigeria with its poor road networks is an inefficient means of mass transit of people and goods. With a goal of a modernized industrialized Nigeria, it is mandatory that Nigeria develops its transport sector.

Land Reforms: While hundreds of billions of dollars have been lost through unused government-owned landed assets, changes in the land laws and the emergence of land reforms will optimize Nigeria's growth through the release of lands for commercialized farming and other large scale business by the private sector. The final result will ensure improvements and boosts to the production and wealth creation initiatives.

Security: An unfriendly security climate precludes both external and internal investment into the nation. Thus, security will be seen as not only a constitutional requirement but also as a necessary infrastructure for the development of a modern Nigerian economy. With its particular needs, the Niger Delta security issue will be the primary focus, marshaled not with physical policing or military security, but through honest and accurate dialogue between the people and the Federal Government.

Education: The two-fold reforms in the educational sector will ensure firstly the minimum acceptable international standards of education for all. With that achieved, a strategic educational development plan will ensure excellence in both the tutoring and learning of skills in science and technology by students who will be seen as the future innovators and industrialists of Nigeria. This reform will be achieved through massive injection into the education sector.

As the preceding shows, agriculture represents the key to achieving these objectives. Of the seven points, three (food security, wealth creation, and land reforms) are directly tied to agriculture, while the remaining four (education, security, transportation, power and energy) have some degree of relationship with the agricultural sectors. Having neglected agriculture since the discovery of oil about forty years ago, coupled with the failures of past socioeconomic programs, a great deal of skepticism exists regarding the nation's ability to muster her abundant natural and human resources to complete the course.

The doubt these skeptics engendered may hold true considering the challenges Nigeria faces that created a litany of past failed programs. These include poor capital and financial infrastructure, political instability, corruption, illiteracy, and inadequate telecommunication facilities, among others.

Despite these daunting challenges, it could be argued that Nigeria has taken the first step in the right direction by recognizing the indispensability of agriculture to its future and attempting to implement an e-agricultural program. For this recognition to translate into tangible results, Nigeria's agricultural policies must be creative and flexible. It is laudable that it has embarked on an e-agricultural policy. There is a need to refine such policies constantly so as to make changes, such as suggestions offered in this chapter and by others. If Nigeria's e-agriculture policy and her wider agricultural policies are to succeed, the government should strive do the following:

- Empower farmers through education to raise their ability to read. Develop training programs in the rural areas to teach rural farmers how to read. The training must include training on how to use some of the new technologies and new farm techniques.
- The content of Nigeria's agricultural policies and e-agricultural programs should be locally generated and reflect the values in the country. Generating local-based content would be easier for rural farmers to learn and implement.
- Telecommunication and media infrastructure must be developed in the rural areas to provide access to rural residents.
- Considering the nation's limited resources, telecommunications services and centres aimed at community use should be

developed. While the government should discourage those who have adequate resources to acquire their own facilities, it must concentrate on providing community-wide facilities. This will help the government to achieve economies of scale to save costs.

- Other infrastructural transformations consist of an availability of electricity, water, roads, and markets. It should be noted that existing facilities should be refurbished and improved instead of embarking on new ones. Efforts to develop new facilities should be based on acute need—this has been a constant source of waste.
- Women make up the bulk of farmers in the rural area. The government should develop programs targeted at helping women to use their farming skills to help themselves and their families. Such assistance may include low interest loans, training, and ownership.
- There must be constant review of e-agriculture and agricultural policies. These reviews should be based on scientific studies that involve farmers. The primary focus would be to develop bottom-to-top, instead of top-to-bottom policies. Using this approach will ensure the farmers are making their contributions and policies reflect their need, instead of the views of government bureaucrats who are not exposed to realities of the group. This is quite important considering the cultural differences in different regions of the country.
- The government should encourage private-government ownership of infrastructure and farm facilities.
- Markets must be developed where rural farmers trade their products. Efforts must be made to ensure that these farmers, who rely on agriculture, must exchange their goods at reasonable prices for them to depend on agriculture for their livelihood. If people realize they can depend on agriculture, more people—especially the younger generation—will turn to farming. The mass exodus to urban cities may decline as agriculture becomes a mainstream activity.
- Nigerian government should not shy away from cloning the policies and programs that have worked in other developing countries when she finds her policy is wanting. They should also seek assistance from international organizations and government as long as that assistance reflects their values and needs.

Although there is a need for a creative and flexible policy, the government must realize the policy must be consistent, but flexible and coordinated. Without a consistent policy, there would be a lack of focus and direction that has plagued past socioeconomic programs. Most of the aforementioned policies would not come to fruition if Nigeria does not reinvent itself in an effort to become more effective. For example, with regard to agriculture, efforts should be made to ensure the various agencies dealing with e-agriculture have a mechanism to coordinate and communicate with each other so they can know what others are doing in order to avoid duplication of services. In conclusion, efforts to ensure the successful implementation of the Nigerian e-agricultural program include continued partnership with local and foreign investors as well as international organizations. Services must be extended to rural residents where most farmers reside. Nigeria must provide an enabling and secure environment by setting up essential legal, institutional, and regulatory frameworks and structures that encourage investors, and facilitate the training of farmers and agricultural workers.

REFERENCES

i4d (2008, April). Anambra government to train youths in agriculture. Retrieved Dec. 6, 2008, http://www.i4donline.net/

i4d (2008, February). Nigeria's agrovision. [Online]. Retrieved Dec. 5, 2008, http://www.i4donline.net/.

Adegite, V. (2006, Sept. 15). *E-Agriculture—Defining a new paradigm for agricultural development*. Retrieved Dec. 5, 2008 http://www.e-agriculture.org/.

Adeniyi, S. (2008a, Aug. 6). WB consultant expresses fear over food crisis despite Nigeria's external reserves. *Nigerian Tribune.* Retrieved Nov. 9, 2008 from http://www.tribune.com.ng/.

Adeniyi, S. (2008b, May 21). 'Without food security, Nigeria can't talk of national security.' *Nigerian Tribune.* [Online]. Retrieved June 1, 2008 from http://www.tribune.com.ng/

Afrol News. (2008, June 7). New media policy being designed in Nigeria. *afrol News.* Retrieved Dec. 5, 2008 http://www.afrol.com/.

Alozie, E. C. (2005). *Cultural reflections and the role of advertising in the socio-economic and national development of Nigeria.* Lewiston, NY: The Edwin Mellen Press.

Aroyo, T. (2006). Promoting the use of information and communication technologies (ICTs) in Nigeria's agricultural extension service. *Moor Journal of Agricultural Research,* 7(1&2), 100–106.

Ayers. S. (2005). *ICT, poverty and the global economy: Challenges and opportunities for developing countries.* Information for Development (InfoDev). Retrieved Nov. 4, 2008 http://www.infodev.org/en/.

Babatunde, R., & Oyatoye, E. (2005, Oct. 11–13). *Food security and marketing problems in Nigeria: The case study of marketing maize in Kwara State.* Presentation at Deutscher Tropentag, Hohenheim. Retrieved Nov. 25, 2008 www.tropentag.de/2005/abstracts/.

Business Day Nigeria/Africa News Network (2007, Nov. 22). Why Nigeria imports $3 billion of food annually. *Business Day Nigeria/ Africa News Network*. Retrieved November 24, 2008 from http://africanagriculture.blogspot.com/

Castells, M. (Ed.). (2004). *The network society: A cross-cultural perspective.* Northampton, MA: Edward Elgar.

Corcoran, K. (2008, March 24). Food prices soaring worldwide. *Associated Press.* Retrieved December 6, 2008 from http://abcnews.go.com/International/

D'Costa, A. (Ed.). (2006). *The new economy in development: ICT challenges and opportunities.* Basingtoke, England: Palgrave Macmillan.

Dada, J. (2007, Jan. 25–27). *Nigeria.* Presented at the inauguration of the Telecentre Network of Nigeria. Retrieved November 25, 2007, http://www.giswatch.org/

eJournalUSA, (2007, September). U.S. food aid: Reducing world hunger. Retrieved Nov. 24, 2007, from http://usinfo.state.gov/journals/

Eribo, F. (2001). *In search of greatness: Russia communications with Africa and the world.* Westport, CT: Ablex.

Ernst, D., & Lundvall, B. (1997). *Information technology in the learning economy: Challenges for developing countries.* DRUID, Copenhagen Business School, Department of Industrial Economics and Strategy/Aalborg University, Department of Business Studies in its series DRUID Working Papers with number 97–12. Retrieved Nov. 4, 2008, http://www.druid.dk/

Guardian.co.uk (2008, May 10). Food: A perfect storm | Leader: The rocketing price of food | Comment is free | *The Guardian*. Retrieved Dec. 5, 2008 from http://www.guardian.co.uk/commentisfree

Gumm, D. (2008, May 8). Nigeria can feed her citizens, become world food exporter—Govt. *Vanguard*. Retrieved Dec. 8, 2008 from http://www.vanguardngr.com/

Hamelink, C. (1997). *New information and communication technologies, social development and cultural changes.* United Nations Research Institute for Social Change. [Online]. Retrieved Nov. 4, 2008, from http://www.unrisd.org/

Idacha, F. (1980). *Agricultural research policy in Nigeria*. Research Report 17. Washington, DC: International Food Policy Research Institute.

International Finance Corporation/World Bank. (2008). *Information & Communications Technology in Africa.* Retrieved Nov. 30, 2008, from http://www.ifc.org/ifcext/africa.nsf/

IRIN (2002, July 4). Nigeria: IRIN Focus on shift towards offshore oil production. Retrieved Sept. 14, 2008 from http://www.irinnews.org and http://www.africaaction.org/

Jallow, B. (2004). Community radio for empowerment and impact. *The Journal of Development Communication, 15*(2), 56–68.

James, A. (2008, April 24). World food crisis hits home. *Seattleepi.com.* Retrieved December 6, 2008 from http://seattlepi.nwsource.com/business/

Minasyan, G. (2006, Oct. 13). *Environment and ICT: "enemies or friends."* Retrieved November 4, 2008, http://www.athgo.org/downloads/position_papers/Minasyan_Gohar

Moemeka, A. (1994). Development communication: a historical and conceptual overview. In A. Moemeka (ed.), *Communicating for development: a new pan-disciplinary perspective,* (pp. 3–12). Albany, NY: State University of New York Press.

Mohammed, I. (2008, May 5). Road map to attaining food security in Nigeria. *Nigerian Tribune.* Retrieved Nov. 24, 2008 from http://www.tribune.com.ng/.

National Telecommunication Policy (2000, September). Federal Republic of Nigeria: National Telecommunication Policy, 2000. Retrieved November 25, 2008 from http://www.bpeng.org/NR/rdonlyres/

News, B. B. C. (2002, Jan. 16). Nigeria's economy dominated by oil. Retrieved Sept. 14, 2008 from http://news.bbc.co.uk/ and http://www.africaaction.org/

News, B. B. C. (2007, July 12). Police stop Nigeria hunger march. Retrieved November 9, 2008 from http://news.bbc.co.uk/2/hi/africa/

News, B. B. C. (2008, Nov. 13). Nigerian satellite fails in space. Retrieved Dec. 6, 2006 from http://newsvote.bbc.co.uk/

Nigeria's Agriculture Sector (2007, May 21). Opportunities in Nigeria's agriculture sector. Retrieved November 24, 2008 from http://agriculturepro.blogspot.com/2007/05/nigerias-agriculture-sector.html/.

Nigerian Embassy Vienna (n.d.). Yar'adua seven point agenda to transform Nigerian. Retrieved December 2, 2008 from http://www.nigeriaembassyvienna.com/YarAduasSevenPointAgenda/

Nigerian Tribune (2008, April 29). Food crisis in Nigeria. Retrieved December 8, 2008 http://www.tribune.com.ng/

Obasanjo, O. (2004, March 8). *Transparency, accountability and good governance through eGovernment*. President Speech. Retrieved from September 14, 2008 http://www.negstglobal.com/sitefiles/Retrieved.

Obi, (2008, October 29). A panacea to global food crisis. *The Tide Online*. Retrieved November 9, 2008 from http://www.thetidenews.com/

Ogen, O. (2007). The agricultural sector and Nigeria's development: Comparative perspectives from the Brazilian agro-industrial economy, 1960–1995. *Nebula 4.1*(March). Retrieved November 25, 2008 from http://www.nobleworld.biz/

Okuneye, P. (2002). Rising cost of food prices and food insecurity in Nigeria and its implication for poverty reduction. *CBN Economic & Financial Review 39*(4). Retrieved November 25, 2008 from http://www.cenbank.org/out/Publications/

Over, E. (2001). *Social justice in world cinema and theatre*. London: Ablex Publishing.

Oxfam (n.d.). World food crisis. Retrieved Dec. 6, 2008 from htttp://www.oxfam.org.uk/Oxfam/

Patterson, C. (n.d.). Development communication. The Museum of Broadcast Communication (MBC). [Online]. Retrieved Nov. 24, 2008 http://www.museum.tv/archives/etv/D/

Pendakur, M., & Harris, R. (Eds.). (2002). Citizenship and participation in the information age. Aurora, Ontario: Garamond Press.

People's Daily Online. (2008, Nov. 7). World Bank to adopt Nigeria's food security policy. English: People's Daily Online. [Online]. Available at http://english.peopledaily.com.cn/ Retrieved Dec. 7, 2008.

Punch On the Web. (2008, Oct. 8). Nigeria, fifth world hunger prone nation—FIIRO DG. Punch on the Web. [Online]. Available at http://www.punchng.com/ Retrieved Dec. 11, 2008.

Robertson, R. (1992). *Globalization: Social theory and global culture*. London: Sage.

Singhal, S., & Domatob, J. (2004). The field of development communication: An appraisal (a conversation with Professor Everett M. Rogers). *The Journal of Development Communication, 15*(2), 51–55.

Tadégnon, N. (2007, Sept. 3). E-agriculture for Togolese farmers. Highway Africa News Agency. Retrieved Dec. 6, 2008 from http://hana.ru.ac.za & http://www.africa.upenn.edu/afrfocus/

Tamukong, J. (2007). *Analysis of information and communication technology policies in Africa*. Retrieved November 25, 2008 from http://www.ernwaca.org/panaf/

The Economist (2008, April 17). Food: The silent tsunami. Retrieved Dec. 6, 2008 http://www.economist.com/opinion/

The Guardian (2008, Dec. 11). Africa, Asia most affected by food shortages, says FAO chief. Retrieved December 13, 2008 from http://www.ngrguadian.com/

The Guardian Online (2001, August 13). Nigeria is world's 10th populous nation, says report. Retrieved August 15, 2001 from http://www.ngrguadian.com/.

The ITU Telecom Africa. (2004, May 4–8). *Has African renaissance begun?* Presented at The ITU Telecom Africa, Cairo Egypt. [Online]. Retrieved November 25, 2008 from http://www.itu.int/AFRICA2004/

The New York Times (2008, April 10). The world food crisis. *The New York Times*. Retrieved November 28, 2008 http://www.nytimes.com/2008/04/10/opinion/

The World Bank (2004). Millennium Development Goals. Retrieved December 2, 2008 from http://web.worldbank.org.

Tradenethome (n.d). *Market information on your mobile*. Retrieved Dec. 6, 2008 from www.tradenet.biz

Vanguard (2008, Sept. 2). FG urges telecom firms to invest more in Nigeria with their profits. *Vanguard*. Retrieved Nov. 25, 2008, from http://www.vanguardngr.com

ENDNOTE

[1] The rate of the Nigerian naira to U.S. dollar at the time of writing is about N130 to $1.

Chapter 4
Newly Emerging e-Commerce Initiatives on the Agricultural Market in Poland

Mieczysław Adamowicz
Warsaw University of Life Sciences (SGGW), Poland

Dariusz Strzębicki
Warsaw University of Life Sciences (SGGW), Poland

ABSTRACT

Establishment of the new institutions that could improve Polish agricultural market was one of the main goals of the Polish government agricultural policy in the period of economy transformation. The project of creating agricultural markets was successful. Several regional wholesale markets and commodity exchanges were established and most of them still function with good performance. These markets are the key important marketing channels for market-oriented farmers in Poland. They can also be seen as a source of electronic commerce innovations on the Polish agricultural market. The aim of the chapter is to present the process of establishment and first experiences of electronic market of agricultural products as one of the new e-commerce initiatives on the Polish agricultural market. The chapter also discusses conditions of the electronic exchange development and its impact on the Polish agricultural market.

INTRODUCTION

The government policy of early 90s in Poland was focused on transforming central planned economy into market economy. One of the most important tasks of the Polish agricultural policy of that time was to establish new market institutions to improve the organization of agricultural market. The project of wholesale markets and commodity exchanges development started at the beginning of 90s. As a result several wholesale markets and commodity exchanges were founded and most of them are operating with a good performance at present. Among others important markets such as Warsaw Commodity Exchange and Wielkopolski Wholesale Market in Franowo were established. For many farmers in Poland these institutions are important marketplaces and constitute the new marketing channels. The institutions are not only important places where farmers can sell their products but they are also the leading sources of e-commerce innovations on the Polish agricultural market. The main e-commerce

DOI: 10.4018/978-1-60566-820-8.ch004

initiatives in Poland undertaken by wholesale markets and commodity exchanges are:

- Activities connected with electronic integration between a wholesale market and a farmer that improve information exchange and enable creating electronic orders, farmers' access to the information on wholesale inventory levels etc.
- Creation of electronic markets, which is the activity of commodity exchanges.

The chapter discusses Warsaw Commodity Exchange e-commerce initiative of establishing public electronic market for agricultural products named e-WGT. We also discuss determinants of the electronic exchange development and its impact on the Polish agricultural market. Statistical data from the Polish Central Statistical Office and information from the interview with the chairman of e-WGT[1] were used in the analysis.

Background

The new media and particularly the Internet have many important features that allow migration of markets to the cyberspace. These unique characteristics are: interactivity, hypertext, multimedia, great information capacity and global reach. Internet is not only a medium but it is also a market (Hoffman & Novak, 1996). Electronic market can be defined as an inter-organizational information system that allows the participating buyers and sellers to exchange information about prices and product offerings (Bakos, 1991). The basic functions of electronic markets are matching buyers with sellers, enabling exchange of information, products, payments and providing market infrastructure (Zwass, 2003).

There are two basic types of electronic Business-to-Business (B2B) markets that are private markets and public markets. Private markets are of the one-to-many type and are usually established by an organization representing buy side or sell side of a market. Public markets are markets of the many-to-many type and are created by a third party enterprise (Turban, 2006).

Electronic markets could be distinguished using many criteria. The market categories include following (Grieger, 2003):

- buyer oriented, seller oriented, neutral markets (market orientation criterion);
- vertical and horizontal markets (product criterion);
- auction, exchange, catalog (pricing mechanism criterion);
- MRO hubs, catalog hubs, yield managers, exchanges (manufacturing vs. operating inputs, spot vs. system sourcing – buyer behavior criterion);
- open, closed markets (market access criterion);
- markets enabling information exchange, negotiation, settlement (transaction phase realization criterion);

According to Turban (2006) electronic markets are characterized by greater information richness, low information search costs, access to more buyers and more sellers, buyers and sellers are able to be in different locations, less information asymmetry between market participants. Bakos (1998) considers market aggregation achieved by more choices, rich and transparent information and lower transaction costs as a very important benefit provided by electronic markets.

Subba Rao et al. (2007) distinguished two different kinds of benefits of using electronic markets. The first one is a market aggregation which creates value for market participants enabling to overcome market inefficiencies such as market fragmentation by offering greater market reach, more market choices, more available information about products and firms, lower searching costs and transparent prices. The second one is a interfirm-collaboration which creates value for market participants enabling improvements of processes in the supply chain. Very often high assets specificity is associated with the collaboration. The collaboration can appear when a firm make strategically important and difficult to manage purchases. Market participants may have different motives in using electronic markets because of the two kinds of benefits that the electronic markets can provide (Rao, Truong, Senecal, & Le, 2007).

Formation of an electronic market can have its sources in market competition. When it is low and it is not difficult to find customers and firms don't want to broaden their market they can just utilize extranets for digital transactions and communication, but when the market becomes competitive and customers groups are difficult to identify, extranets would be replaced with digital markets because they are able to reach more customers with lower costs (Dou & Chou, 2002).

The key to the electronic exchange success is to attract to the market many buyers and sellers to achieve sufficient market liquidity, that ensures to facilitate the price discovery (Chambers, Hopkins, Nelson, Perry, Pryor, Stenberg & Worth, 2001). The more buyers and sellers are using electronic exchange the more attractive the electronic market is for market participants and the more income generated for the market owners.

When discussing the development of agricultural electronic exchanges it is important to underline the specificity of e-commerce development in agribusiness. Among many universal barriers to e-commerce development (such as difficulties with the Internet access, difficulties in evaluation of the future benefits from investment in EC, lack of e-commerce usage by business partners) there are also some barriers that are especially significant and characteristic in agribusiness like complexity of some agricultural product or importance of face-to-face transactions (Leroux, Mathias, Wortman, 2001). High description complexity products are more likely to be exchanged through hierarchical coordination than market coordination because of high transaction costs connected with the exchange of complex description (Grieger, 2003). Many of farmers consider that relationship strengthening via the Internet is difficult (Boehlje, 2000).

Electronic exchanges coordinate transactions for standardized commodities. It could be considered as a flaw because agricultural markets are moving away from a situation where agricultural transaction feature non-differentiated products (Chambers, Hopkins, Nelson, Perry, Pryor, Stenberg, & Worth, 2001). Perfect information transparency and competitive bidding among suppliers cause also the situation in which buyers get lowest prices but profit margins of suppliers are approaching zero. Besides, factors such as quality, timing, long lasting strategic relations with suppliers are often more important for buyers than price on the market (Wise & Morrison, 2000).

However examples of developing countries prove that farmers using electronic agricultural markets can benefit from market aggregation and matching that is a very important benefit of using exchanges and auctions and helps to overcome market fragmentation. For instance in India, the newly emerging electronic exchange promise to give Indian farmers better possibilities for selling their crops, to reduce the intermediaries, and help farmers to find better prices. A large number of Indian farmers sell to few buyers, so the buyers can dictate prices. The new electronic market in India is an agricultural commodity spot market with national reach. The market provides platform where farmers can sell at the best rate and buyers can buy at the most competitive rate. Thanks to the electronic exchange farmers can achieve stronger hand seeing bidders for their products than simply nearest local buyers (Kaur, 2007). Another case from Philippines shows that electronic market called B2Bpricenow.com is growing more and more popular among farmers because it helps them to omit middleman and get better prices for their products (Rimando, 2008). Both in India and in Philippines the electronic markets are supported both by government agricultural agencies and NGO's such as farmer's cooperatives

Table 1. The percentage of the households having access to the Internet in big cities and rural areas in Poland 2006. http://www.stat.gov.pl

	2004	2005	2006
Big cities	34	40	45
Rural areas	15	19	25

Source: GUS Polish Central Statistical Office. Information Society. Research results in 2004-

and associations. To encourage farmers to use electronic market some efforts such as trainings in electronic market usage or free participation on the electronic market are made.

The Polish agricultural market is characterized by many small market players and many fragmented markets. Poland is one of the major agri-food producers in the European Union, but the production is very fragmented and the number of farms that sold on the market more than 50% of their production in 2007 was about 1,48 million farms (GUS 2008). In the period of central planned economy most of the farms in Poland were individual farms and very often they sold their products on the small, local marketplaces. Thus prices on the Polish agricultural market were created by a market mechanism since a long time.

Poland joined EU in 2004. Following successive reforms agricultural policy of EU has changed in the direction of less market protection of agricultural products. The vast majority of support paid to EU farmers now comes in the form of direct payments which are "decoupled" from production (and linked to high standards of environmental care, animal welfare and public health). Decoupled payments leave farmers free to base production decisions on the market, and therefore do not distort trade under WTO rules. Probably further reforms will continue the direction of changes aimed at more liberal trade. Thus, Common Agricultural Policy is becoming more focused on market needs. Market information have stronger and stronger impact on farmers production decisions.

Internet Usage in Poland

The degree of Internet usage in Poland plays a key role in electronic exchanges development. According to statistical data the percentage of households having access to the Internet in Poland is increasing dynamically (Table 1).

In 2007 more than a half of the Polish households possessed computers (54% of households), 41% of households had access to the Internet and 30% of households had a broadband access to the Internet. There was also a fast growth in the number of people buying products via the Internet and an average value of the purchases increased. There was 12 – 16% growth in the number of people between 16-74 years old that buy via the Internet (GUS 2008). The percentage of Internet access on rural areas in Poland will increase rapidly because of planned wireless technology implementation.

There is a growing number of agricultural information portals in the Polish Internet. Some of them are evolving in the direction of electronic markets. These electronic markets are rather catalogs of buy and sell offers (ads) of various products connected with agricultural sector. "Giełda Rolna" (www.gieldarolna.pl) is an example of this kind of agricultural portal. In 2007 the Web site was visited by about 4000 internet users daily and about 100 offers were added daily to the catalog of offers.

Many Polish farmers sell and buy products using popular electronic auction services like Allegro www.allegro.pl (which is the Polish equivalent of e-Bay). On Allegro auctions, beside common products, farmers buy and sell also agribusiness products like machines, fertilizers and pesticides. Auction Web sites become more popular among farmers. These web sites cause a growing popularity of auction and dynamic pricing which was not popular in Poland before the Internet era and was not used in agriculture at all.

The Profile of e-WGT

E-WGT JSC was established by Warsaw Commodity Exchange JSC. It is an electronic, public agricultural market (e-market) of the type many-to-many. Warsaw Commodity Exchange was established in 1995. Initially the commodity exchange served as a market for cash transactions of agricultural products. In 1999 futures contracts were introduced. In 2001 Warsaw Commodity Exchange undertook first activities of using the Internet in purpose of trading goods. In 2005 division of WGT created two companies:

- WGT which specializes in the futures contracts.
- e-WGT which specializes in the cash transactions.

The main goal of the e-WGT is to organize a convenient electronic marketplace and to attract as many agricultural market participants as possible. The main groups of the electronic market participants are agri-food producers, food processing enterprises, and traders. The main products that are traded on this exchange are grain, livestock, butter and sugar.

E-WGT consists of the three main parts:

- Information portal – in which information and news on agricultural market are published.
- "Table of offers" – which is an easy tool to search and browse buy and sell offers using many searching criterion. It also enables to make transactions within the electronic exchange.
- Auction – which is the part where auctions are conducted.

Within the electronic market we can distinguish a few different categories of the market participants:

- Not registered market participants – who have access to the information portal and can browse buy and sell offers.
- Registered participants – who are able to use all the tools of the "table of offers" part along with the "safe transaction" tool.
- Ordinary members – who have all the abilities of the registered participants and can use the auction part of the market. Every farmers and other market participant can become an ordinary member and the application for the membership is not complicated.
- Public members – who have all the abilities of ordinary members and can make transactions representing other market participants. Only those who run a business can become the public members.

There are two main transaction safety solutions on the electronic exchange e-WGT:

- Clearing margins deposits that market participants are required to maintain on their accounts. These deposits are set by a seller and are equal for sellers and buyers. Electronic system is checking automatically the value of the deposits. Only the proper deposit value enables to take part in the auction.
- "Safe transaction" tool which enables security of payments resulting from transactions. This is achieved by using Settlement Office of the e-WGT. Before performing delivery, purchaser pays the value of the commodity to the e-WGT clearing house, which is transferred to the selling side only

when the delivery is accepted by the buyer without any complaints.

The basic sources of income for e-WGT are transaction fees based on the value of the transaction. The higher the value of the commodity, the higher the total transaction fee levied. On e-WGT sellers and buyers pay transaction fee that is 0,3% of transaction value for each transaction side.

E-WGT could be also described using above mentioned criteria of electronic markets classification. The e-WGT is a public neutral market because it is driven by a third party. It is also a market for agricultural products and therefore, it is an industry specific vertical market, in the contrast to horizontal markets that facilitate sale and purchase of goods used by a range of industries. E-WGT enables two pricing mechanisms which are catalog and dynamic pricing of auction. By using the criterion of buying behavior the e-WGT should be classified to the category of exchanges as a market of spot sourcing of manufacturing inputs that are raw materials that go into the products of processors. The next criterion is an access to the market for potential market participants. e-WGT is an open market because access to the market is open for everybody who follow market's rules.

Electronic markets differ from each other in the scope of transaction phases that are realized. Every electronic market is enabled to realize at least one of the three transaction phases. e-WGT supports information, negotiation and settlement transaction phases but each of the phases is supported to the certain degree, in a simple way and with the basic functionalities which are:

- information phase is supported by the possibility to search buying and selling offers and the access to a simple market information portal
- negotiation phase is supported by the solutions like chat rooms, internet communicators that are provided within the electronic market
- settlement is supported by secure payment solutions, but transportation of the commodities is not supported by the electronic market.

There are not many additional services and opportunities for information sharing and collaboration between market participants within the electronic market. Support for the e-WGT development is provided by government policies and producer associations (e.g. the case of the pig market).

Poland is one of the biggest pig livestock producers in the EU following Germany and Spain. Pig production is an important source of income for many farms in Poland. The market of pigs in Poland is characterized by cyclical, seasonal, and economic fluctuations. These market conditions make pig production in Poland very risky for farmers because of difficulties in price predictions. Livestock prices fluctuations make the income of many farmers in Poland very unstable. The level of market organization measured by the share of producers' groups and cooperatives is very low and producers have low bargaining power. The livestock price fluctuations are very strong in Poland comparing to the other EU members. The Polish system of livestock trade has not changed since many years and buyer is a dominant transaction side. Participation on the electronic exchange could change this disadvantageous market situation for farmers. Polish Ministry of Agriculture, Polish Association of Pig Producers and e-WGT answering the need for creation of electronic market that will enable pig producers price negotiation and equilibrium price creation on the market, signed an agreement in 2008. In the agreement they decided to undertake the activities that will help in electronic market development. A large electronic exchange could establish new market conditions for pig producers. The activities undertaken by the sides within the agreement framework include the following:

- Organizing trainings and instructions of e-WGT usage for farmers and other market participants.
- Providing sufficient technical and logistical resources for the trainings.
- Encouraging farmers to take part in the electronic trading.
- Promoting the idea of electronic exchange of livestock.
- Advertising e-WGT market using their own media such as Web sites, brochures, bulletins, and periodicals.

Conditions of the e-WGT Development

Electronic exchange e-WGT is a unique in Poland electronic market where farmers can sell their products. The market was established by an organization that had been specializing in the agricultural products trade, which is Warsaw Commodity Exchange (Warszawska Giełda Towarowa WGT). This is an important reason of good recognition and trust towards the market among farmers and other agricultural market participants. All the cash transactions of WGT were handed over to the newly created electronic exchange, what caused its liquidity at the beginning of its operation. These were very important factors of the electronic exchange success. Currently (11.2008) the e-market has about one thousand registered participants and members and the number is growing dynamically. As the chairman claims the present number of buyers and sellers on the market generates a right income for the market to exist and to develop.

E-WGT aims continually at increasing number of its participants. Advertisements of e-WGT appear in agricultural periodicals. Information connected with the e-market operation are transmitted via such popular programs of the Polish public TV as agricultural "Agrobiznes" and the main news program "Wiadomości". Thanks to the information campaign most of the farmers in Poland are conscious of the e-market existence and possibilities of the Internet usage to sell and buy agricultural products. However, as the chairman claims, direct marketing is the most important tool of promoting the e-market. Employees of e-WGT try to get directly to the farmers to explain how the e-market works and to give instructions and trainings on the electronic exchange usage.

Government agricultural policy plays an important role in the e-market development. The process of organized markets creation started at the beginning of 90s. Some of the markets were privatized with time. The establishment of e-WGT is a continuation of the efficient market development that started in 90s. Although e-WGT is a private market already, Polish Ministry of Agriculture still propagate the idea of an efficient agricultural electronic market development, and actively participates in the trade agreements connected with the electronic exchange development.

In the chairman's opinion there are no competitors of e-WGT in the sense of e-marketplaces where farmers can sell their products. There are emerging some agricultural Web sites in the Polish Internet that are mainly information portals and catalogs of sell and buy offers of broad range of products connected with agribusiness and these ads are not numerous.

The chairman considers two main factors that are attracting farmers to the electronic exchange which are "convenience" and "safety". Convenience means that the e-market participants have access to many potential buyers and sellers from various regions of Poland and they are able to compare prices in a short time. The e-market members are also able to participate in a dynamic process of price creation by selling and buying on an auction. In Poland auctions were never used in trading agricultural products. The e-market popularized this form of making transactions in Polish agriculture.

Safety is the second important factor of e-WGT attractiveness. When using the "safe transaction" tool buyers don't have to pay directly on the sellers accounts. The money are transferred by e-WGT to the sellers only when all the transaction conditions are fulfilled and completed. Also the system of deposits on the e-WGT auctions makes the transactions safer.

In the chairman's opinion the most important barriers to using the e-market by farmers include the following:

- Psychological barrier to using the Internet as a tool for selling agricultural products.
- Lack of trust to make transactions via the Internet.
- Farmers' attachment to the places they have been selling so far.

And much less important are barriers like:

- Costs of transportation.
- Lack of the Internet access.
- Unknown potential buyers and sellers.
- Difficulties with meeting product quality standards.
- Difficulties with appropriate product description in an offer.

Psychological barrier of using the Internet as a way of selling agricultural products is particularly important, especially in case of elderly farmers. The e-WGT is most often used by young farmers. Using electronic market and electronic auction in particular requires some computer skills and computer dexterity. Besides, for elderly farmer's auction is a new way of making transactions. Young farmers are accustomed with bidding because they often use popular auction Web sites. The e-WGT employees very often observe that elderly farmers use the e-market with their children's help. In the chairman's opinion a big scale of selling agricultural products via the Internet requires a new generation to enter the market.

As the chairman claims the Internet will have a strong impact on the agricultural products trade in Poland. Farmers who sell relatively big quantities of agricultural products on the market will get from e-market such benefits as:

- Broader market with new customers and suppliers.
- Lower transaction costs that can be achieved by shorter time of searching for market information and improving communication among market participants.
- Better prices for their products.
- Possibility of participation in the price creation.
- New distribution channels.

In the opinion of the chairman electronic markets probably will not change the trade system of the small agricultural market participants. For farmers who offer only small quantities of agricultural products on the market it is not profitable to sell or buy on an electronic market because transportation costs are too high and the time to learn how to use an e-market is too long therefore they prefer to sell their products to the local buyer getting lower, not negotiated prices.

The e-WGT managers plan to add new product markets to the offer of the electronic exchange. They plan for instance to start with the seed market within the e-WGT in the near future.

CONCLUSION

In the period of economy transformation in Poland it was a very important goal of agricultural policy to create new market institutions that would improve agricultural market organization in Poland. As a result of the policy wholesale markets and commodity exchanges were created. These markets have become important institutions for marketing channels development and price creation on the Polish agricultural market. The markets also undertake initiatives of innovation in trade such as e-commerce of agricultural products. These initiatives are continuation of the process of establishing efficient agricultural markets that started at the beginning of 90s. e-WGT is a unique public electronic market in Poland where farmers and other agri-food market participants can sell and buy agricultural products. The e-market is developing and the number of its participants and transactions is growing fast.

There are two main determinants of the electronic exchange development. The first is the type of market owner and the second is the specificity of Polish agricultural market. e- WGT was created by an institution that had been specializing in agricultural commodity trade. This institution was able to secure proper liquidity for the e-market at the beginning of its operating. Specificity of the Polish agricultural market cause that the idea of a public electronic market meets the needs and requirements of the Polish agricultural market featured by a great number of geographically dispersed market participants. Polish farmers perceive the opportunities that public e-market can give to them. They can search for the new buyers or suppliers within one virtual place and they can negotiate prices and participate in auctions.

However electronic markets are not popular among Polish farmers yet. It is expected that electronic agricultural markets will become more popular when younger generation of farmers appears on the market. Also improvements in information technologies will probably diminish such barriers of EC development like the importance of face to face contacts when making transaction and the agricultural product complexity.

We also underline a great role of agricultural policy in the electronic public markets creation. The e-WGT market was established as a continuation of the changes on the Polish agricultural market that begun in 90s. Currently this market has no competitors; therefore we can assume that there would be no electronic market of agricultural products in Poland if not the initiative of Warsaw Commodity Exchange. The e-market creation required capital investment but also farmers' trust to the institution that created the market was also very important. The e-WGT example proves that development of the electronic market is an evolutionary process and the changes of the market are the result of the continuous electronic exchange adjusting to the agricultural market needs and requirements.

REFERENCES

Bakos, Y. (1991). A strategies analysis of EM. *MIS Quarterly*, *15*(4), 295–310. doi:10.2307/249641

Bakos, Y. (1998). The emerging role of electronic marketplaces on the Internet. *Communications of the ACM*, *41*(8), 35–42. doi:10.1145/280324.280330

Boehlje, M. (2000). Critical dimensions of structural change: policy issues in the changing structure of the food system. In *Proceedings of the American Agricultural Economics Association Preconference Workshop*, Tampa, FL, July 29, 2000. Retrieved from http://www.agbioforum.org

Chambers, W., Hopkins, J., Nelson, K., Perry, J., & Pryor, S. Stenberg, & P., Worth, T. (2001). *E-Commerce in United States Agriculture*. ERS White Paper. http://www.ers.usda.gov.

Dou, W., & Chou, D. (2002). A structural analysis of business-to-business digital markets. *Industrial Marketing Management*, *31*, 165–176. doi:10.1016/S0019-8501(01)00177-8

Grieger, M. (2003). Electronic marketplaces: A literature review and a call for supply chain management research. *European Journal of Operational Research*, *144*, 280–294. doi:10.1016/S0377-2217(02)00394-6

GUS (Polish Central Statistical Office). (2007). Information Society. Research results of 2004-2006. Retrieved from http://www.stat.gov.pl

GUS (Polish Central Statistical Office). (2008) *Characteristics of agricultural holdings in 2007*. Retrieved from http://www.stat.gov.pl

GUS (Polish Central Statistical Office). (2008). *The usage of information technologies in households in 2007*. Retrieved from http://www.stat.gov.pl

Hoffman, D., & Novak, T. (1996). Marketing in hypermedia computer-mediated environments: conceptual foundation. *Journal of Marketing*, *60*(July).

Kaur, G. (2007). *From APMCs to electronic markets*. Retrieved from http://www.indiatogether.org/2007/jul/agr-spotex.htm

Leroux, N., Wortman, M., & Mathias, E. (2001). Dominant factors impacting the development of business-to-business (B2B) e-commerce in agriculture. *International Food and Agribusiness Management Review*, *4*, 205–218. doi:10.1016/S1096-7508(01)00075-1

Rao, S., Truong, D., Senecal, S., & Le, T. (2007). How buyers' expected benefits, perceived risks, and e-business readiness influence their e-marketplace usage. *Industrial Marketing Management*, *36*, 1035–1045. doi:10.1016/j.indmarman.2006.08.001

Rimando, L. (2008). *Electronic Market For Farmers*. Retrieved from http://www.b2bpricenow.com/pr/ WhatIsB2B.htm

Turban, E. (2006). *Electronic Commerce 2006: A Managerial Perspective*. Upper Saddle River, NJ: Pearson Prentice Hall.

Wise, R., & Morrison, D. (2000). Beyond the Exchange: The Future of B-to-B. *Harvard Business Review*, (November – December): 86–96.

Zwass, V. (2003). Electronic Commerce and Organizational Innovation: Aspects and Opportunities. *International Journal of Electronic Commerce*, *3*, 7–37.

ENDNOTE

[1] The interview was conducted by the authors with the e-WGT chairman prof. Michał Jerzak on 12 November, 2008. Information from the interview were used to write the "Conditions of the e-WGT development" part of the chapter.

Section 2

E-Agriculture Development:

Conceptual Frameworks

Chapter 5
The Development of e-Agriculture in Sub-Saharan Africa:
Key Considerations, Challenges, and Policy Implications

Blessing Mukabeta Maumbe
Eastern Kentucky University, USA

ABSTRACT

The rapid growth of Information and Communication Technologies (ICT) has led to new opportunities to improve food and agricultural production, processing, distribution, and marketing functions in Sub-Saharan Africa (SSA). Such ICT-led transformations are expected to provide vast social and economic benefits to poor agricultural communities thus help uplift their standards of living. The process of how ICT should be applied in agriculture to raise living standards of millions of poor Africans is not yet well understood. Therefore, there is a need to deepen our understanding of ICT deployment and the socio-economic benefits expected from their application in African agriculture. Historically, some technologies failed in Africa and parts of Asia because of inadequate attention to context specific issues, irrelevance, and relatively prohibitive costs. In that regard, this chapter describes a framework for sustainable e-agriculture development in SSA. The proposed framework is based on three related models; (i) e-agriculture service delivery (ii) ICT development and diffusion pathways, and (iii) e-information flow and e-content development landscape. In order to facilitate the effective diffusion and adoption of e-agriculture, a set of "preconditions" and "e-value creation" opportunities are assessed. The identified preconditions help to filter out "irrelevant" ICT, and "e-value creation" facilitates use of context specific and demand- driven e-innovations in agriculture. The chapter identifies and discusses ICT illiteracy, ICT policy gaps, infrastructural deficiencies, and poverty as key challenges affecting the future success of e-agriculture in SSA. The chapter recommends the development of e-policies and e-strategies on e-content, e-trust, e-security, and e-value addition to promote sustainable e-agriculture development on the African continent.

DOI: 10.4018/978-1-60566-820-8.ch005

INTRODUCTION

The rapid diffusion and adoption of information and communication technologies (ICT) around the world has led to a growing desire to understand clearly how these modern technologies should be applied in key economic sectors in a developing country context. There is a growing realization that ICT are essential for accelerating socio-economic development and the modernization of agriculture in Sub-Saharan Africa (SSA) in the 21st century. The use of ICT to promote agricultural development in SSA offers tremendous socio-economic benefits and new opportunities to transform the livelihoods of poor agricultural communities and society in general. Some of the expected benefits from ICT are food security arising from productivity gains in crop and livestock enterprises, better access to national and global agricultural markets, improvements in rural financial service delivery, reduction in transaction costs, faster communication methods, e-health services for farmers, and the provision of accurate, reliable, and timely information for farm household decision making. The use of ICT in agriculture and rural development (i.e., e-agriculture) will make a significant contribution towards reducing poverty and malnutrition on the continent (Bertolini, 2003). Therefore, the diffusion and adoption of ICT offers a renewed promise to improve SSA's prospects to achieve the Millennium Development Goals (MDGs).

The benefits of ICT applications in the financial (e.g. e-banking, m-banking, etc.) and education sectors (e.g. online learning, distance education, NEPAD e-school initiative etc.,) are relatively well known. What remain unclear are the potential benefits of ICT use in agriculture and rural development in SSA. Although ICT use in the agricultural sectors of most developed nations started more than a decade ago (e.g. precision farming, e-auctions, computerized record keeping, online input procurement services offered by dealerships etc.), it has been lagging behind in developing countries. The past five years has seen a surge in interest to explore potential opportunities offered by ICT to smallholder farmers, agribusinesses, emerging rural entrepreneurs, governments, and non-governmental organizations (NGOs) in developing countries. In that regard, this chapter describes a framework for e-agriculture development in SSA. The chapter proposes three main inter-related frame-works; (i) an e-agriculture service delivery model, (ii) ICT development and diffusion pathway model, and (iii) an e-agriculture information flow and e-content development landscape. The development and implementation of these frameworks requires certain preconditions and e-value strategies to enhance the sustainability of e-agriculture.

In the past, the world experienced the euphoria associated with new technologies that did not live up to their initial high expectations. Today, ICT stand at a similar crossroad where societal expectations of the socio-economic benefits may be higher than what is practically possible. Some green revolution technologies failed in Africa and parts of Asia because of failure to articulate context specific issues, irrelevance, and prohibitive costs (Atkins and Bowler, 2001). The challenge therefore, is to design innovative strategies that enable the development of e-agriculture in order to avoid some of the common technological pitfalls of the 20th century which now provide important lessons for e-agriculture. In that respect, establishing certain preconditions for ICT use in agriculture, delineation of e-services, development of affordable technological platforms or diffusion pathways, and identification of tangible benefits are necessary but not sufficient conditions for successful e-agriculture development in SSA. In addition to that, there is need to aggressively promote e-value creation in e-agriculture service delivery across the agricultural value chain. Ultimately, the major driving force behind long-run demand for e-agriculture services will be end user e-accessibility and e-satisfaction. Putting the end user needs at the center of e-agricultural development is pivotal for its future success.

Definition of e-Agriculture

Before discussing the proposed framework for e-agriculture development, it is important to develop a working definition for e-agriculture. E-agriculture is an evolving field and so is its definition. In this chapter, e-agriculture is defined as-- *the application of modern information and communication technologies (ICT) to agriculture and rural development with the goal to improve human livelihoods by reducing poverty and hunger among the rural communities.* According to the United Nation's Food and Agriculture Organization (FAO) definition; *e-Agriculture is an emerging field in the intersection of agricultural informatics, agricultural development and entrepreneurship, referring to agricultural services, technology dissemination, and information delivered or enhanced through the Internet and related technologies. More specifically, it involves the conceptualization, design, development, evaluation and application of new (innovative) ways to use existing or emerging ICT (*www.e-agriculture.fao.org*).* The term "e-agriculture" can be used interchangeably with "digital agriculture" which comprises both "electronic" agriculture and "mobile" agriculture. To fully understand e-agriculture, it is important to indentify the target end users who are mostly agricultural communities including government policy makers, researchers, agribusinesses, rural traders and entrepreneurs, NGOs and smallholder farmers.

E-agriculture lies at the intersection of an ICT revolution, input procurement, agriculture production and food security, food marketing management, rural development, poverty and hunger reduction, and e-innovations along the food and agriculture value chains. E-agriculture thrives under different socio-economic, technological, political, cultural and legal environments. Similarly, e-agriculture revolves around diverse multimedia tools or channels (e.g. computers, mobile phones, etc.), strategic partners (e.g. information technology firms, service providers, e-content developers etc.) and the policy environment. The future of e-agriculture will depend upon the development of a sound enabling environment and that includes a robust research and development agenda. In addition, the ability of policy makers, researchers, industry practitioners, subsistence farmers and other key stakeholders to identify and derive tangible benefits or "e-satisfaction" from e-agriculture will shape its long term development trajectory in SSA. Further, understanding the "push" and "pull" factors for e-agriculture is critical for its sustainable development, deployment and uptake in SSA. Furthermore, the pace and pathway upon which e-agriculture should evolve is also crucial given the diverse resources endowments and challenges that face SSA countries and the need to minimize poor judgment in the development process.

This chapter is divided into six major sections. The next section describes the framework for e-agriculture development in SSA which is divided into four main phases. This is followed by a description in section 3 of the e-value considerations in e-agriculture and an overview of selected innovative ICT applications in agriculture globally. Section 4 explores some of the key challenges facing e-agriculture development in SSA. Finally, section 5 concludes this chapter with key insights and policy considerations for sustainable e-agriculture development in SSA.

E-AGRICULTURE DEVELOPMENT FRAMEWORK FOR SUB-SAHARAN AFRICA

The proposed framework identifies four inter-related phases of e-agriculture development in SSA. These phases are the establishment of e-agriculture pre-conditions, e-agriculture service framework, ICT diffusion pathways and benefit streams and, mapping an e-agriculture information flow and e-content development landscape. These four phases are summarized in figure 1 below and discussed in the following sections. The first phase describes the pre-conditions for ICT use. Basically, this could take the form of a needs assessment for different ICT uses in agriculture. The second phase describes the core e-services required for the development of e-agriculture. The establishment of core e-services highlights the need for priority setting in e-service development. The third phase examines the ICT investments, technology channels, stakeholder interactions, and associated socio-economic benefits. The fourth stage underscores the importance of an e-agriculture information flow and e-content development landscape. The framework gives an overview of the model features, key elements, their interconnectedness, as well as the potential scale and scope of e-agriculture requirements. Although these phases are discussed separately below, it is important to recognize the interconnectedness and the need for a balanced focus on the core issues covered in all the four phases. Given the integrated nature of these phases, the absence of one implies that e-agriculture will be poorly conceived. More importantly, sustainability issues should be accorded focused attention throughout the four phases to ensure the long term success of e-agriculture.

Phase 1: Establish e-Agriculture Preconditions

The diffusion of ICT should commence by addressing four fundamental questions that should set the preconditions for e-agriculture development. Neglecting to ask the fundamental questions identified in figure 2 below could lead to a potential failure of ICT use in agriculture in the long run. The first critical question to address is: *Why should farmers use ICT?* The tendency "to run with Joneses" in this era of ICT diffusion is inevitable (Maumbe, et. al, 2008). Discussions about "leapfrogging" ICT tends to overshadow the much needed debate about whether or not ICT are really necessary. Due to the failure to ask the why question, most poor farmers in SSA and the developing world could end up using "irrelevant" technology. It is important to first define the tangible benefits likely to accrue to farmers from the proposed ICT applications in agriculture. If prospective ICT uses do not generate significant benefits to the target farmers or end users, it is considered unnecessary. The second important question to ask is: *What kind of ICT should be recommended for use by poor farmers?* The ICT revolution has unleashed numerous kinds of technologies ranging from the traditional radio, television, faxes and pagers to the more modern office computers, lap-tops, personal digital assistants (PDAs), 3G mobile phones, i-phones, blackberries and other smart phones. Given the proliferation of new ICT, deciding on the kind of modern technology to use in agriculture has become even more challenging and complex. The third question to address is: *Where should ICT be applied in the agricultural sector?* Governments, agribusinesses and farmers have different technological and information needs, and so it is important to understand clearly which part of the agricultural sector has such needs. The demand for ICT is influenced by (i) nature of the agricultural enterprise (e.g. domestic versus exported orientation) (ii) stage of food and agriculture supply chain (i.e. input, production, distribution, and marketing), (iii) geographic location (rural versus urban), (iv) type of ICT (i.e. mobile versus fixed, etc.) available, (v) level of ICT literacy levels among different actors involved in agriculture and (vi) affordability issues among others. The multiplicity of ICT application areas, diverse user needs, wide choices in ICT tools, and differences in functional literacy levels emphasizes the importance of the need to understand precisely where to deploy ICT before proceeding blindly with e-agriculture. The basic guiding principle therefore should be to apply ICT

Figure 1. Summary of the phases of e-agriculture development in Sub-Saharan Africa, 2009.

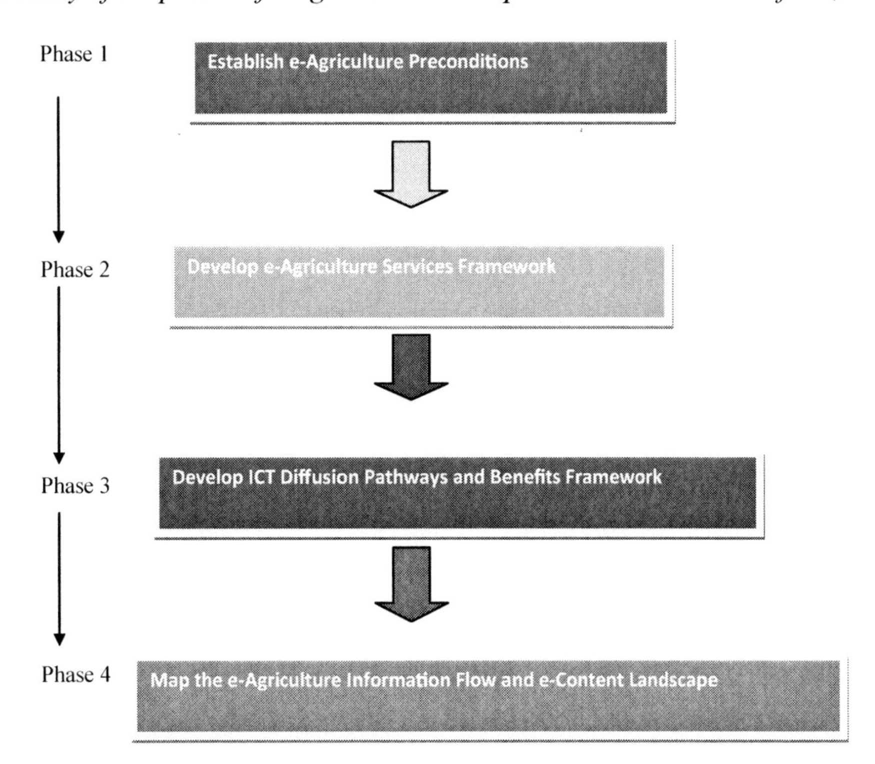

where they can generate the greatest return to the farm business. The final question is: *How should ICT be applied in agriculture?* This question is intertwined with the foregoing "where" "what" and "why" questions. For instance, today's ICT connectivity comes in various formats, WiFi, WiMax, GSM and VSAT just to name a few. It is therefore essential to address the issue of how to deploy ICT for agriculture development ahead of time. The type of ICT channel to be utilized is vital in determining the cost efficiencies of the technology. Smallholders are more likely to adopt e-agriculture activities if the ICT cost is right and the benefits are clear. If the proposed ICT are unaffordable and will not simplify or improve current agricultural practices, then their use becomes "irrelevant". Failure to effectively assess these "e-agriculture pre-conditions" could result in low ICT penetration rates in agriculture, fewer opportunities to increase agricultural production, and heightened risk of food insecurity, malnutrition and worsening poverty. The "e-agriculture pre-conditions" should be used to judge or pre-qualify the initial viability of proposed e-agriculture development projects in SSA and in other developing countries. Asking the right questions is bound to put e-agriculture on a path of future long-term success.

Phase 2: Framework for e-Agriculture Service Development Model

The e-agriculture service development model in Figure 3 describes an overview of the core e-services that should comprise e-agriculture development in SSA. These services have been divided into several categories namely, input supply, production, processing, distribution and marketing and e-value creation. Although the framework appears "static", it is important that it be responsive to changing needs of end-users. The dynamic element of the framework will

Figure 2. The pre-qualifying questions for e-agriculture development, 2009.

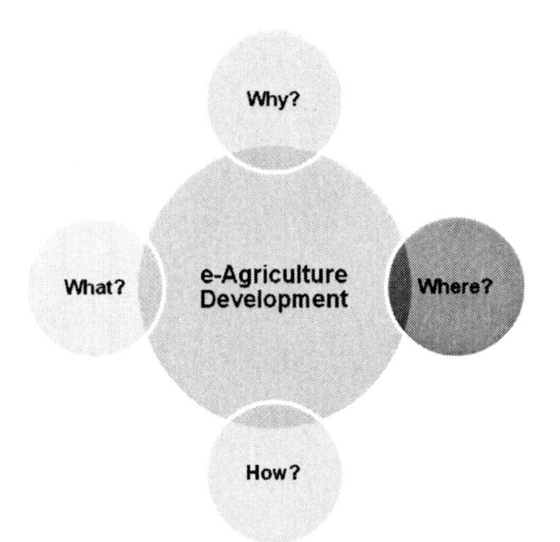

form the basis for a sustainable e-agriculture development and is expected to emerge from the e-value creation process. The e-value-creation service category will generate innovative uses of ICT or new service offerings that will provide key agricultural stakeholders the opportunity to wean services that are "obsolete" or redundant and select the ones that are new and "add value" to their e-agriculture development process. Each stage of the food and agricultural supply chain is characterized by a list of core e-services for its smooth operation. Although in the diagram the different e-services have been grouped under each stage of input supply, production, farm processing, distribution and marketing, and e-value creation, these services are interrelated.

At any one time, a farmer taps into e-services from each of these core areas depending on the level of specialization or diversification of their production enterprises. More specialized agricultural enterprises tap into e-services from predominantly a single core area. In contrast, more diversified enterprises (i.e. via vertical, horizontal or conglomerate integration) tap into the different core e-service areas identified above. The demand for e-services used in agriculture depends on the types of ICT at the deposal of the farmer and his or her level of ICT functional literacy (e.g., ability to operate the computer). Those farmers who have access to multiple ICT such as mobile phone, computers and PDAs are able to use a combination of e-services (e.g. web-based marketing, electronic inventory controls, mobile banking services etc.) while those with limited ICT access use only a limited set of e-services.

In Figure 4, the service menu for the core e-agriculture service areas identified above are highlighted. To ensure e-inclusion and e-accessibility in e-agriculture, it is important to deploy various ICT such that farmers, rural agribusinesses and other users do not face undue restrictions to key e-services they need to conduct their operations. The development of agro-portals is a significant step in providing one stop shop with diverse e-services for agricultural communities in a given country. Globally, the organizational structure of agriculture is changing and e-service providers face the constant challenge to repackage their service menu to suit new forms of organizational arrangements in agriculture. The new organizational forms include cooperatives, local and global franchises, strategic alliances, joint ventures, and producer bargaining associations. Internet service providers (ISP) and other technology firms should

Figure 3. E-agriculture service development model, 2009

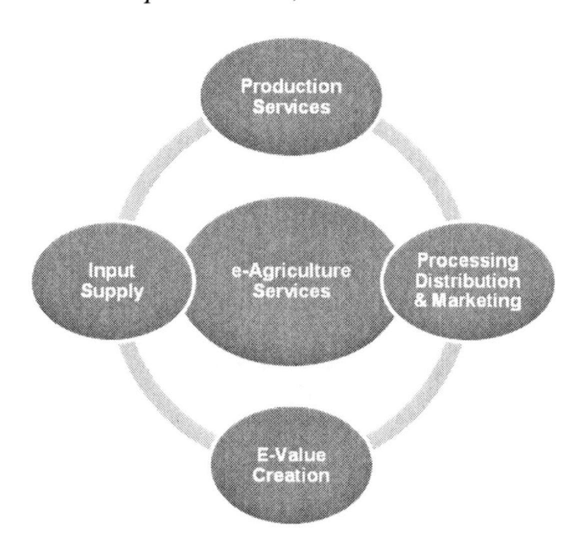

anticipate the diverse e-service needs of these various agricultural communities. The unique needs of individual farmers, contract farmers, agricultural cooperatives, and even large agribusiness firms will vary accordingly.

The varying levels of functional literacy among the different agricultural stakeholders in SSA are a major challenge to e-agriculture development. This is particularly important given that most farmers in remote rural areas may not understand the ICT and let alone the services that can be derived from the new technologies. Addressing such issues of "unconscious ignorance" proactively through training and awareness campaigns will help optimize the use of e-services by the various stakeholders in SSA.

Phase 3: e-Agriculture Technology Development and Diffusion Pathways

Understanding the benefits that could be derived from ICT use in agriculture is vital to their effective utilization. As already mentioned in phase 1, it is important to identify the specific benefits that are expected to accrue to the agricultural communities (e.g. farmers) from ICT utilization. After tangible benefit identification is completed, it is essential to match these benefits with the most appropriate ICT delivery channels. The identification and selection of the most cost effective ICT channels that have greater penetration into the target community should form a central part of the deployment criteria. A fundamental guiding principle is to deploy ICT in ways that maximize benefits that accrue to poor farmers. In other words, use of ICT that bring only limited benefits to agricultural communities should be avoided. Figure 5 below highlights the need to align the ICT investments, e-channels, business relationships, and the expected economic and social benefits. The key socio-economic outcomes are food security, employment creation, poverty alleviation, rural development, democratic participation, provision of farmer advisory services, and development of a knowledge society. The priorities accorded to these socio-economic benefits will vary from country to country in SSA. After the identification of tangible expected benefits and their classification into specific groups based on national priorities, the next step involves a systematic process to align ICT channels with the benefits to ensure effective e-service delivery to the target agricultural communities. Equally important is the need to create

Figure 4. Description of service menu for e- agriculture development, 2009

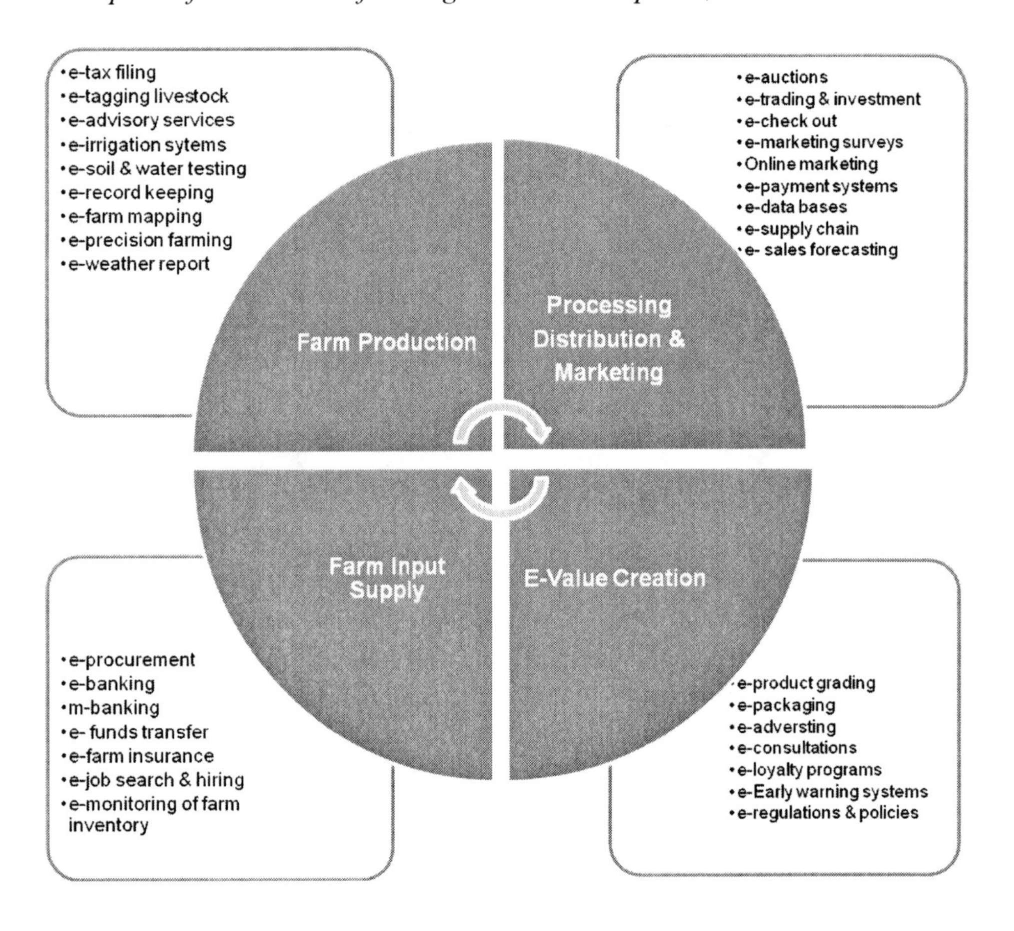

an enabling ICT policy environment to facilitate the process of matching technology choices to benefits. Table 1 provides a broad description of the potential benefits that could arise from using ICT in the agricultural sector. These benefits can be divided into four main categories; economic, social, institutional and organizational, and political and legal depending on whether the individual, community or country is the target beneficiary.

Phase 4: e-Agriculture Information Flow and e-Content Development Landscape

Matching technology channels to expected agriculture community benefits will not deliver sustainable e-agriculture without the relevant e-content to service the e-channels. This is particularly a major issue in most African countries with diverse languages where most people may not necessarily understand English e-content. Mapping the e-agriculture information flow and e-content development landscape is one of the most challenging ingredients for the development of successful e-agriculture. The various stages through which e-agriculture information flows are illustrated in Figure 6 below. The e-agriculture information landscape shows all the key stages that information should permeate. The e-agriculture information content must be balanced at the global, regional, country, village and down to the farmer level. Insufficient information availability at one level could easily suffocate the development of e-agriculture at other levels. The generation of new information to improve e-agriculture should not be top down, but bot-

Figure 5. ICT development and diffusion pathways, support policies and potential benefits, 2009

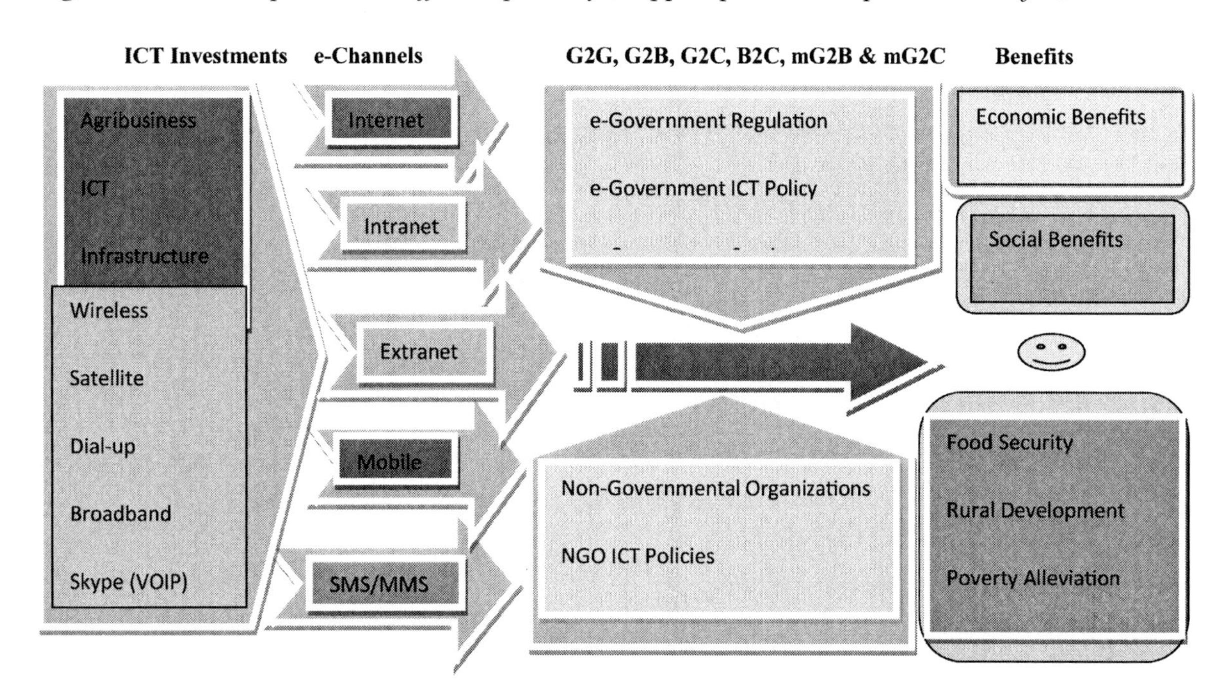

tom up, consultative and interactive across the different stages.

According to figure 6, key information nodes should be identified for sound development of e-agriculture. These key nodes are located at the global, regional, national, district (or municipal), village, local and individual farmer level. Some examples of the key actors that serve as e-agriculture information nodes at the global level include the Food and Agriculture Organization (FAO), International Fund for Agricultural development (IFAD), the World Summit for Information Society (WSIS), the International Center for Agricultural Research (CGIAR), International Telecommunication Union (ITU), and global agribusiness research and development firms. These organizations are actively involved in the systematic collection, processing and dissemination of e-agriculture content and the promotion of relevant technologies around the world. The key regional institutions that are major contributors to the development and diffusion of e-agriculture content include the African Virtual University (AVU), Africa Information Society Initiative (AISI), the International Society Technologies for Africa (IST), and the SADC Agriculture and Natural Resources program (e.g. SADC Gene Bank). In addition, entities such the Open Society Initiative for Southern Africa, Open Society Initiative for West Africa (OSIWA) and *Kwetu* in East Africa also play a vital role in the provision of online resources that are useful for e-agriculture development.

Within individual countries specialized Agricultural Universities on the continent (e.g. Sokoine University of Agriculture in Tanzania), Departments of Agriculture at various national universities and the Institute for Development and Planning (lead e-curriculum development institute) have the responsibility to provide locally relevant and culturally balanced e-content required to support the development of e-agriculture portals in SSA. These nodes also play a secondary key role as facilitators for the transfer of e-agriculture content to African farmers, who are the ultimate beneficiaries of e-agriculture. These farmers

Table 1. Socio-economic benefits from e-agriculture development in Sub-Saharan Africa, 2009.

Potential Benefits	Description of potential benefits and opportunities
Economic Benefits	
Agricultural productivity & efficiency	Improved profitability, food security, poverty alleviation, e-food banks, electronic land registration and administration, e-resource mapping, food ration calculators, e-irrigation systems, soil and water testing sensors, e-procurement systems, e-tax filing, e-consultations and advisory services, e-newsletters etc.
Rural banking	Improved rural financial service delivery, e-banking service, m-banking service, e-deposits, electronic fund transfer transactions, online investment opportunities, credit scoring techniques, rural remittances etc.
Market development	Marketing information services, market access and market coordination, identity preservation, online marketing services, e-auctions, elimination of middlemen, reduction of marketing margins, e-price reporting, e-databases etc.
Rural development	Rural info-structure development and employment generation. Online employment registration sites etc.
Social Benefits	
Networking	Rural networking, partnerships, and e-agriculture information sharing
Social capital	E-collaboration and social capital development, e-conferences, etc
ICT Literacy	e- Agriculture training programs, e-advisory and support services etc.
Regulatory	Participatory e-regulations and e-strategies development etc
Institutional & Organizational Benefits	
Research and Training	e-Agriculture research and training opportunities
Security Institutions	e-Agriculture and ICT security infrastructure development institutions
Content development	e-Agriculture content development institutions
Financial Institutions	Development of rural micro-finance service institutions for e-agriculture
Political/Legal	
Empowerment	Empowerment of women in agriculture
Knowledge Development	Development of a knowledge society especially in rural areas
e-Democracy	Rural grassroots community participation in democratic process

should not be viewed as passive beneficiaries but equal participants in the e-agriculture knowledge generation process. Without the necessary interaction and broad-based consultation among strategic partners at all the key nodal points of the information flow landscape, e-agriculture development will remain stunted. Moreover, any attempt to implement traditional top-down approaches in the development and diffusion of e-agriculture information will be "dead on arrival," as one of the cornerstones for the sustainable development of e-agriculture is mutual collaboration and broad-based consultation. In other words, the e-content should address the needs and concerns of the end users, as it is their satisfaction that will guide future demand for e-agriculture content and related services.

The development of viable and effective

Figure 6. The e-agricultural information flow and e-content development landscape, 2009

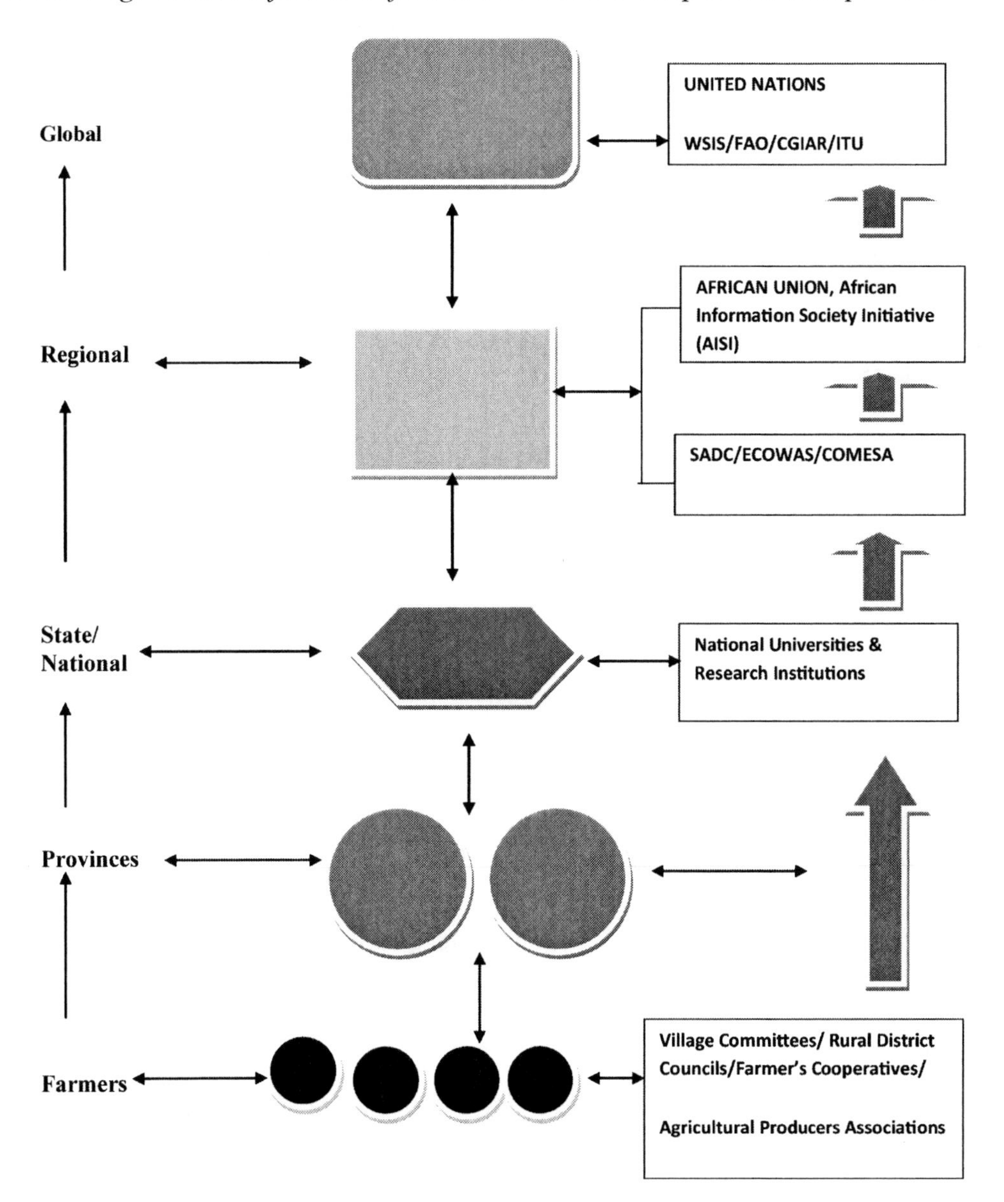

e-agriculture information architecture requires quality information. The information quality should be evaluated using the criteria of relevancy, reliability, adequacy, timeliness, accuracy, and dependability. In addition to the foregoing quality criteria, e-inclusivity and e-accessibility in terms of whether rural women and also disabled members of the agriculture community are able to utilize the e-agriculture information for decision making just like their able-bodied counterparts is of paramount importance and ought to be addressed proactively. Additional key quality assessment criteria gaining momentum include e-security, e-trust and use of standard operating procedures across back-end e-information systems that are compatible across government departments, national, regional and global institutions. The development of an e-information flow and e-content landscape is the lifeblood of e-agriculture, and without such investments, e-agriculture success will be limited.

Table 2 highlights the key stakeholders that are crucial in holistic e-agriculture development in SSA. It is important to not only identify the different stakeholders but to also coordinate their specific roles to avoid unnecessary duplication, develop, store, and disseminate the much needed e-content at all levels. The stunted development of e-agriculture content at one of the key strategic levels will lead to poor coordination and harmonization of e-agriculture development initiatives on the continent. A well developed e-agriculture information flow and e-content landscape makes it easy to identify any blockages, eliminate duplication, and share crucial e-content, and develop an institutional framework for information quality assessment. The importance of mapping the information flow landscape and e-content development needs (i) helps in coordinating and integrating e-agriculture development at national, regional, and global levels and (ii) it minimizes the fragmented e-agriculture development that has characterized e-agriculture development on the African continent (Aregu et. al, 2008).

e-VALUE CREATION IN e-AGRICULTURE DEVELOPMENT

Meaning of e-Value Creation in Agriculture

This is the value-addition brought about by the use of ICT in agriculture. The use of ICT does not necessarily guarantee value to the user as functional literacy and user-friendliness affects e-value derived. E-value is perceived as "additional gain" or benefit derived from using the ICT and it can be expressed as marginal utility or satisfaction obtained by the end user. Such utility is obtained at various stages across the agricultural value chain where ICT is deployed. The four sources of e-value creation are (i) farmers (ii) agribusiness industry (and ICT service providers), (iii) government extension workers and policy makers and (v) non-governmental organizations (i.e., private researchers, consultants, foundations, non-profit organizations etc). Each of these stakeholders play a pivotal role in promoting value creation in e-agriculture.

There are a number of ways in which e-value creation arises in e-agriculture. First, e-value actually occurs when new agricultural uses are identified for existing technology. Second, e-value creation is generated when new ICT are rolled out and new users or new markets developed. According to Bryceson, (2006), the need for value creation in an electronically enabled e-world is equally important as it is in none electronically enabled world. The author indicates that e-value arises from reduction in information asymmetries between buyers and sellers arising from ICT usage, online product bundling, and the use of lock-in strategies to encourage repeat transactions through e-loyalty programs. Other possibilities for increasing e-value are reduction in information search costs, e-trust, e-content development, speed, accuracy, and timeliness of information flow up and down the supply chain, and simplicity of transactions (Bryceson, 2006; Maumbe et al, 2006). The possibilities for e-value creation are immense, and increased competition across the value chain will also drive new innovations in e-agriculture.

Consequences of Lack of e-Value Creation

The likely impact of paying poor attention to e-value creation is poor diffusion and adoption of ICT in agriculture. End user satisfaction is compromised as e-services are partially provided. E-value creation generates new demand for e-agriculture services and it promotes sustainable development of e-agriculture. The more e-value creation opportunities there are, the greater the

Table 2. Global, regional and domestic institutions for e-agriculture development, 2009

Organization/Institution/Resources	Description
Global Institutions	Food and Agriculture Organization (FAO), International Fund for Agricultural Development (IFAD), International Telecommunications Union (ITU), Center for International Agricultural Research System (CGIAR)
Global e-Information Resources	AGORA, AGRIS, TEAL, CTA, FARA etc
Continental Institutions	Africa Information Society Initiative (AISI), African Virtual University, IST Africa Conference
Continental Resources	Open Society Initiative for Southern Africa (OSISA), Open Society Initiative for West Africa (OSIWA), *Kwetu*
Regional Institutions	Southern African Center for Cooperation in Agricultural Research (SACCAR), Association for Strengthening Agricultural Research in East and Central Africa (ASARECA)
Regional Resources	Southern African Development Community (SADC) Plant Genetic Resources Center (SPGRC)
Domestic Institutions	National Universities, Universities of Agriculture, National Agricultural Research Councils, Ministries of ICT, Ministries of Agriculture and Rural Development, Ministries of Livestock Development, Non-governmental Organizations (NGOs).

likelihood of success of e-agriculture. Increased isolation could result when global e-supply chains are integrated to reduce transaction costs and increase product marketing efficiencies without similar developments taking place in SSA.

Figure 7 provides a basic model for e-value creation in e-agriculture. The diagram identifies the key actors that drive e-value creation in e-agriculture. The e-value creation process operates on multiple frontiers. Conceptually, the space between the actors or stakeholders is the enabling environment for e-value creation in e-agriculture. It is important to note that, e-value creation does not occur without dedicate institutional arrangement for generating assets such as e-security, e-trust, e-policies, and ethics among others (see Figure 8). Sound research and development processes results in cutting edge technologies and e-innovations in e-agriculture. An enabling environment that engenders the development of e-innovations, e-security and e-trust assets require a policy framework that guides ICT deployment in the economy. Crafting sound e-policies and e-strategies will help SSA countries articulate the rules and regulations that promote the development, diffusion and adoption of best practices in e-agriculture. With e-value creation, e-agriculture development will be more sustainable and opportunities derived will continue to expand into the future.

In agriculture, the application of ICT has been growing at terrific pace and new and innovative approaches have been developed. As shown in Table 3, the use of ICT in agriculture in different countries cut across the entire agriculture supply chain. Examples of key innovations in e-agriculture are found in countries as diverse as Greece, India, Thailand, Senegal, and Zimbabwe to name a few. Some of the latest applications include input procurement (e-orders), e-consultation and e-advisory services, irrigation e-monitoring systems, food ration calculators, seed tracking systems and e-resource mapping, regional plant genetic resources network or gene-banks and rural digital banking services (i.e. e-banking and m-banking etc). The specific ICT intervention varies from real time communication, real time decision making, and real time access to key strategic information.

Figure 7. e-value creation structural organization in e-agriculture

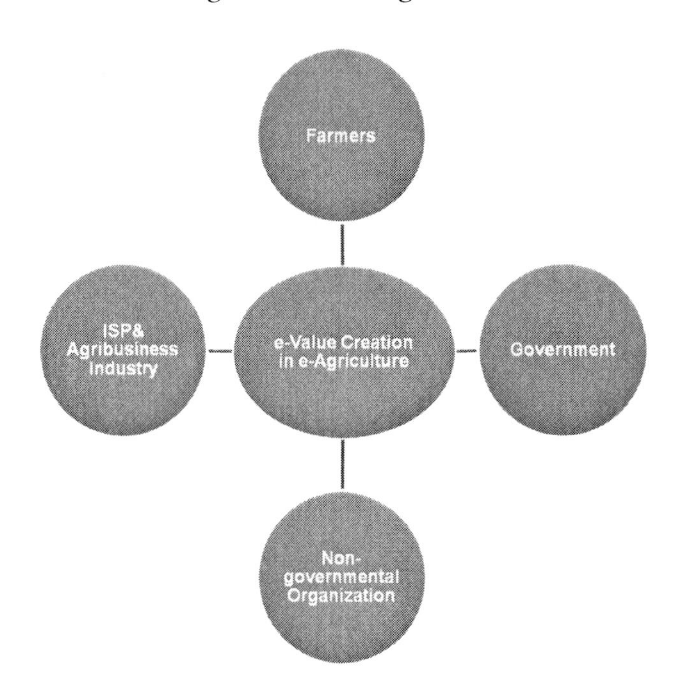

Figure 8. e-value creation in e-agriculture in Sub-Saharan Africa, 2009

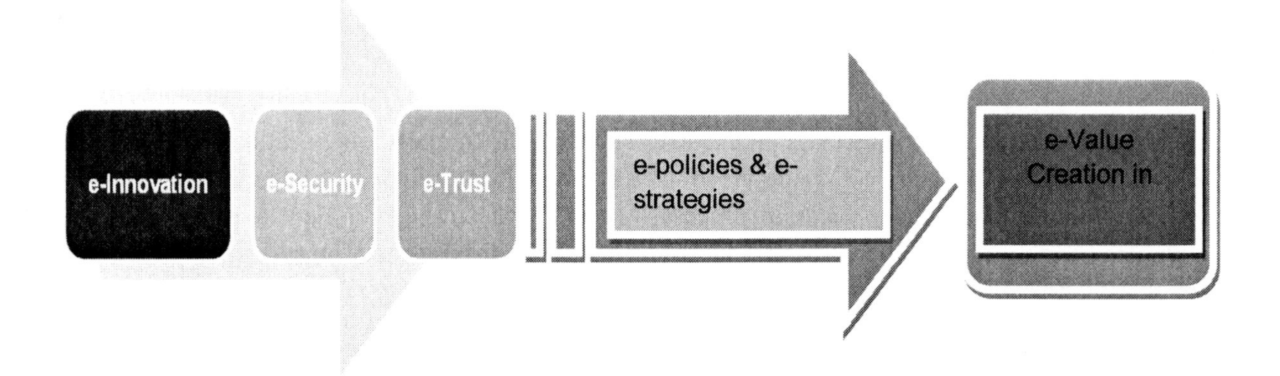

Real time strategic information includes input and product pricing data, weather reports, pest and disease surveillance reports, market assessments such as availability of buyers (i.e., market demand conditions), and changes in product availability (i.e., market supply conditions).

In marketing, new applications have mushroomed and these include electronic sales forecasting, new online payment systems, (e.g. pay pal etc.), and use of e-kiosks in rural areas. What is also clear is that apart from individual country initiatives, there seems to be concerted effort to promote growth and development of e-agriculture at the regional (e.g. SADC Plant Genetic Resources Center (SPGRC) Network) and global levels (e.g. FAO e-agriculture initiative). The sustainable application of modern ICT in agriculture and rural development depends to a large extent on the level and degree of coordination, integration and even standardization of technological processes.

Table 3. Selected global innovative ICT applications in agricultural development, 2009

Type of ICT Use in Agriculture	Development Sector	Applications	Organization/Region/Donor
e-Bario Project Online Community	Rural development Handcrafts making	E-marketing of barrio (fragrant) rice, e-repository for cultural knowledge.	UNIMAS, Marudi District & Bario Communit, Comserv, Telekom Malaysia
Smart cards	Rural banking services Rural insurance services	Micro-deposits Smaller ATM Banking for the poor & migrant workers	ICIC bank, International Finance Corporation (IFC) in India
e-auction software	Tea auction centers Tea sales governance	e-auctions for tea trading	NSE IT Limited, Tea Trade Association of Cochin, India
Online agro-technology Internet Kiosks	Targets rural farmers Certificate and diploma level training	E-learning modules in tissue culture, farm management, and marketing techniques	University of Sri Lanka, Colombo, Sri Lanka & Sri Lanka Government's Distance Education Modernization Project (DEMP) funded by Asian Development Bank grant.
Rural e-Gov Project	Rural SMEs access to e-government services	Rural e-Gov Observatory E-learning E-content development	European Union, Greece etc.
Consultative Group on International Agriculture Research (CGIAR) web-based learning resources	Online agriculture resources repository Global network of organizations	Online learning resources (OLR) for agriculture	CGIAR's ICT-KM Program & European ARIADNE Initiative
Rural Radio Stations	Radio stations for farmers	Broadcasts on weather, latest technology, and pest outbreaks to promote agriculture development.	Birsa Agricultural University (BISA) & Agriculture Technology Management Agency
Mobile Reuters Market Light (RML)	Links farmers to markets	Mobile phone based service. Market information service. Updates on pricing, agricultural news and weather.	Global Reuters Market Light Initiative 7 Government of India
Vintage Point Network	Serving farmers' crop management needs using a paid-subscription web presence.	Web-based farm management advice. Provision of precision farming information, weather forecasts, and financial records	John Deree, Special Technologies Group
Wireless soil sensors	Farm productivity Soil and water management	Collect data on soil moisture, temperature and nutrient.	Iowa State University
Type of ICT use in Agriculture	**Development Sector Targeted**	**Applications**	**Organization/Region**
Smart Grid Energy Internet	Energy conservation Emission reduction Climate change	Smart building designs, smart logistics, smart electricity grids, smart industrial motor systems, tele-working, video-conferencing, e-paper & e-commerce.	The Climate Group & Global e-Sustainability Initiative (GeSI).
www.e-agriculture.org	Agriculture and rural development	Interactive exchange of opinions & good practices. Promote sustainable agriculture & food security through increased use of ICT.	United Nations, Food and Agriculture Organization (FAO) & Global Community of Expertise Initiative.

continued on following page

Table 3. continued

Type of ICT Use in Agriculture	Development Sector	Applications	Organization/Region/Donor
SADC Plant Genetics Resources Centre (SPGRC) Regional Network	Regional Plant Genetic Resources-Genebank	Coordination of SADC plant genetic resources Long-term storage of plant genetic resources Immediate use in crop improvements	Southern Africa Development Community (SADC) Plant Genetic Resources conservation and safe preservation
Mobile Technologies	Livestock production system	Livestock production management Linking farmers to markets	The Cyber-shepherd Program, Senegal
Thailand Marketing Information System	Agricultural Marketing	Real time marketing information dissemination, GIS, etc	Thailand
e-Hurudza	The 5 Agro-ecological Zones in Zimbabwe	Improvement of farm management and production yields through e-consultation and e-advisory services	Government of Zimbabwe
National Spot Exchange Limited (NSEL) for direct marketing of oilseeds, pulses and spices.	Transformation of rural economy	e-trading platform National level electronic spot market for agricultural commodities. Development of an India common market	Joint venture of Financial Tech India Limited (FTIL), Multi-Commodity exchange (MCX) and National agriculture Cooperative Marketing Federation of India Limited (NAFED)
Mobile phones for rice farmers. IMUS's Agriculture Network in Philippines.	Delivery of rural farm services to rice farmers	Assessment of ICT needs of underserved rural communities. Provision of readily access information across the agriculture value chain.	Swedish Program for Information and communication Technology for Developing Countries (SPIDER).
PARMA Knowledge Network (PARMa KN)	Global network of agri-food professionals	State of the art research Information for business	City of Parma, Italy.

It is important to note that the pace at which ICT are integrated into the national school education curriculum is pivotal to the future success of e-agriculture development in SSA. The educational curriculum in some SSA countries has been infused with ICT through the NEPAD e-school initiative, and teaching subjects on e-agriculture will benefit its development substantially. The coordination and integration of e-agriculture developments at the national, regional and global levels as demonstrated by the information flow and e-content development landscape will spur e-agriculture development worldwide.

Despite the early advances that have been made in e-agriculture development, there are numerous hurdles that still need to be surmounted and these are discussed in the next section.

KEY CHALLENGES FACING e-AGRICULTURE DEVELOPMENT IN SUB-SAHARAN AFRICA

The widespread use of ICT in African agriculture faces a number of challenges. These challenges arise from the diverse nature of African countries, and in particular the unique socio-economic conditions of smallholder farmers in general. The common challenges discussed in the section below are, poor diffusion and awareness raising,

ICT skills shortage in agriculture, wide diversity of ICT to learn, diverse languages and cultures, lack of robust e-agriculture curriculum on the continent, digital divide and rural infrastructural gaps, riskiness of agriculture, poverty dilemma, and lack of enabling policy environment.

Poor ICT Diffusion and Awareness

Most SSA countries have relatively low level of e-readiness despite the fact that Internet use in Africa has been rising at more that 600 percent between 2000 and 2007. Despite such high penetration rates, majority of the people are unaware of the Internet and have difficulties conceptualizing benefits from a resource they do not fully understand or know. Rapid uptake of e-agriculture depends on ICT infrastructure, expanding knowledge about the benefits from e-services, e-collaboration and the use of e-communication strategies. Placing the rural people at the centre of ICT revolution increases community awareness, empowerment and social participation. This requires optimizing the use of different ICT channels such as Internet, kiosks, mobile phones, podcasting, iTVs, and PDAs to enhance awareness, social inclusion and penetration.

ICT Skills Shortage in Agriculture

Most African countries face the stark reality of a serious shortage of requisite skills in ICT. The future success of e-agriculture development in Africa will depend to a large extent on the availability of critical mass of ICT literate individuals especially in rural areas. High staff turnover in the public sector in countries such as South Africa create a new twist to ICT development as implementation gaps arise when e-champions relocate from public to private sectors or from one province to another.

Wide Diversity of ICT to Learn

Learning how to use new ICT technologies is a major challenge in Africa. The availability of various ICT channels is both an opportunity and a constraint. It is an opportunity in that it affords the agricultural community a wide range of technology choices. However, it is a constraint as it forces users to learn how to use the ever increasing diverse technologies at the same time. Learning the computer takes time, so does the use of new software. When new technological gadgets are added on the list of what farmers should understand, this creates literacy challenges. Effective use of ICT requires thorough understanding of how they work or function. The rapid introduction of new modern technologies with powerful applications in e-agriculture such as 3G phones, i-phones, black berries, PDAs and other smart phones to name a few simply means additional time will be required to train local farmers on how to use the western technologies. The functional literacy challenge is complicated by the numerous innovative ICT applications constantly unleashed on the global market annually.

Diversity of Languages and Cultures on the African Continent

Africa has more than 50 different countries and each country has several languages and dialects. For instance, South Africa has more than 11 different languages. Making the ICT available in local languages enhances its uptake and use. However, developing new or converting the old e-content available mostly in English into the various languages spoken in different African countries and regions or districts within countries is a major undertaking. Yet failure to make the modern technologies available in useable formats or languages is a major disincentive to their adoption in the long run. The diverse culture also presents challenges people expect to see a balanced representation of indigenous and cultural knowledge in the available e-content.

Lack of a Robust e-Agriculture Curriculum

The sustained development of e-agriculture requires the effective development and integration of e-agriculture in the school, college and University curriculum. Meaningful application of ICT in agriculture does not arise unless if it is supported by the development of experts that will provide leadership to the growing industry in the 21st century. Although e-content in agriculture has been developed by several organizations including the Food and Agriculture Organization (FAO), Southern Africa Development Community (SADC), and the International Research System (CGIAR), the formal education system in African countries is yet to adopt e-agriculture curriculum, or at least offer specialized degree programs in this area with such a growing importance.

The Digital Divide and Rural Infrastructural Gaps

The problem of the digital divide in SSA has been well documented. At the global level, the digital divide is still pervasive. According to the International Telecommunications Union (ITU), only 18 percent of the global population has Internet access. While the digital divide is shrinking, ITU estimates that about 1 billion people majority of who depend on agriculture for their livelihood, still lack connection to any kind of ICT. The digital divide is particularly a major problem for e-agriculture development in SSA because farmers reside in remote rural locations that are poorly serviced by ICT infrastructure. For e-agriculture to have positive developmental impact on the continent, it is critical that issues of access to ICT are resolved. One way to minimize the negative impact of the digital divide is for African governments to invest heavily in the development of Internet Service Provision and the necessary support infrastructure for broadband, satellite, dial up, and mobile services including the towers for relaying the cell-phone signals in rural areas. A major commitment is required in sustained human capital development such as webmasters, cell-phone and computer technicians, content development specialists, civil and electrical engineers to champion the spread of the ICT revolution in the rural areas.

The Risk of African Agriculture and its Implications on e-Agriculture Development

African agriculture faces diverse production and business risk. Production risks arise from fluctuations in weather patterns, drought, floods, outbreaks of pests and diseases, while business risk is associated with fluctuations in factor and product markets including interest rates. African agriculture is also characterized by some of the most severe droughts recorded in human history. Since smallholders dominate African agriculture, their ability to cope with production and business risk is limited. Failure of their enterprises tends to overshadow the success that might arise from utilization of ICT. It is also important to understand that even if seasons are normal, ICT have the inherent risk that comes from inefficiency of Internet Service Providers (ISPs), erratic supplies of electricity, and technological malfunction. The riskiness of agricultural enterprises and that associated with specific technology failures imposes challenges to the success of e-agriculture on the African continent.

Lack of Enabling Policy Environment

The absence of systematic ICT policy development remains a major constraint on the continent (AISI, 2008). Most African countries do not have ICT policies suggesting that current e-agriculture initiatives may not necessarily be supported by clear priority setting, and development of e-strategies and procedures. An enabling policy environment assist in the development of regulatory frameworks, institutional reforms and the crafting of tax incentives for ICT venture capitalists and infrastructure development. The development of e-agriculture in a policy vacuum makes it difficult to secure the intended benefits in the long-run.

The Poverty Dilemma for e-Agriculture Development

Poverty is both an opportunity and challenge for e-agriculture development on the continent. The opportunity is that innovative applications of ICT help minimize the impact of poverty on the continent through increased food production, improved access to markets, reduction of information search costs, better decision making, and rural employment generation. The challenge is that majority of the farmers do not afford the state of the art ICT and affordable ones may offer only basic functions which may not lead to meaningful transformation of their livelihoods. The problem of poverty is worsened by failure to articulate the rationale and benefits of ICT deployment for e-agriculture among improvised communities in Africa. Finally, the question of providing Internet to communities without access to basic services such as water, health and even education raises ethical and developmental questions. Nonetheless, the fact that ICT application in agriculture has the potential to reduce poverty and facilitate the achievement of the MDG is not disputable.

CONCLUSION

The chapter described the growing importance of e-agriculture development in Sub-Saharan Africa (SSA). The rapid development and diffusion of ICT has transformed agricultural development by lowering transaction costs, enhancing production technologies, and improving access to both local and global agricultural markets. Most developing countries have adopted ICT in their agricultural sectors, and some of the innovative uses are showing potential signs of raising the standards of living of the people by reducing the impact of poverty.

The chapter proposed a framework for e-agriculture development. It highlights four phases that should guide the systematic deployment of ICT in agriculture in SSA. These phases are; the establishment of preconditions for e-agriculture development, e-service delivery model, ICT development and diffusion pathway, and e-agriculture information flow and e-content development landscape. The chapter discussed the importance of e-value creation in developing a sustainable e-agriculture sector and provides selected case studies of successful e-agriculture initiatives around the world.

The chapter outlined some of the specific challenges that should be addressed by African governments in order to maximize the success of e-agriculture in their respective countries. The pace and pathway of e-agriculture development will vary from country to country depending on resource endowment, human capital and ICT investment levels and policy. The chapter provided important lessons about the (1) critical consideration in e-agriculture development (2) the key challenges to address when developing sustainable e-agriculture in SSA.

Policy Implications

From the chapter, it is important to observe that the growth of e-agriculture requires that SSA countries should develop policies, strategies and support institutions to enable ICT use to flourish in agriculture and rural development. These interventions should include the following;

- National e-policies and e-strategies to promote e-agriculture development and incentives for e-value creation.

- Establish and strengthen institutions that develop e-agriculture information and build culturally relevant and context specific balanced multi-lingual e-content.
- Development of e-agricultural markets, agricultural market information system, and e-agricultural services at all the different stages of the food and agricultural value chain.
- Promote policies that boost regional economic integration and incentives for global agricultural trade e.g. e-security, e-trust issues for regional information databases, etc.
- Develop and integrate e-agriculture curriculum in national higher education institutions.

The future success of e-agriculture on the continent of Africa requires the development of locally relevant e-content, deployment of suitable affordable ICT, improvements in rural ICT infrastructure, and the identification of perceptible benefits arising from the use of ICT. Government policy makers should craft e-policies, e-strategies, and programs that stimulate e-agriculture investments. In that regard, the development of comprehensive demand driven e-services is vital for e-agriculture's future success. Finally, it is critical that governments should try and invest in sustained e-innovations to enhance end user satisfaction, expand opportunities of ICT utilization in agriculture and rural development, and contribution towards raising the standards of living of majority poor smallholders in Sub-Saharan Africa.

REFERENCES

Aregu, R., Bagaya, M., & Nerbonne, J. (2008). An ICT-Based Digital Content Information Access Framework in Developing Countries: the Case of Agricultural Informatics Access and Management in Uganda. In *Proceedings of IST-Africa Conference*, Windhoek, Namibia.

Atkins, P., & Bowler, I. (2001). *Food in Society: Economy, Culture and Geography*. London, UK: Hodder Arnold.

Bertolini, R. (2003). *Making Information and Communication Technologies Work for Food Security in Africa*. Washington, DC: International Food Policy Research Institute.

Braga, F. & Baker, G. A. (2008). Parma Agrofood Research Management Knowledge Network: PARMA KN. *International Food and Agribusiness Management Review, 11*(3).

Bryceson, K. P. (2006). *'E' Issues in Agribusiness: The 'What', 'Why' and 'How.'* Gateshead, UK: Athenaeum Press.

Economic Commission for Africa. (2008). *Africa Information Society Initiative: A Decade's Perspectives*. Addis Ababa, Ethiopia: Author.

Kapange, B. W. (2008). Using ICTs for Managing Network Genebank: The Case of SADC Plant Genetic Resources Centre. In *Proceedings of IST-Africa Conference*, Windhoek, Namibia.

Maumbe, B. M., Owei, V., & Alexander, H. (2008). Questioning the Pace and Pathway of e-Government Development in Africa: Case of South Africa's Cape Gateway Project. Government Information Quarterly, 25.

Maumbe, B. M., Taylor, W. J., Erwin, G., & Wesso, H. (2006). E-Value Creation for Government Web Portal in South Africa. In A.Tutnall (Ed.), *Encyclopedia of Portal Technologies and Applications*, Hershey, PA: IGI Global.

Songan, P., Yeo, A. W., Hamid, K. A., & Jayapragas, G. (2006). Implementing Information and Communication Technologies for Rural Development: Lessons Learnt From the eBario Project. In *International Symposium on ICT for Rural Development*, Faculty of Computer Science and Information Technology, UNIMAS, Kota Samarahan, Sarawak, Malaysia.

Chapter 6
Diffusion and Dissemination of Agricultural Knowledge:
An e-Communication Model for Rural India

I.V. Malhan
University of Jammu, India

Shivarama Rao
Wipro Technologies, India

ABSTRACT

Almost sixty five percent of Indian population is engaged in agriculture that contributes to food security of the world's second largest populated country. Though agriculture sector shares 26 percent of GDP, this sector is very crucial for the sustainable growth and development of India. The emerging agricultural challenges demand information intensive agriculture work and applications of state of the art knowledge to enhance agricultural productivity, but non-accessibility of information and subsequently awareness and knowledge gaps that exist in this sector, enormously affect agricultural productivity. Efforts are being made for e-communication of information in rural India. This chapter portrays such efforts of public and private sectors, pinpoints the problem areas for accessibility of latest agricultural knowledge and suggests an e-communication model suitable for transfer of agricultural knowledge in the rural areas of India.

INTRODUCTION

The world is facing twin challenges of economic growth and food and nutritional security of its inhabitants. Knowledge for enhancing agricultural productivity exists but is confined to limited pockets of populations. Many poor people do not know how to access it and use it to the best advantage of their limited land holdings. They therefore continue to live in misery and deprivation. "Of the world's 1.09 billion extremely poor peoples about 74 percent (810 million) live in marginal areas and rely on small scale agriculture for their livelihoods" (Bage, 2005). A number of persons form such farming communities are illiterates and are undernourished. This seriously affects their capacities to access knowledge and even work effectively. They are the producers but not the adequate consumers of food because of poverty. "Most of the 842 million undernourished people in the developing world

DOI: 10.4018/978-1-60566-820-8.ch006

today are from farming families in developing countries" (Prolinnova, n.d.).

Increasing the agricultural productivity with the application of state of the art agricultural knowledge is not only essential for feeding the growing global population but is also linked to improving the economic conditions of people and overall development and growth of the world economy. "Agriculture is a vital development tool for achieving Millennium Development Goal that calls for halving by 2015 the share of people suffering from extreme poverty and hunger. That is the overall message of 2008's World Development Report (WDR)....The report provides guidance to governments and the international community on designing and implementing agriculture-for-development agenda that makes a difference in the lives of hundreds of millions of rural poor" (World Development Report,2008). There is a great potential for boosting the economy by enhancing the purchasing power of rural populations. This is possible by increasing productivity through knowledge intensive agriculture, focusing development activities in the rural areas through appropriate technologies that will further help in creation of jobs and movement of goods and services in rural areas.

Management of agricultural resources including the agricultural knowledge, building farmers' capacities and removal of bottlenecks for good governance of agriculture at all levels is essential for increasing the contribution of the agricultural sector in the growth of economy. "To-day, rapidly expanding domestic and global markets; institutional innovations, finance, and collective information technology offer exciting opportunities to use agriculture to promote development. But seizing these opportunities will require the political will to move forward with reforms that improve the governance of agriculture" (World Development Report, 2008).

Agricultural Challenges and Opportunities in India

"India is endowed with a lot of water, sunshine, many rivers and good knowledge about agriculture. This needs to be better planned to improve yields, cold chains and exports of packaged and fresh products" (I-watch). India has a very rich biodiversity and is home to thousands of rare herbs and other plants. The country has about 127 different agro-climatic zones. Because of different climatic conditions at various geographical areas, one can find sandalwood trees in Karnataka, mango trees in Uttar Pradesh, apple trees in Himachal and Kashmir, coconut trees in Kerala and tea plants in Darjeeling and Assam and saffron flowers at Bhaderwah. India's traditional system of medicine namely Ayurveda is also mainly based on medicinal plants and products derived from plants. "Up to 80,000 crores can be earned from exports of medicinal plants"(Pimbert & Tom, n.d.). A program has been launched in Himachal Pradesh on August 3, 2008 to make this hill state the herbal state by encouraging every rural family to plant a medicinal plant on its land. The medicinal plants were provided by the forest department free of cost.

The first level of Green revolution helped India to reach the level of food sufficiently from a situation of food scarcity. The next level of Green revolution demands strategic management of natural resources, bridging the knowledge gaps and incessant flow of latest agricultural knowledge to rural areas. Enhancing agricultural productivity is important but effective management of agricultural produce in terms of storage, processing and marketing is equally important. "India produces the maximum of fruits and vegetables in the world about 55 million and 75 million tons per year and 40% is wasted. What are we doing about it? Should we let it rot? Or should we let others pick it up and export or market in India" (I-watch). There is a lot of potential for setting up vegetable and fruits based food processing industries in this country.

India's major agricultural problem is awareness gaps and subsequently the knowledge gaps that exist between this nation and other parts of the world and within the country, is a great challenge for facilitating the next level of Green revolution. "Given our current per hectare output, there is scope for at least a 50% increase in yields. That if we just use the technology currently available in the country.... Statistics bear this out. In rice for instance, per hectare in 2006 – roughly double the average our farmers were getting in 1960's according to data from the international Rice Research Institute (IRRI). But increase pales in comparison with yields farmers are getting in China, where the figure for 2006 stood at 6.26t/ha. Even the average for Asia as a whole is 4.7t/ha, almost 25% better than that in India" (Bhattacharya, 2008). The situation is not different for other crops. "In wheat, the gains have been a bit more impressive – yields have almost tripled since 1960's to 2.6 t/ha. But here too it is far below the output in China (4.1 t/ha) and almost half of what Europe produces (5+ t/ha). The stagnation in food-grains output is a public policy failure that can be reversed says Swaminathan who is credited with fathering the Green Revolution in this country" (Bhattacharya, 2008).

Several problems are cropping in the way of increasing crops yield. The ruthless use of fertilizers and pesticides increased the soil salinity and there are now initiatives for organic farming. Bumper crops are devastated at some places by floods whereas drought conditions prevail at other places. There is thus need for more effective management of water resources and irrigation systems in the country. The rise in temperature levels due to global warming is also affecting the agricultural productivity in India. Decreasing level of landholdings due to population growth, the rapid expansion of real estate are other factors telling upon the Indian agricultural productivity. At the individual farmer's level interrupted power supply, exploitation by money lenders, increasing cost of agricultural inputs, timely non-availability of fertilizers and quality seeds at some places is seriously affecting the productivity. Hurdles on routine work due to one reason or another and profits in comparison to efforts lead to lack of motivation among farmers. Agricultural productivity of India is in fact the combined balance sheet of productivity of individual farmers. The mantra for increasing crop yields is to render help to every farmer by imbedding latest agriculture knowledge and best practices in his work and fields.

The transfer of latest agricultural knowledge from labs to land is one of the neglected dimensions of Indian agriculture. "We have a huge gap between what is produced on research stations and demonstration fields and the average actual production. And that gap can be up to 200%. This means that at least in theory there's a potential for doubling yields if the recommended practices and crop varieties are followed" (Bhattacharya, 2008). Creation of new agriculture knowledge is thus not as much a problem as its communication to intended users. "According to latest National Sample Survey Organization report, 60% of farmers in India have no access to agriculture technology. Disturbed by this report, the Union Agriculture Ministry is now considering a proposal to revamp Indian Council of Agriculture Research (ICAR)... in a bid to bridge the gap between technology generation and technology dissemination" (Suryammurthy,2005).

A survey of hundred farmers each conducted in the Ludhiana district of Panjab and the Jammu district of Jammu & Kashmir indicates that even there are differences in crop yields in the same area for the same crop among some farm houses primarily due to knowledge gaps among farmers. Some farmers are able to create more economic value from their fields because of more awareness. "I planted the *Parmal* variety [of rice] in my fields last year while my neighbor Jasmeet Singh opted for *Muchhal* while I got a price of Rs. 750 per quintal for my produce, Jasmeet's crop fetched him a rate of Rs. 2200", said Joginder Singh, a farmer from Jeewan Nagar village of Sirsa"(Sushil,2008).

Presently the challenge for Indian agriculture is to transform Indian farmer's vocation to practice agriculture for subsistence to agriculture for profit. India is a welfare state and looking at the plight of farmers, Government is making every possible effort to help these communities. The Government of India, to tackle the debt problems of farmers and to prevent the agrarian crisis decided to free the farmers from their bank loans by keeping a budgetary provision of Rs. 5,00,000 million in the last year budget"(Economic Times, 2008). The best way to prevent Indian farmers from getting mired in economic problems and instead make profits is to help them to get access to high quality and tested agricultural knowledge that can help to increase their crop yields and create more economic value from their fields and efforts. Some individual initiatives for profitable farming ventures do exist here and there but there is no quick adoption and absorption of technologies that exist to increase yields and ensure more profits for farmers." Mohammed Mushtaq, a farm expert has helped as many as 265 farmers to start green house technology in Bilaspur, Hamirpur and Mandi regions of Himachal Pradesh. He has set up his own green house in 30,000 square yards in Morni hills in Sirmur district, where from he expects to earn 5 lakhs a day in 10 years time from now and about 10 – 15 farmers are consulting him daily"(Dhaliwal,2006). The agricultural societies, the farmer's associations, state and central government departments also make efforts to help farmers to shift to better methods of production and cash crops. For instance, "the National Horticulture Board in Collaboration with the state horticulture department had initiated efforts to promote strawberry cultivation in Jammu" (Khajuria, 2008).

All such combined efforts are not adequate enough for the large number of farmers that do farming in India. "The extension personnel of the Department of Agriculture disseminated the technological messages to the farmers manually. Through this approach information has not been able to reach majority of the farmers who have spread across the whole country. This gap remains a challenge for the extension system even today. ... Farmer's needs are much more diversified and the knowledge required to address them is beyond the capacity of the grass root level extension functionaries (Sharma, 2003). Further adoption of new technologies not merely depends on how many farmers are reached for transfer of knowledge but also how many farmers clearly understood the advantages of it and do the farmers approached really rely on and adopt with confidence the innovative approaches communicated by the extension workers? The agricultural information accessibility is just one part of the problem, making sense out of it and properly putting the acquired knowledge to practice in fact only help to serve the desired purpose of communication. The part of the knowledge understood is in fact is the knowledge communicated especially in the illiterate farming communities.

Knowledge and information is not the prerogative of agricultural scientists and extension workers only, a number of farmers having perceptible minds learn from their own experiences and innovate. "Darshan Singh Tibba, from Hayatpura village near Machhiwara has proved it right.... Some of his innovative methods of farming have now been adopted by scientists at Punjab Agricultural University (PAU)" (Bhakoo, 2008). Innovations in agriculture take place with a blend of experimentation and experiences conveyed. Flow of information and ideas from fields to research institutions is as important as transfer of knowledge from lab to land. Efforts have been made by the public and private sectors to create modern infrastructures for instant communication of information in rural areas. A country like India in fact requires continuing both the traditional approaches for transfer of agricultural knowledge and adopting modern methods for facilitating access to available knowledge and best practices.

Communication Infrastructure: Public and Private Sector Initiatives

Besides the traditional agriculture extension departments that reach up to farmers, the ICT based infrastructure has been created and is being expanded in India for communication of agricultural information. The agriculture extension requires to be revamped to make it more farmer related, information technology driven, integrated, accountable and employment oriented. The Department of Agriculture of the Government of India launched a scheme during (2005–06) that provides support to extension programs of various states for undertaking extension reforms. The effective communication of agricultural knowledge also depends upon what we actually have for communication and what potential impact the content has on improving the crop production, cutting costs and putting more profits in the farmer's baskets. If there is something potential and good for farmers, the later themselves also make efforts to reach up to the knowledge source.

Though agricultural knowledge is communicated generally by informing some farmers personally regarding better seed qualities, pest control, demonstrating better implements and innovative methods, the modern communication infrastructure do play a crucial role to reach up to huge number of farmers. The Government of India has set up the world's largest satellite based communication network namely National Informatics Centre Network (NICNET) which connects all the state capitals and districts of India with a super computer at Delhi. This network carries content of government information including information pertaining to the Ministry of Agriculture. NICNET has also established community information centers and thousands of web pages of information are hosted on this network. The National Informatics Center that manages and run this network is planning to develop a separate portal on each crop. The Department of Agriculture and Co-operation, Ministry of Agriculture, Government of India has initiated e-Governance programs such as ICT infrastructure for headquarters, networking of directorates and field units, development of Agricultural informatics and Communications (DAIC) and Kisan Call Centers, Agricultural Resources Information System (AgRIS) in eight pilot districts, Agricultural Marketing Information Network, Agricultural Census, Registration of Pesticides. "Government of India has decided to launch a Central Sector Scheme titled, 'Strengthening/Promoting Agricultural Informatics & Communications'. Of which one component is AGRISNET. The Objectives of AGRISNET is to provide improved services to the farming community through use of ICT.... Funds under AGRISNET will be provided to the State/UT governments or to autonomous bodies identified by the state/union territory governments on at the basis of specific project proposals submitted for the purpose"(Dept of Agriculture & Cooperation, a).

A National e-Governance Plan in Agriculture is in the process for e-governance of activities and programs of the agriculture sector. "Phase 1 of the plan identified the prioritized list of services to be provided to the farming community and list of processes requiring re-engineering for the purpose. The phase 1 of NeGP is nearing completion. The phase 2 will define the role private sector and civil society" (Dept of Agriculture & Cooperation). This e-governance plan when fully operational will help in just in time transfer of information, better coordination of activities and programs pertaining to agriculture and facilitate more awareness of ongoing work.

The process of economic liberalization has not only increased foreign investments in India but also brought new kinds of corporate work culture at some places as it also encouraged foreign firms to invest and improve infrastructure though for enhancing their own business operations and expanding their customer bases. This motivated the Indian private sector and public sector companies especially in the telecommunication sector to make more investments and because of growing competition set the process of frequent reduction in the charges of their services. Flexible regulations and private investments in telecommunication have also shaken the public sector telecommunication company from slumber to ameliorate its facilities, expand its infrastructure, reduce service charges and add value to its services. The tariff wars between telecommunication companies have brought telephony in India to affordability level for a large number of people and created an environment of completion to add unique value to their services that definitely helped to improve the quality of telecommunication services. The enhanced communication is subsequently helping in economic growth of the nation through diffusion of knowledge and best practices.

The rate at which the mobile telephony is expanding in India is fastest in the world. The telephone companies are even collaborating with other service providers, to enhance the olevel of their services. "Telecom major Bharti Airtel Limited and multi-state farmers' fertilizer cooperative Indian Farmers Fertilizer cooperative Limited (IFFCO) launched their rural joint venture initiative IFFCO Kisan Sanchar Ltd. The two promoter companies say joint venture company will provide a major boost to Indian agriculture by placing the power of telecom services in the hands of the farmers so that he may access information vital for enhancing farm productivity" (Bharti Airtel, 2008), As the mobile telephone network is now available in rural India and even in difficult terrains, this will facilitate quick transfer of information to farmers and just in time communication among farmers. The Bharti Airtel and IFFCO joint venture will facilitate enhanced communication among farmers and quickly provide requisite information to farmers. "The Joint Venture Company will offer products and services specifically designed for farmers through IFFCO – run societies in villages across the country . . . On offer will be affordable mobile handsets bundled with Airtel mobile connection. The farmer will also get access to a unique VAS platform that will broadcast five free voice messages on mandi prices, farming techniques, weather forecasts, dairy farming, animal husbandry, rural health initiatives and fertilizer availability etc. on a daily basis. In addition, the farmers will be able to call a dedicated helpline, manned by experts from various fields, to get answers to their specific questions" (Bharti Airtel, 2008).

The internet based services are also making inroads in Indian villages through various initiatives. Indian Tobacco Company (ITC)'s Agribusiness division has set up e-choupals in several Indian villages to help farmers to have access to information of their concerns and interests. "Launched in June 2000 'e-choupals', has already become the largest initiative among all Internet-based interventions in rural India. 'e-choupal' services today reach out to more than 4 million farmers growing a range of crops-soyabean, coffee, wheat, rice, pulses, shrimps in over 40,000 villages through 6450 kiosks across 8 states (Madhya Pradesh, Karnataka, Andhra Pradesh, Uttar Pradesh, Maharashtra, Rajasthan, Uttaranchal and Tamil Nadu)"(About e-choupal). Besides, helping the farmers to provide information of their interest in their local languages the ITC use this network for their own business for procurement of agriculture produce and selling goods and services to farmers. This network is being continuously expanded "The ITC is targeting to set up 20,000 e-choupals in one lakh villages by 2010" (Srinivasin, 2005).

The ITC's e-choupals is not the only corporate sector ventures that reach up to Indian villages through ICT networks. "In order to impact both livelihood opportunities and living standards of rural communities 'I-Shakti-an IT based rural information service has been developed by the Hindustan Levers Ltd to provide information services to meet rural needs in agriculture education, vocational training, health and hygiene. The promise of I-Shakti model is to provide need based demand driven information and service across a large variety of sectors that impact the daily livelihood opportunities and living standards of village community" (HLL Shakti,2008).

There are several other initiatives for online transmission of agricultural information in India "The wireless in Local Loop (WLL) technology developed by IIT Chennai has helped in providing the internet connectivity to 250 community Kiosks that offer these services to over 700,000 people in rural India"(Gaur,2003). EID – Parry also used WLL technology in Cuddalore districts of Tamilnadu for village Kiosks. Gyandoot, net is another attempt to set up rural information facility by the district administration of Dhar, Madhya Pradesh "The experience of "Gyandoot" indicates that the village information Kiosk can be self sustainable enterprises with potential to provide jobs for two young rural people at each Kiosk" (Sharma, 2003).

The major hurdle in providing e-agriculture services in Indian villages is frequent interruptions in power supply in some areas or non-availability of electric power in several villages. Of the 5.93 lakh villages in India over one lakh are still without electricity. The inhabitants of villages without electricity also need information for their work "Researches at the Kanpur- Luknow Lab of Media Labs Asia, based at IIT, Kanpur, have created "Infothela" – a mobile pedal-driven visit geared to bring the benefits of Internet, telephones and fax to villages where there is no land telephone and no electricity. The rickshaw-mounted PC is driven by a bank of batteries, which are charged by the dynamo action of the pedal. And to provide the Internet-connection, the media Lab Asia has created a 75 km long corridor between Kanpur and Lucknow using WIFI technology which was also unlicensed by the Government in recent months" (Kanpur-Luknow,2003).

Though a variety of technologies are available and efforts are being made both by the public and the private sectors to reach up to the Indian villages with information for developing the rural sector for their own reasons, the major impediment for transfer of agriculture knowledge is illiteracy and lack of information literacy even among literate population. The problem in India is not information availability or even information accessibility but that of getting understanding of information conveyed and putting knowledge to work. India needs an e-communication model where the technology may reach every village and agricultural information is transferred in an interactive mode and in a friendly way. It is not possible to set up state of the art infrastructure for knowledge transfer in every village. The government should develop interactive multimedia kits and some printed documents in local languages that can be kept at public libraries. For instance Nalluswamy Anandaraja, a farmer's son begged the young scientist award for developing farmer friendly interactive multimedia compact disc and testing its effectiveness in the transfer of farm technology. "He prepared a multimedia CD to educate farmers in three villages in Thoundamuthur block of the Coimbatore district of Tamilnadu state on reasons for eryophyte mite reducing the yield of coconuts. His finding was that the knowledge of the pest and its management among farmers went up from below 2 percent to over 50 percent" (Kumar, 2007). Such efforts for development of multimedia kits should be encouraged for all kinds of crops in various regions of India.

Farmers should also be informed and motivated through mobile teleconference facilities taken to fixed locations and at fixed time schedule serving a cluster of villages. The National Institute of Agricultural Extension Management, Hyderabad, popularly known as Manage is providing mobile video conferencing facility to farmers in their own environments. This facility is uplinked to satellite through a motorized antenna. "Farmers and farming communities are having a direct two way video communication with the district officials, technical resource persons and the agricultural experts. A team of technical experts on information and communication technology coupled with social resource persons from extension are facilitating communication and learning in these sessions" (MANAGE). Though the Internet based facilities are making in-roads in several Indian villages, mobile telephony, television and cable networks are popular sources for communication and information accessibility. Efforts should be made for developing high quality channels for communication of agriculture knowledge through television and cable services and developing mobile telephone based agricultural information services.

The strategies for e-communication of agricultural knowledge will also depend upon place to place within India. Though several villages in India do not have electricity, the state of Punjab has cent percent electrification. A random survey of hundred farmers in the Ludhiana district of this state revealed that 17 of these farmers even had the Internet connectivity at their homes. Similarly, though India is still grappling with the problem of illiteracy, the state of Kerala has hundred percent literacy rate. This state is now moving fast to train one member of every family in e-literacy. If this model is replicated by the other states, it will help to solve several problems associated with the transfer of agriculture knowledge and facilitate knowledge based agricultural work in India.

CONCLUSION

A country like India has unique problems for diffusion and dissemination of agriculture knowledge. Best practices and knowledge resources that can enhance agricultural production do exist in the country. There is no adequate use of such knowledge and practices due of various reasons. Because of wide spread illiteracy, the extension departments need to make extra efforts to demonstrate to farmers how to imbed new knowledge in their work to accomplish better crop yields and more profits. Knowledge gaps continue to exist as extension departments with their limited resources and infrastructure are able to reach only limited farmers. Should India ameliorate the traditional extension programs or develop and expand electronic communication networks for transfer of agricultural knowledge to farmers? A country like India requires doing both. The National agriculture policy emphasizes the use of ICTs for more rapid development of agriculture in India. The National Knowledge Commission (NKC) set up by Government of India also looked into issues related to agriculture research and extension to enhance facilities for creation and dissemination of scientific knowledge. The Department of Agriculture and Cooperation is taking care of plan for e-governance of agriculture.

There are several ongoing programs and appropriate technologies exist for modernization of agriculture. Efforts are also being made both by the public and private sectors to reach up to Indian villages with the expansion of their computer networks. What is now required along with further expansion and modernization of facilities is consolidation of such efforts and synergy in various initiatives for facilitating more effective utilization of the infrastructure created or being created. At the same time we also require to initiate massive programs for capacity building for our extension workers and farmers for faster diffusion of knowledge and absorption of appropriate technologies in the interest of better performance in the agriculture sector. The state of Kerala has taken a lead for capacity building of people.

Though the Internet based services have reached several Indian villages, cable, television, radio and mobile telephones are most popular media for farmers to access and exchange information. There is also urgent need to create content of need based agriculture knowledge in local languages and disseminate it keeping in view the local culture and social ecology. Regular demonstrations and training of farmers largely help in transfer of agricultural knowledge because when the farmers are demonstrated new innovations and trained in handling new technologies for better productivity and profitability, they automatically get integrated in their work.

REFERENCES

Bage, L. (2005). *Statement delivered on the launch of MDG Report, 18 January, 2005*. Retrieved February 1, 2009, from http://www.ifad.org/events/mdg/ifad.htm

Bhakoo, S. (2008). Even scientists recommend his farming methods. *The Tribune, May 14, 128*(133), 5.

Bharti Airtel set up joint venture with IFFCO to provide rural mobile phone services, (n.d.). Retrieved February 1, 2009, from http://www.domain_b.com/companies/companies_b/bharti tele_ventures/20080502_joint.

Bhattacharya, A. (2008). India's food grain yield half of China's; scope for at least 50% increase in production. *Times of India, Wednesday, April 2, 7*.

Dept of Agriculture & Cooperation, Ministry of Agriculture, Govt. of India (n.d.). Retrieved from http://agricoop.nic.in/policyincentives/BRIEF%20ON%20 AGRISNET.htm Dept of Agriculture, Ministry of Agriculture, Govt. of India, (n.d.). Retrieved January 3, 2009, from http://agricoop.nic.in/Annualreport0607/INFORMATION%20TECHNOLOGY.pdf/

Dhaliwal, S. (2006). Helping farmers to change their fortune. *The Tribune, October 12, 2006, 126*(283)

e-choupal. Retrieved January 1, 2009, from http://www.echoupal.com/frontcontroller.ech

Economic Times (2008, February 29). Rs 644 crore for agriculture insurance scheme. Retrieved December 29, 2008, from http://economictime.indiatimes.com/articles show/2826829.cms

Gaur, R. K. (2003). Rethinking the Indian digital divide; present state of digitization in Indian management libraries. In T. A. V. Murthy, (Ed.) *Mapping technology on libraries and people* (p. 108). Ahmedabad, India: Inflibnet.

HLL Shakti changing lives in rural India. (n.d.). Retrieved December 1, 2008, from http://www.hllshakti.com/sbcms/templ.asp?pid=46802246

i-watch a wake up call for India. (n.d.). Retrieved February 1, 2009, from http://www.wkeupcall.org/employment/improving business.php.

Kanpur-Luknow media lab @ IIT Kanpur media lab Asia (2003). Retrieved November 12, 2008, from http://www.iitk.ac.in/MLAsia/infothela.htm

Khajuria, R. K. (2008). Strawberry growers laugh their way to bank. *The Tribune*, April 17,2008.

Kumar, B. A. (2007). Technology transfer paramount. *The New Indian Express*, Chennai, January 04.

MANAGE. (n.d.). *Making inroads into rural India through mobile VSAT videoconferencing van*. Retrieved November 17, 2008, from http://www.manage.gov.in/managelib/NewEvents/mobile%20VSAT.htm

Pimbert, M. P., & Tom, W. (n.d.). *Prajateerpu; a citizen jury- scenario on food and farming futures for Andhra Pradesh*. Retrieved January 13, 2009, from http://www.farming solutions.org/pdf.db/EPWPrajateerpu.pdf.

Prolinnova (n.d.). Retrieved January 13, 2009 from http:/www.oaklandinstitute.org/?g=node/view/159

Sharma, V. P. (2003). *Cyber extensions: connecting farmers in India-some experiences*. Retrieved January 13, 2009 from http://wwwgisdevepoment.net/pdf/i4d003.pdf.

Srinivasin, L. (2005). *HLL, ITC draw up two prolonged strategies to woo customers*. Retrieved February 1, 2009, from http:/www.ictportal.com/newsroom/press_apr_05.htm/p5.

Suryammurthy, R. (2005). ICAR revamp to benefit farmers. *The Tribune, September 27, 2.*

Sushil, M. (August 1, 2008). Farmers dump parmal for muchhal. *The Tribune, August 1, 2008, 7.*

World Development Report. (2008). *Agriculture for Development.* Washington, DC: The World Bank.

Chapter 7
Using ICT to Resolve the Modernization Paradox for Rural Communities in Africa:
A Theoretical Exploration and Conceptualization

Dawn Hinton
Saginaw Valley State University, USA

Joseph Ofori-Dankwa
Saginaw Valley State University, uSA

ABSTRACT

Rural communities are being heavily influenced by the ongoing modernization process taking place in all African economies and nations. Theoretically the modernization process is intended to help lead to an increase in the economic well being of the citizenry. However, one of the unanticipated outcomes of continuing urbanization and modernization, particularly for rural communities would be the loss of local social relations within such communities. This is similar to what happened in the Western context, where modernization, in the form of industrialization resulted in the loss of social relationships and increasing sense of alienation as cities formed. There is therefore a very real fear that in the African context, the ensuing modernization will result in a paradox where modernization may lead to an increase in economic well-being, but have the unintended consequence of increasing alienation and reducing the sense of community that exists in rural villages. The purpose of this chapter is two-fold. First, the authors theoretically explore the possibility of using Information and Communication Technologies (ICT) to develop a sense of community in rural villages and thus offset and mitigate the more negative aspects of the modernization process. Second, they propose a way to conceptualize this potential paradox by integrating the well established sociological concepts of Gemeinschaft (community) and Gesellschaft (individualism) with current paradox models of diversity and similarity curves. Such an approach has pedagogical utility in helping to describe and explain the modern paradox confronted by most of the African countries.

DOI: 10.4018/978-1-60566-820-8.ch007

INTRODUCTION

Countries in Africa are going through major economic and socio-cultural transitions associated with the modernization process. This modernization however produces a classical paradoxical situation where while, in some instances, it may result in the in increasing economic well being of farmers, it unfortunately may have several unintended consequences such as loss of sense of community, and increasing feelings of alienation and anomie. This is historically consistent with what happened in the Western economies, where the modernization process and its resultant industrialization has resulted in both economic wellbeing of their citizenry and an increasing loss of sense of community and relationships, particularly in larger urban cities.

The purpose of this essay is three fold. First, we argue for using Information and Communication Technologies (ICT) to develop community and thus leading to a situation where increasing levels of modernization would result in both increasing levels of economic well being and decreasing levels of alienation. We will however point out that, ICT should in no way be considered a panacea as its design and implementation, in the African context, is likely to face several daunting challenges and problems. Second, using the economic model of supply and demand as a heuristic, we will propose a way to conceptualize this paradox and its resolution by integrating the well established sociological concepts of Gemeinschaft (community) and Gesellschaft (individualism) (Tonnie, 1957) with current paradox models of diversity and similarity curves (Ofori-Dankwa & Julian, 2002, 2004). Finally we highlight the pedagogical implications of this essay.

Two caveats are however in order. First, while we recognize that the main focus of this book is on agriculture, we set out specifically to present a theoretical essay focusing on the broad effects of modernization on the rural communities in Africa, taking into account while there is some evidence of urban agricultural enterprises (e.g. Ezedinma & Chukuezi, 1999), that to date, most of the agricultural production in Africa comes from the rural communities. Second, though we generalize about the effects of modernization in Africa, we are very much aware that Africa is by no means monolithic and have distinct communities and regions. The trends that we highlight however have broad and general applicability.

Modernization and Loss of Community in Africa: A Modern Paradox

African countries are slowly going through the developmental process, in part spurred by modernization process. Indeed, many developing countries in Africa are experiencing some of the implication of modernization, in some ways are similar to the process that the US and other Western economies experienced them during the industrial revolution.

From a sociological perspective, the industrial revolution and its modernization process lead to rapid social changes. One of the earlier thinkers and scholars of the consequences of this urbanization process was Wirth (1938) who suggested that urbanization, if unchecked could result in social disorganization and the disintegration of the sense of community. Wirth (1938) pointed out that as people began to move into urban communities from rural places they dissolve intimate ties that were established in rural communities and fail to make these ties in the new urban communities. These ties are difficult to make in urban areas, according to Wirth (1938), because "diverse population elements inhabiting a compact settlement thus tend to become segregated from one another in the degree in which their requirements and modes of life are incompatible with one another and in the measure in which they are antagonistic to one another" (p. 15). Hence, as people gather in these large, dense and heterogeneous areas it becomes difficult, if not impossible, for intimate ties and relationships to develop among different groups. "Personal disorganization, mental breakdown, suicide, delinquency, crime, corruption, and disorder" (p. 23) are the results of urbanization. Tonnie (1957) also describes in detail how the modernization process is impacting the traditional sense of community (Gemeinschaft) and slowly shifting communities towards a sense of individuality and high level of self-centeredness (Gesellschaft) that he characterized as dysfunctional.

Interestingly, several decades later, sociological scholars are still pointing to the potential negative consequences associated with unfettered and unchecked urbanization and modernization (Wuthnow, 1998; Putman, 2000). Putnam (2000) extols the virtues and bemoans the ongoing loss of what he termed "social capital", and its resultant effects on the development of our communities. Social capital, for Putnam (2000), refers to "connections among individuals—social networks and the norms of reciprocity and trustworthiness that arise from them" (p. 19). He attributes this decline in social capital to a combination of things including suburbanization, the pressures of time and money and other generational changes. He suggests that social capital is beneficial in that it "makes us smarter, healthier, safer, richer, and better able to govern a just and stable democracy" (p. 290). Indeed, both Wuthnow (1998) and Putnam (2000) suggest that the connections that individuals make are not the same as have been made in the past. According to Wuthnow (1998), "these connections are often looser than was true in the past" (p. 7).

Within the different nation states in Africa, one can discern a similar process in play. Slowly the modernization process is losing the strong communal ties that bind its citizenry. Particularly in the rural areas this modernization process is taking its toll, in part because there is an going movement from rural areas, where typically most of the agricultural sector is housed to the urban areas (Tacoli, 1998; Fall, 1998). For example, Fall (1998) documents how traditional kin structures are slowly being replaced by new "solidarity networks" arising out of both formal and informal work and religious associations. These new associations fray the old established traditional networks and resulting in fewer visits by rural migrants and increasingly undertaking only occasional participations in family ceremonies (Fall, 1998). The net result of this ongoing process is in essence creating a paradox where modernization may lead to an increase in economic well-being, but has the unintended consequence of increasing alienation and reducing the sense of community that exists in rural villages.

Several other scholars (e.g. Wuthnow, 1998, Gans, 2005), however, suggest that the consequences of modernization may not always and necessarily have to be as dire as Wirth (1938) predicts. Wuthnow (1998) argues that in the wake of the modernization process, individual's involvement in communities should be conceptualized not as simply declining but rather as changing. Gans (2005) is critical of Wirth's view and suggest that the modernization process, typically associated with the increasing size, density and heterogeneity of resulting urban cities do not necessarily mean that the total loss of primary group ties. Nor does it mean that relationships within such communities are irreparably and adversely impacted. Gans suggests that there are other cultural and social structural factors that either exist or can be developed that will positively influence the relationships that develop in urban areas. Thus, in essence, while Gans acknowledges the potential loss of some primary group ties, he nevertheless argues that individuals living in urban communities develop alternative relationship ties that are based on quasi primary groups.

With respect to these alternative relationships, Wuthnow (1998) suggests that it is possible for individuals to join organizations or interact with individual that meet their narrow specific needs as opposed to involvement in communal relationships with broad far reaching goals. Consequently, Wuthnow (1998) suggests that such relational changes may not necessarily result in the aforementioned negative effects. Indeed Wuthnow (1998) points out that we can't view these loosing of relational ties outside of the social structural context, because "this argument does not take adequate account of broader changes in social institutions. By ignoring these larger changes, it assumes that loose connections are the cause of social problems, rather than recognizing that loose connections and social problems are both results of other developments" (p. 222)

Thus, while our primary concern is how communities can loss relational and primary group ties as the result of the rapid social change associated with the modernization process, it is clear that by taking adequate steps, these potentially negative aspects of the modernization process may be mitigated. *Indeed, ironically, it may be through one of the manifestations of the modernization process (ICT), that these more negative and dysfunctional aspects of the modernization process may be mitigated.* From the perspective of the individual citizen, the potential for ICT to enhance the democratic process is sometimes referred to as *e-participation* (Amoretti, 2006). Adopting, extending and adapting this definition to our study, we think of e-community as centering on the extensive and effective contribution of ICT to increase the sense of community in rural areas.

In general, the potential for ICT to increase the sense of community with the rural villages of a nation will depend on two distinct variables—the government and individual citizenry (Yildiz, 2007; Fountain, 2001; Gil-Garcia & Pardo, 2005).

Using ICT to Mitigate the Negative Effects of the Modernization Paradox

The term "Information and Communication Technology" (ICT) refers to emerging technology revolving around the increasing availability and use of the Internet, personal and organization-wide computer systems and global wireless and satellite systems. The result of the ICT revolution is an increasing level of local and global connectivity. Several scholars point to ICT as a potentially powerful source for democratic governance and citizenry participation (Evans & Yen, 2005; Evans & Yen, 2006; Amoretti, 2006; Layne & Lee, 2001; Schelin, 2003; Fountain, 2001). In a similar vein, we think that ICT can be a potentially powerful source of generating, sustaining a sense of community in rural African settings.

In Africa, there has been a more liberal policy to improve the infrastructure of its existing ICT system and to attract investors. There is also the increasing trend of privatization and liberalization of the telecommunication industry in several countries in Africa (Wright, 2004). Several African countries have begun to introduce and implement ICT and its related policies (Haruna, 2003). A leading advocate of African ICT is the African Information Society Initiative (AISI) that was set up by the Economic Commission for Africa (ECA) which advocates the development and implementation of national information and communication infrastructural (NICI) plans. This is part of a trend noted by Juma (2005) where developing countries are beginning to work together to meet their technological needs. By the end of the year 2000, thirteen countries in Africa had NICI plans in place, while ten were actively designing them.

Without a doubt, the ICT capabilities of firms in Africa are significantly weaker than those of the industrialized countries such as Britain and Japan, and also the newly industrialized countries in Asia and Latin America, thus revealing a major development constraint (Lall, Navaratti, Teitel & Wiggnaraja 1994). In the same way that ICT, if unattended can increase the technological divide between more- and less-developed countries (Ifinedo, 2005; Singer et al, 2005; Wright, 2004), ICT can also increase the technological divide of communities that exist within the African nations. Referencing a 2003 United Nations Crossroads study, Evans & Yen (2006) compares and ranks global regions in terms of e-government readiness—Africa placed last. A similar study arrives at the same conclusion (Ifinedo, 2005).

Ghana's ICT policy is representative on ongoing efforts in different African countries. Ghana's ICT process was undertaken under the auspices of the African Information Society Initiative. The process, termed "ICT for Accelerated Development," identified the key role that ICT can play in furthering national economic development if these technologies are effectively utilized (www.ict.gov.gh). In August of 2002, the National ICT Policy and Plan Development Committee in Ghana highlighted a three-phased process based on the AISI/ECA methodology. In 2005, the National Telecommunication Policy was presented. It had the vision of creating a "true Information Society" which would provide "the greatest opportunity for economic growth, social participation, and personal expression."

The Advantages of Internet Communities

The potential advantage of participating in Internet communities is the potential to share information beyond your individual sphere of influence. Wellman (2001) suggest that Internet users can rely on the contacts of others when seeking needed information and that this expands the circle of contact for Internet users. Thus, it is likely that "Friends forward communications to third parties and, in so doing; they provide indirect contact between previously-disconnected people who can then make direct contact. When one's strong ties are unable to provide information, one is likely to find it from weak ties" (Wellman, 2001, p. 246). Within Internet communities strong ties are developed, those connections that are based on your direct circle of friends, are supplemented by weak ties (Granovetter, 1983), those relationships that are based on your friends circle. This type of arrangement has created a network that is socially and culturally heterogeneous. These relationships that develop via online communities can be beneficial to farmers in developing countries, in that they have access to a world of information that extends far beyond their sphere of interaction.

Although the creation of Internet communities is based on the culture and social structures of those who participate in these communities, they are not homogeneous. "Despite the Internet's potential to connect diverse cultures and ideas, people are drawn to online communities that link them with others who share common interests or concerns" (Wellman, 2001, p. 246). Many are members of these communities based on a common interest which transcends social demographic characteristics, and this can be empowering for members of lower-status (Mele, 1999; Wellman, et al., 1996). In addition, the idea that these groups have knowledge and information that may be beneficial to others can increase the self-esteem of its participants.

Agro-Based Advantages of ICT

An added benefit of ICT is that it can be helpful in developing the agro-base and the agro-business sectors of indigenous African economies (Badu, 2004). The possibility of substantially revitalizing the indigenous agricultural marketing system through the use of e-commerce practices is important (Entsua-Mensah, 2005). This is important because about substantial portions of the gross domestic product of African countries is agrarian-based, and about most of the population are located in rural areas. Such revitalization relies on ICT being conceptualized as a process that would lead to a reduction of transaction costs, reduction of market uncertainty, the expansion of markets, and the overall implementation of market-oriented institutional changes (Bertolini, et al, 2001). For example, vendors in African markets can utilize mainly cell phones to discuss fluctuating market prices, delivery and payment schedules, and the prices of competitors' goods (Bertolini et al, 2001).

Further, potential global ICT linkages beginning to provide jobs and employment opportunities to developing countries could be steered towards the agriculture sectors of the African economies (Brah, 2001; Juma, 2005; Zachary, 2002; Zachary, 2003). In addition, there is tremendous potential of ICT in effective management of existing forests in Africa (Sraku-Lartey, 2003; 2006a; 2006b). For example, the development of a computerized management information system for the forestry sector and its associated training programs will improve the skills of forestry information managers in Africa and the African continent will clearly benefit from the potential roles and advantages that such skilled personnel bring to the forestry management process (Sraku-Lartey, 2003; Sraku-Lartey, 2006a). A step in this direction is the Global Forest Information System (GFIS) spanning Ghana, Senegal, Gabon, Zimbabwe and Kenya, in which an integrated forest data management system was developed and personnel from the different countries were trained in its utilization (Sraku-Lartey, 2006b).

Role of Government in Developing ICT Related Communities

There is a relatively broad consensus among scholars about the use of information technology in government for the promotion of democratic values and through providing citizens with easy access to political information and increased mechanisms for participation (Gil-Garcia & Pardo, 2005; Gil-Garcia & Luna-Reyes, 2003). We think that this can be expanded to include generating a sense of community through the use of ICT. This focus we call e-community.

There are several things that governments in Africa can do to assist this drive. First, governments in Africa have to come up with easy ways to rapidly educate rural communities about ICT technologies. For example, government could facilitate, encourage or set up low level technology fairs to educate the broad mass of its citizens about ICT and its potential. Vesley (2003), for example, describes the technology fair set up by Ghana's African Information Technology Exhibition & Conference (AITEC) in Accra. Such technology fairs have great potential for increasing the awareness of the citizenry and businesses about ICT and its advantages. Indeed such technology fairs could be made mobile, and thus could be moved from rural community to another and thus help inform and educate a large segment of the rural population.

Second, governments in Africa could help set up, in rural areas and villages "information kiosks" that contain a cell phone, radio, television, and computer, easily bringing important information to secluded villagers (Mayur & Daviss, 1989). The concept of information kiosks is broadly similar to the ICT–hub model that Jacobs & Herselman (2005) proposes. Along these lines, an ongoing project in Bangladesh that is worthy of note, by Nobel laureate Mohammad Yunus, founder of the now famous micro-lending Grameen Bank involves placing a cell phone in each of the 65,000 villages in Bangladesh. An additional illustration of this point is the recent decision by Nigeria to provide a significant number of "$100 laptops" to a large segment of their school populations, and to teach them how to effectively use these for educational and civic communal purposes. This program, if effectively implemented, could serve as a major catalyst for advancing ICT use.

Third, and linked to the above, government-sponsored dialogue could utilize ICT to enhance e-community. We think that existing local government structures could be adapted to facilitate the development of e-community (Clossson, et al., 2002; Evans & Yen, 2005; 2006; Owusu, 2005). African governments should therefore develop the appropriate structures and mechanism that will help increase the levels of dialogue through, for instance, provide access to expression mediums (websites, phones, short message services (SMS), online videos, online chats, and etcetera) that are specifically geared to the encourage dialogue and interaction between farmers. To provide another example, there could be the development of a training program aimed at teaching farmers the fundamentals of blogging is another potential way public participation in governance can be increased. In addition, the continued introduction of computerization in the educational system, and a drive to provide the Ghanaian citizenry with access to low-cost computers and Internet connection, are all ways in which government-fostered dialogue can be enhanced.

Role of Educational Institutions

There is a significant drive to increase ICT-related skills and expertise in African, and several public and private institutions have been set up to achieve this end (Mangesi, 2007). For example, with respect to the role of ICT in education within Ghana, describes several initiatives and ongoing programs. Example of these ICT initiatives and projects include: expanding the deployment of ICT in both secondary and tertiary institutions, promotion of distance learning, the establishment of high speed ICT infrastructures, awards for teachers who excel in the utilization and teaching of ICT, and the establishment of several ICT-focused youth clubs (Mangesi, 2007).

A good example of the role of educational institution is illustrated by the Ghana Telecom University is a publicly funded tertiary institution focusing specifically on ICT and inaugurated in August of 2006. The university is affiliated with the Kwame Nkrumah University of Science and Technology and DePaul University in the U.S., and is planning to set up more centers of telecommunication and information nationwide (Mangesi, 2007). Plans are also underway to introduce online distance education in Ghana. The Ghana Telecommunications Training Center also provides several ICT-related workshops and seminars and trains more than 2,500 people a year mostly from the Ghanaian telecommunications sector (Mangesi, 2007).

Role of Non Governmental Institutions

In the African context, one should also not underestimate the potentially powerful role that NGO can play in the promotion of public awareness and education regarding e- community issues and implications (Closson, et al, 2002; Bingham et al, 2005). For such initiatives to succeed, there is the need to effectively use the more decentralized and non-governmental organizations (NGOs) (Closson, Mavima & Siabi-Mensah, 2002) . Several scholars point to the potential use of NGOs in Europe to increase the utilization of ICT, and thus serve as a tool for democracy enhancement (Taylor & Burt, 2005; Bingham, 2005). For example, voluntary sector organizations, through their independent gathering and dissemination of relevant information, can be helpful in developing e-community by providing easy access to in-house information (Taylor & Burt, 2005). Also, by providing hot links to websites of other organizations that are the sources of relevant and related information, knowledge and expertise, they assist the citizenry in engaging in the democratic process, usually making sure that this can be done with ease and immediacy, and typically engaging traditionally marginalized sectors and communities of a nation in the democratization process.

Using ICT to Sustain Relational Ties: Promising Signs

Several scholars indicate the increasing availability of ICT in Africa (Zachary, 2002; 2003; Zelnick, 2000; Obeng, 2003). To the extent that ICT becomes more readily available to the broad mass of rural citizens in Africa, the e-community process is greatly enhanced. In addition, increasing availability of ICT in Africa means that the technological divide is slowly narrowing (Zachary, 2002). Furthermore, mobile phones are becoming increasingly accessible to all Africans, and thus can be seen in a sense as a technology-equalizing agent (Zelnick, 2000). Thus, while the task of introducing ICT to rural areas in Africa might be difficult, we only need to take a look at the very successful introduction and utilization of mobile telephone technologies through out both rural and urban economies in Africa to realize the potential for new technologies to be effectively used. Indeed, the increasing ubiquity of cellular phones in Africa, a fact assisted by the continuing drop of their purchase prices points to their potential for helping to bridge the technological divide.

Potential Challenges

While we have identified and suggested the potentially powerful role that ICT can play in increasing the sense of community among social units that are being impacted by modernization, we wish to point out that ICT should not be viewed as a panacea. There are, in reality several major challenges and problems that African countries face as they try to use ICT to generate a sense of community in rural areas. First, some scholars that the increasing use of the internet may actually lead to social alienation and may produce loneliness and depression (e.g., Nie & Erbring, 2000; Kraut et al.,1998). This fear has however been mediated by research which indicates that those who frequent online communities also maintain their offline social ties and relations as well (Wellman 1996; Baym 1998; DiMaggio et al., 2001). Second, a major challenge comes from the fact that there is a substantial lack of requisite human capital and skilled ICT personnel (e.g. Zachary, 2003; Binns et al, 2005). Another major ICT challenge in Africa stems from the inadequacy of current information, communication and technological infrastructure which results in frequent breakdowns in network connections, and equally frequent power outages and vulnerable to potential virus attacks (Hinson, 2005; Adjei, 2004; Adjei & Ayernor, 2005). Yet another major challenge arises because the implementation of ICT initiatives, will run into this major mindset divide. By definition, ICT implies a highly decentralized, fast-paced, and continually evolving mindset. Yet, several scholars note the major cognitive challenge associated with the bureaucratic and centralized mindset that African government institutions often possess, and the mindset of citizenry empowerment that is associated with the decentralized institutional frameworks that they are supposed to institute (Mensah, 2005; Owusu, 2005; Haruna, 2003). Finally, finance is another area in which ICT-related challenges are encountered in the African context (Hinson, 2005; Binns et al, 2005; Vesley, 2003). Information communication technology is not inexpensive or readily available, especially in resource-constrained economies in Africa (Bertolini et al, 2001; Kwapong, 2007).

SUMMARY AND CONCLUSION

Given the above, ICT-related initiatives and programs that encourage and promote increased levels of democracy in developing countries in Africa are important, yet oftentimes underestimated, tools that assist with national economic development. Our recommendation is that governments, in an effort to balance the impact of increasing technology through modernization, develop targeted plans that would provide training and inexpensive access to the Internet for rural communities (Wellman, 2001). Thus farmers and rural dwellers are not likely to suffer the loss of community ties if given access to the technology that make online communities possible. We suggest that this corresponding increase in community, through Internet technology, would help offset the potential loss of personal ties and relationships that are likely to be associated with the modernization process. Indeed, several scholars suggest that ICT has the added benefit of making it possible for more rapid and sustained development of economies without the draining and negative aspects of the industrialization process (e.g. Mayur & Daviss, 1998; Jacobs & Herselman, 2005).

It is important to recognize the numerous challenges and problems that are likely to be associated with the development and effective implementation of an agricultural related ICT policy in Africa that helps to both modernize and minimize potential alienation (Alhassan, 2005; Alhassan, 2007). We have identified above a potential problem that is associated with an increase in modernization in the agricultural sector of developing countries in Africa. Specifically, we suggest that as there in an increase in modernization, there is the potential for a loss of sense of community and its attendant increase in levels of alienation and anomie. This also has the tendiency to be associated with increase levels of depression, anxiety and other neoro-pathological symptoms.

To counter this potential problem, we suggest that ICT as a potential avenue for addressing this problem. While this has potential, it is not without its own problems as well, especially in the African context. We identify several of these potential problem areas but point to promising trends in Africa that appear to be addressing these problem areas.

Depicting the Modernization Paradox Using Sociological and Management Conceptual Tools

We have presented the potential problem associated with the Modernization paradox and have highlighted the use of ICT as its potential resolution. We now turn to existing scholarly sociological and management tools to help us to conceptually and graphically represent this modernization paradox and its potential resolution. We use the notion of gemeinschaft and gesellschaft from the sociology literature and the notion of diversity and similarity curves from the management literature. Such a depiction has several advantages. It enables the relative ease in terms of pedagogy. The use of such graphs represents powerful and useful tools in helping teachers to teach about and graphically represent this on-going modernization phenomenon. Further, similar to the utility of graphs and curves, such as the supply and demand curves graphs from the economic discipline, it enables students, policy makers and also researchers to more fully and easier understand the dynamic implications of increasing modernization process.

Gemeinschaft and Gesellscahft: Sociological Conceptual Tools

The scholarly work of Tonnies (1957) has been very influential in establishing the concepts of gemeinschaft and gesellschaft in the sociology literature, and thus has long informed discussions on urban and community sociology. With respect to *gemeinschaft*, Tönnies (1957) suggest that "all intimate, private, and exclusive living together is understood as life in Gemeinschaft" (p. 33). Not only is gemeinschaft characterized by living together but also by the consensus that keeps us together as a totality. This consensus is based in unity, which according to Tönnies (1957), is characterized by "the closeness of blood relationship and mixture of blood; physical proximity; and for human beings, intellectual proximity" (p. 48). All things within gemeinschaft are held in common, including common evils, common friends, and common enemies. Ultimately, "the relationship between . . . the community and its members, is based not on contracts, but upon understanding, like that within the family" (p. 59). It is the commonalities among members of the community that facilitate the development of meaningful relationships within the community. Hence, community for Tönnies involves not only a spatial dimension but more significantly the social relationships among those members. Within gemeinschaft individuals "remain essentially united in spite of all separating factors.

The orientation of *gesellschaft* on the other hand is viewed as substantially different. Tönnies (1957) suggests that gesellschaft is when communities are "essentially separated in spite of all uniting factors" (p. 64). Tönnies (1957) suggest that within gesellschaft, most interactions are motivated by individual profit seeking behavior and that the relationship is so toxic that people in such communities "are forced to crowd each other out or to trip each other up. The loss of one is the profit of the other, and this is the case in every individual exchange" (p. 77).

Tönnies (1957) classical work seemed to conceptualize gemeinschaft and gesellschaft as a continuum, with communities moving from gemeinschaft towards gesellschaft. However, more recent scholars (e.g. Bender, 1978; Keller, 1998; Bell and Newby, 1972; Christenson, 1984) argue is that both aspects of gemeinschaft and gesellschaft can and do exists in all communities. Gemeinschaft and gesellschaft should therefore be better described as two distinct patterns of relationships and social bonds that can co-exist in societies. For example, Gemeinschaft and Gesellschaft have been described as different "forms of human interaction" (Bender, 1972, p. 33). Christensen (1984) supports Benders assertions in suggesting that "while there may be a shift toward more Gesellschaft types of social relationships in modern society, contemporary American should reflect complex forms of Gemeinschaft as well as Gesellschaft rather than a continuum of Gemeinschaft to Gesellschaft" (p. 162).

Diversity and Similarity Curves: Current Paradox Management Tools

The original formulation of the Diversity and Similarity Curves (DSC) was used to reflect the implications of workforce diversity and value congruence implications (Ofori-Dankwa & Julian, 2002). It was subsequently proposed as an approach to graphically represent diversity and similarity as a paradox and to capture different social science paradoxes such as work and play, and novelty and continuity (Ofori-Dankwa & Julian, 2004).

The diversity curve was conceptualized as curvilinear, slopes downward and represents the extent of one aspect of demographic diversity in an organization. The similarity curve is also curvilinear but upward sloping and represents the extent of value congruence around important values in an organization (Ofori-Dankwa & Julian, 2004). The DSC model can be used to graphically exhibit how diversity and value congruence can produce both positive and negative organizational outcomes (Ofori-Dankwa & Julian, 2004). Ofori-Dankwa & Julian (2004) suggests that "the DSC model's relatively dynamic and inherently robust nature allows for cross-disciplinary explorations of the interaction of dual paradoxes in the social sciences" (p. 13).

Integrating GM/GS and DSC to Reflect the Modernization Paradox

This study seeks to pair the gemeinschaft and gesellschaft concepts (from sociology) with DSC model (from management), and thus provide an opportunity to simultaneously explore the potential implications of modernization on both the economic well being and social relational ties of African communities. The pairing of these two conceptual frameworks from different disciplines also enables us to conceptually show how by increasing the use of ICT, the sense of community of rural African communities can be increased.

In applying the DSC model, economic well being is located on the X-axis and social isolation is located on the Y-axis. The X-axis represents the potential range of a community's economic well-being and starts with low levels to the left and high levels to the right. The Y-axis represents the potential range of social isolation within a community and starts at the bottom with low levels of social isolation and high levels of social isolation at the top.

There is substantial similarity in conception between the nature and implications of gemeinschaft and that of the concept of value congruence that underpins the similarity curve idea. Consequently, we argue that similar to the arguments detailed in

Figure 1. Diversity and similarity curves as Gessellschaft and Gemeinschaft curves

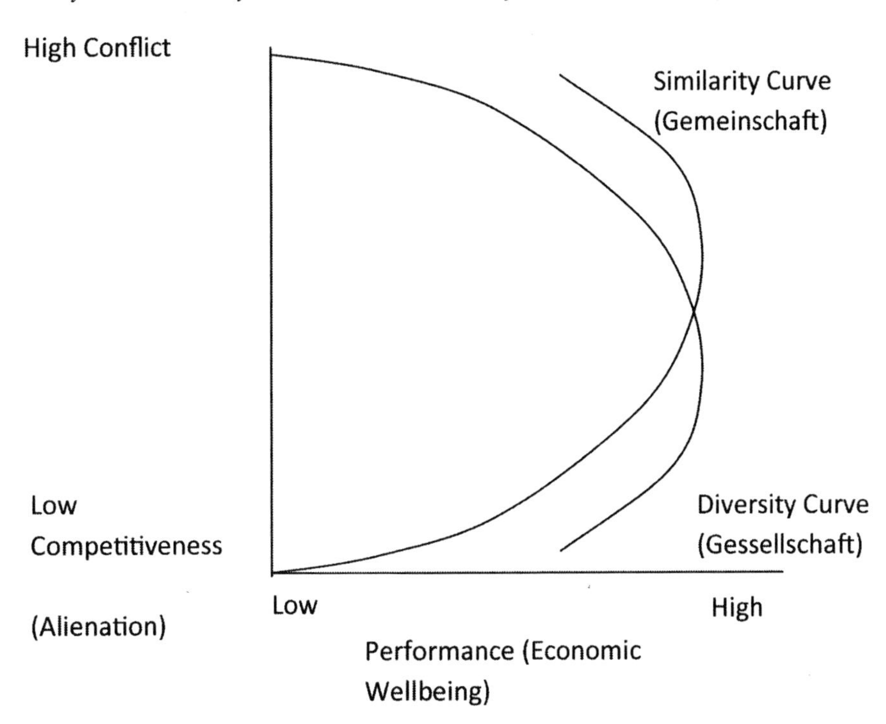

deriving the similarity curve (Ofori-Dankwa and Julian, 2004), we conceptualize the gemeinschaft curve as curvilinear, upward sloping and representing the extent of intimate social relationships in a given community. There is also substantial similarity between the nature and implications of gesellschaft and the idea of demographic diversity. Thus, we also argue that the gesellschaft curve is also curvilinear, downward and this represents the extent of gesellschaft-like relationships in a community. The intersection of the gemeinschaft and gesellschaft curves is associated with a certain level of economic well-being, depicted on the X-axis, and a certain level of social alienation, depicted on the Y-axis (See Figure 1).

Effects of Increases in Levels of Modernization

To depict the potentially paradoxical effects of modernization on a given African community, we can turn to Figure 2. Let us assume that a highly traditional community has an initial level of gesellschaft (D1) and an initial level of gemeinschaft (S1). In this case the intersection of gemeinschaft and gesellschaft correspond with low level of alienation (CO0) and also low levels of economic well-being (CR0). As modernization takes place, it has the potentially to increase levels of gesellschaft. As our figure 2 depicts, assume that the level of gemeinschaft remains the same (S1), then increasing levels of gessellschaft associated with increasing modernization will be represented by (D2). Its interaction with (S1) would be associated with higher levels of economic well being (CR1) than initially but with increasingly high levels of alienation (CO1). However as we consistently have higher levels of modernization without a corresponding mechanism to increase in a sense of community, this can result in the level of gessellschaft shifting to (D3). This would then be associated with very high levels of alienation (CO2) and a reduction in the level of economic wellbeing (CR2).

Figure 2. Implications of outward shifts of Gesellschaft (diversity curves) due to modernization with Gemeinschaft (similarity curve) not shifting

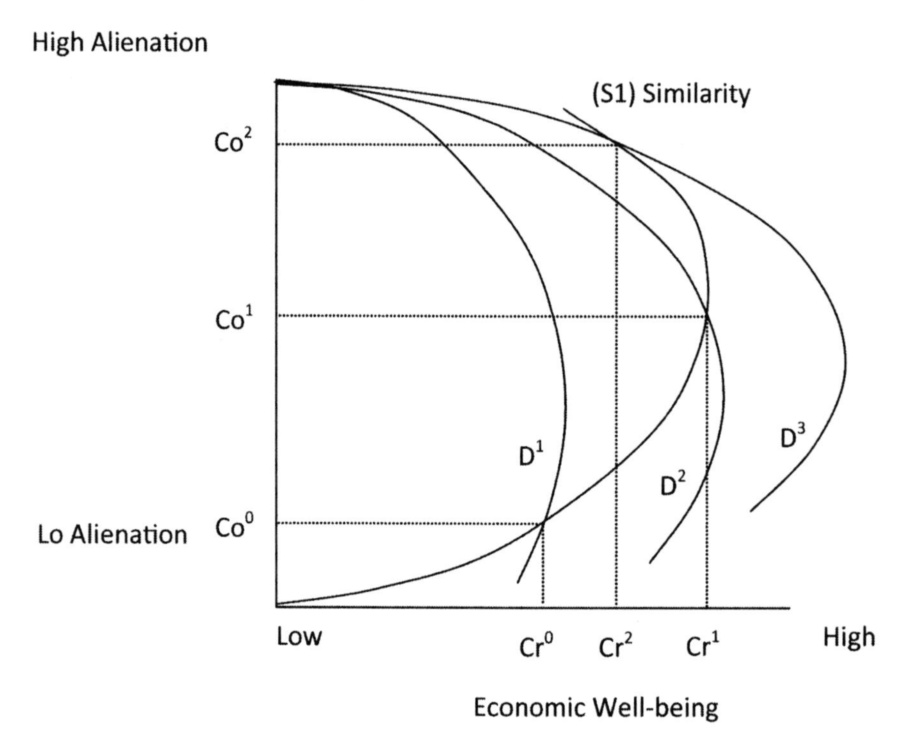

The scenario above suggests that a substantial increase in gesellschaft, with no change in gemeinschaft, can have dire effects on communities by increasing the level of social isolation and decreasing the levels of economic well-being.

Effects of Using ICT to Increase Community When Modernization is Increasing

Finally, to depict the effects of using ICT to increase community to resolve the modernization paradox, we turn to Figure 3. Assuming that there is an initial level of gemeinschaft (S1) and an initial level of gesellschaft (D1). If what sociologists such as Tönnies (1957) suggest is true then increasing levels of modernization will lead to increases in gesellschaft represented by D2. If the level of gemeinschaft in a society remains the same (S1), its interaction with high levels of gessellschaft will result in high levels of alienation (CO1) and nto much increases in levels of economic well being (CR1). However, if the increase in gessellschaft (D2) is associated with a corresponding increase in gemeinschaft (S2), this results in moderate levels of alienation (CO2) and high levels of economic well-being (CR2).

The above, thus represents an ideal situation when in discussing the modernization paradox, it potentially negative aspects are mitigated by having a situation of both high levels of gemeinschaft and gesellschaft. High levels of modernization provide individuals within a community substantial access to external resources. Indeed, as Portes and Sensenbrenner (1993) have suggested, the longer communities resist economic mobility through external means, the greater the chance that there will be little advancement in the community. However, encouraging the modernization process while increasing the sense of community through ICT use makes it possible particularly for communitie's individual members to tap into the

Figure 3. Implications of outward shifts of Gemeinschaft (similarity curves) through ICT as outward shifts of Gessellschaft (diversity curves) occur due to modernization

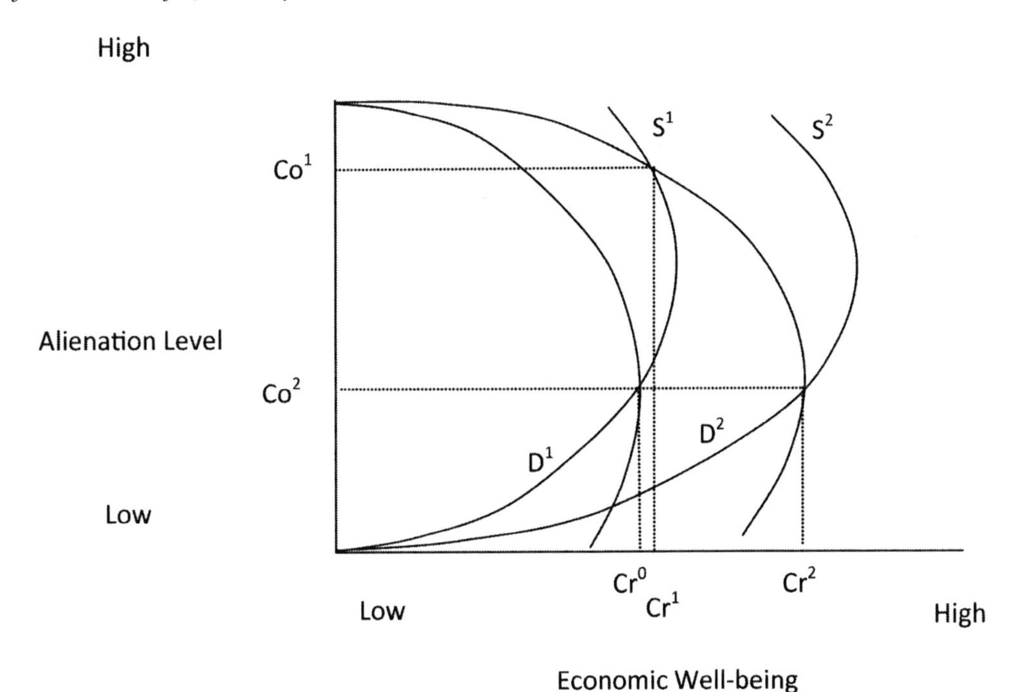

potentially powerful advantages associated with modernization without forcing them to "break their ties with their community" (Portes and Landolt, 1996, p. 4).

It is important to caution that we are definitely not advocating a whole sale focus on increasing the level of community and a whole sale stoppage of the modernization process. Indeed scholars such as Woolcock (1998) have described the disadvantages of social capital associated with the excess of community characterized by "fierce ethnic loyalties and familial attachments that members are discouraged from advancing economically, moving geographically, and engaging in amicable dispute resolution with outsiders" (p. 171). As Portes (1998) suggests, "the same strong ties that bring benefits to members of a group commonly enable it to bar others from access" (p. 15). Communities placing constraints on members by suggesting that interaction with outsiders should be limited encourage prejudice against non-members. Paradoxically, this lack of external contacts can prevent the success of business initiative by members (Brint, 2001; Portes, 1998; Portes and Sensenbrenner, 1993). Thus, similar to the notion of groupthink at the group/team level of analysis, an 'excess of community' does not allow for financial and economic development, and to the contrary may undermine and potentially destroy a community's development potential.

In other words, it is by combining both the advantages of having high levels of community and also high levels of modernization, that communities can reap the advantages of both situations. As the early segment of this paper suggest, in seeking to increase the "sense of community within communities", governments, NGO and other community institutions may increasingly have to turn to ICT as the new wave of the future

Concluding Remarks

This essay suggests that as developing countries increasingly modernize their economies, it is likely to result in increases in the economic well being of the country and its farmers. There may however be an unintended consequence where the modernization process is also likely to be an increase in the level of alienation and a reduction in the sense of community within these communities. We propose ICT as a potential avenue to help increase the sense of community within communities, reduce the level of alienation and thus offset the negative aspects of the modernization process. We also suggest that this can be undertaken by the provision of cheap inexpensive access to ICT by governmental and non-governmental and educational agencies and institutions within these countries.

This essay also integrates the traditional and well established sociological concept of Gemeinschaft and Gesellschaft with more modern paradoxical concepts of diversity and similarity curves (from the management field) to explore and conceptually demonstrate how, through the use of ICT, the negative and dysfunctional aspects of increasing levels of modernization can be mitigated. There are several implications of this essay.

First, for African nations, the increasingly global and interrelated nature of world wide economies means that there is an inevitability associated with the modernization process. There is also the reality that there are several negative and dysfunctional aspects associated with the modernization process. If history is a guide, one of the potentially dysfunctional aspects of modernization is the high level of alienation, anomie arising from the accompanying loss of communal relationships. Consequently, it is important to begin to think about ways and means to effectively manage this inevitability. As we suggest, ICT, though not a panacea, represents one of the approaches to help offset the negative aspects of modernization.

Secondly, there will be a pedagogical need to depict and reflect this modernization paradox and its complex implications. We think that similar to the supply and demand heuristic used in the field of economics to explain different economic conditions such as prevailing level of wages and labor, our proposed model will be useful for professors and researchers seeking to describe, explain and graphically represent the modernization paradox and its implications and complexities.

Finally, while we have argued for the use of ICT mechanisms to help increase the sense of community and to combat the potential for alienation that is associated with modernization, we wish to note that we are NOT against the modernization process. Indeed there is a down side of social capital that is developed within the context of indigenous communities, if they discourage modernization and only encouraging limited outside contact. Such approaches results in excessive community pressures and can be viewed as forms of social control that attempt to enforce conformity.

REFERENCES

Adjei, E. (2004). Retention of medical records in Ghanaian teaching hospitals: Some international perspectives. *African Journal of Library Archives & Information Science*, *14*(1), 37–52.

Adjei, E., & Ayernor, E. T. (2005). Automated medical record tracking system for the Ridge hospital, Ghana Part 1: Systems development and design. *African Journal of Library Archives and Information Science*, *15*(1), 1–14.

Alhassan, A. (2005). Market valorization in broadcasting policy in Ghana: Abandoning the quest for media democratization. *Media Culture & Society*, *27*(2), 211–228. doi:10.1177/0163443705050469

Alhassan, A. (2007). Broken Promises in Ghana's Telecom sector. *Media Development*, *3*, 45.

Amoretti, F. (2006). *The digital revolution and Europe's constitutional process. E-democracy between ideological ad institutional practices.* Paper presented at the VII Congresso Espanol De Ciencia Politica Y De La Administration, Grupo De Trabajo 9 Communicacion Politica.

Amoretti, F. (2007). International Organizations ICTs policies: E-Democracy and E-government for political development. *Review of Policy Research*, *24*(4), 331–344. doi:10.1111/j.1541-1338.2007.00286.x

Badu, E. E. (2004). Academic library development in Ghana: Top managers' perspectives. *African Journal of Archives and Information Science*, *14*(2), 93–107.

Baym, N. K. (1998). The emergence of On-Line Community. In S. Jones (ED), *Cybersociety 2.0: Revisiting computer-mediated communication and community* (pp. 35-68). Thousand Oaks, California: SAGE Publications, Inc.

Bell, C., & Newby, H. (1972). *Community Studies: An Intriductionto the Sociology of the Local Community.* New York: Praeger Publishers.

Bender, T. (1978). *Community and Social Change in America.* New Brunswick, NJ: Rutgers University Press.

Bertolini, R., Dawson Sakyi, O., Anyimadu, A., & Asem, P. (2001). *Telecommunication Use in Ghana: Research from the Southern Volta Region.* Working Paper, University of Ghana- Center for Development Research, Bonn University.

Bingham, L. B., Nabatchi, T., & O'Leary, R. (2005). The new governance: Practices and processes for stakeholder and citizen participation in the work of the government. *Public Administration Review*, *65*(5), 547–558. doi:10.1111/j.1540-6210.2005.00482.x

Binns, T., Kyei, P., Nel, E., & Porter, G. (2005)... *Africa Insight*, *35*(4), 21–31.

Brah, K. (2001). Ghana goes for IT lead. *African Business, July/August, 25.*

Brint, S. (2001). Gemeinschaft Revisited: A Critique and Reconstruction of the Community Concept. *Sociological Theory*, *19*(1), 1–23. doi:10.1111/0735-2751.00125

Christenson, J. A. (1984). Gemeinschaft and gesellschaft: Testing the spatial and communal hypothesis. *Social Forces*, *63*, 160–168. doi:10.2307/2578863

Closson, R. B., Mavima, P., & Siabi-Mensah, K. (2002). The shifting development paradigm from state centeredness to decentralization: What are the implications for adult education? *Convergence*, *35*(1), 28–42.

DiMaggio, P., Hargittai, E., Neuman, W. R., & Robinson, J. P. (2001). Social Implications of the Internet. *Annual Review of Sociology*, *27*, 307–336. doi:10.1146/annurev.soc.27.1.307

Entsua-Mensah, C. (2005). Revitalizing the indigenous agricultural marketing system in Ghana through the e-commerce project: A performance appraisal. *IAALD Quarterly Bulletin*, *50*(3), 141–147.

Evans, D., & Yen, D. C. (2005). E-government: An analysis for implementation: Framework for understanding cultural and social impact. *Government Information Quarterly*, *22*, 354–373. doi:10.1016/j.giq.2005.05.007

Evans, D., & Yen, D. C. (2006). E-government: Evolving relationship of citizens and government, domestic and international development. *Government Information Quarterly*, *23*, 207–235. doi:10.1016/j.giq.2005.11.004

Ezednma, C., & Chukuezi, C. (1999). A comparative analysis of urban agro enterprises in Lagos and Port Harcourt, Nigeria. *Environment and Urbanization, 11*(2), 135–144. doi:10.1177/095624789901100212

Fall, A. S. (1998). Migrants' long distance relationships and social networks in Dakar. *Environment and Urbanization, 10*(1), 135–145. doi:10.1177/095624789801000104

Fountain, J. E. (2001). *Building the virtual state: information technology and institutional change*, Washington D.C: Brookings Institution Press.

Gans, H. J. (2005). Urbanism and suburbanism as ways of life: A reevaluation of definitions. In J. Lin & C. Mele (Eds.). *The Urban Sociology Reader* (pp. 42 - 49). New York: Routledge.

Gil-Garcia, J. R., & Luna-Reyes, L. F. (2003). Towards a definition of electronic government: A comparative review. In A. Mendez-Vilas, et al. (eds.), *Techno-legal aspects of information society and new economy: An overview*. Extremadura, Spain: Formatex Information Society Series.

Gil-Garcia, J. R., & Pardo, T. (2005). E-government success factors: Mapping practical tools to theoretical foundations. *Government Information Quarterly*, 187–216. doi:10.1016/j.giq.2005.02.001

Granovetter, M. (1983). The strength of weak ties: A network theory revisited. *Sociological Theory, 1*, 201–233. doi:10.2307/202051

Haruna, P. F. (1999). *An empirical analysis of motivation and leadership among career public administrators: The case of Ghana*. Unpublished doctoral dissertation, University of Akron, Ohio.

Hinson, R. E. (2005). Internet adoption among Ghana's SME nontraditional exporters. *Africa Insight, 35*(1), 20–27.

Ifinedo, P. (2005). Measuring Africa's e-readiness in the global networked economy: A nine-country analysis. *International Journal of Education and Development using ICT, 1*(1), 53-71.

Jacobs, S.J. & Herselman, M.E. (2005). An ICT-hub model for rural communities. *International Journal of Education and Development using ICT, 1*(3), 57-93.

Juma, C. (2005). The way to wealth. *New Scientist, 185*(21), 15–21.

Keller, S. (1998). The American dream of community: An unfinished agenda. *Sociological Forum, 3*, 167–183. doi:10.1007/BF01115289

Kraut, R., Lundmark, V., Patterson, M., Kielser, S., Mukopadhay, T., & Scherlis, W. (1998). Internet paradox: A social technology that produces social involveemnt and psychological well-being? *The American Psychologist, 53*(9), 1017–1031. doi:10.1037/0003-066X.53.9.1017

Kwapong, O.A.T. (2007). Problems of policy formulation and implementation: The case of ICT use in rural women's empowerment in Ghana. *International Journal of Education and Development using ICT, 3*(2), 1-21.

Lall, S., Navaratti, G. B., Teitel, S., & Wiggnaraja, G. (1994). *Ghana under structural adjustment*. New York: St Martin's Press.

Layne, K., & Lee, J. (2001). Developing fully functional e-government: A four stage model. *Government Information Quarterly, 18*, 122–136. doi:10.1016/S0740-624X(01)00066-1

Mangesi, K. (2007). *ICT in education in Ghana, Survey of ICT and Education in Africa: Ghana Country Report*. Retrieved from www.Infodev.org/ict4edu-Africa

Matei, S., & Ball-Rokeach, S. J. (2001). Real and virutal social ties: connections in the Everyday lives of seven ethnic neighborhoods. *The American Behavioral Scientist, 45*(3), 550–564.

Mayur, R., & Daviss, B. (1998). The technology of hope: Tools to empower the world's poorest peoples. *The Futurist*, *32*(7), 46–51.

Mele, C. (1999). Cyberspace and disadvantaged communities: the internet as a tool for collective action. In M. Smith & P. Kollock (eds.), *Communities in Cyberspace*. New York: Routledge.

Mensah, J. V. (2005). Problems of district medium-term development plan implementation in Ghana. *International Development Planning Review*, *27*(2), 245–270.

Nie, N., & Erbring, L. (2000). *Internet and Society (Preliminary report)*. Palo Alto, CA: Stanford Institute for the Qualitative Study of Society.

Obeng, K. W. (2003). Ghana pursues justice and development through computer training. *Choices (New York, N.Y.)*, *12*(4), 20–21.

Ofori-Dankw, J., & Julian, S. D. (2004). Conceptualizing social science paradoxes using diversity and similarity curves model: Illustrations from the work/play and theory novelty/continuity paradoxes. *Human Relations*, *57*(11), 1449–1477. doi:10.1177/0018726704049417

Ofori-Dankwa, J., & Julian, S. (2002). Toward diversity and similarity curves: Implications for theory, research and practice. *Human Relations*, *55*, 199–224. doi:10.1177/0018726702055002183

Owusu, G. (2005). The role of district capitals in regional development. *International Development Planning Review*, *27*(1), 59–89.

Portes, A. (1998). Social Capital: Its Origins and Applications in Modern Sociology. *Annual Review of Sociology*, *24*, 1–24. doi:10.1146/annurev.soc.24.1.1

Portes, A., & Landolt, P. (1996). The downside of social capital. *The American Prospect Online.*

Portes, A., & Sensenbrenner, J. (1993). Embeddedness and Immigration: Notes on the Social Determinants of Economic Action. *American Journal of Sociology*, *98*(6), 1320–1350. doi:10.1086/230191

Putnam, R. D. (2000). *Bowling Alone*. New York: Simon & Schuster

Schelin, S. H. (2003). E-Government: An Overview. In G.D. Garson (ed) *Public Information Technology: Policy and Management Issues,* (pp. 120 - 137). Hershey, PA: Idea Group Publishing.

Singer, P. A., Salamnca-Buentello, F., & Daar, A. (2005). Harnessing nanotechnology to improve global equity. *Issues in Science and Technology*, *4*, 57–64.

Sraku-Lartey. (2006). Developing the professional skills of information managers in the forestry sector in Africa. *IAALD Quarterly Bulletin*, *51*(2), 75–78.

Sraku-Lartey, M. (2003). The role of information in decision making in the forestry sector: Developing a computerized management information system (MIS) for forestry research activities in Ghana. *IAALD Quarterly Bulletin*, *48*(1), 105–108.

Sraku-Lartey, M. (2006). Building capacity for sharing forestry information in Africa. *IAALD Quarterly Bulletin*, *51*(3), 186–190.

Tacoli, C. (1998). Rural-urban interactions: a guide to the literature. *Environment and Urbanization*, *10*(1), 147–166. doi:10.1177/095624789801000105

Taylor, J., & Burt, E. (2005). Voluntary organizations as e-democratic actors: political identity, legitimacy and accountability and the need for new research. *Policy and Politics*, *33*(4), 601–616. doi:10.1332/030557305774329127

Tönnies, F. (1957) *2002. Community and Society*. Trans. and ed. by Charles P. Loomis. East Lansing, MI: The Michigan State University Press.

Vesely, M. (2003). New technology for an old continent. *African Business*, July, 20-21. Website: http://www.ict.gov.gh

Warschauer, M. (2003). Demystifying the digital divide. *Scientific American*, *289*(2), 42–47.

Wellman, B. (2001). Physical place and cyberplace: The rise of personalized networking. *International Journal of Urban and Regional Research*, *25*(2), 227–252. doi:10.1111/1468-2427.00309

Wellman, B., Salaff, J., Dimitrova, D., Garton, L., Gulia, M., & Haythornthwaite, C. (1996). Computer networks as social networks; Collaborative work, telework, and virtual community. *Annual Review of Sociology*, *22*, 213–238. doi:10.1146/annurev.soc.22.1.213

Wirth, L. (1938). Urbanism as a Way of Life. *American Journal of Sociology*, *44*, 1–24. doi:10.1086/217913

Woolcock, M. (1998). Social Capital and Economic Development: Toward a Theoretical Synthesis and Policy Framework. *Theory and Society*, *27*(2), 151–208. doi:10.1023/A:1006884930135

Wright, B. (2004). Telecoms around the continent. *African Business*, May, 16-17.

Wuthnow, R. (1998). *Loose Connections: Joining Together in America's Fragmented Communities*. Cambridge, MA: Harvard University Press.

Wuthnow, R. (1998). Loose Connections. Cambridge, Massachusetts: Harvard University Press

Yildiz, M. (2007). E-government research: Reviewing the literature, limitations, and ways forward. *Government Information Quarterly*, *24*(3), 646–665. doi:10.1016/j.giq.2007.01.002

Zachary, P.G. (2002) Ghana's digital dilemma. *Technological Review, July/August*, 66-73.

Zachary, P. G. (2003). A program for Africa's computer people. *Issues in Science and Technology*, (Spring): 79.

Zelnick, N. (2000). Colonialism? Not again. *Internet World*, *6*(18), 15.

Section 3
E-Government Country:
Case Studies

Chapter 8
Connected Government for a Developing Country Context:
An Assessment of the Extent of Inter-Departmental Integration for Selected Government Departments in Zimbabwe

Ephias Ruhode
Cape Peninsula University of Technology, South Africa

Vesper Owei
Cape Peninsula University of Technology, South Africa

ABSTRACT

An information society begins with a connected government and ICTs are the bedrock and founding pillars of such societies. To assist public administrators think beyond traditional e-government, this study describes a concept of connected government, whose philosophy rests on the integration of back-end processes that facilitate collaboration among government agencies. This article describes a case study of five government-owned organizations in a developing country environment where even the basic e-government services are barely available. This study was carried out to determine the extent of integration within and across government agencies, with the aim of stimulating some thinking within and among government managers and administrators, around the possibility that a connected government can indeed be established in a developing country setting. The study exposes shortcomings to inter-departmental integration not only of the organizations under investigation, but also of other similar enterprises in developing countries within the same context. The paper concludes by proposing a set of recommendations toward diffusing connected government applications for inter-organizational collaboration.

INTRODUCTION

Information and Communication Technologies (ICTs) have been harnessed in many governments throughout the world in order to improve seamlessness and service delivery to citizens, business community and other stakeholders. However, the European Commission (2003) report the emergency of 'islands' of government that are frequently unable to interoperate due to fragmentation resulting

DOI: 10.4018/978-1-60566-820-8.ch008

from uncoordinated efforts at all levels of public administration. Information sharing across government agencies provides new opportunities to enhance governance, which can include improved efficiency, new services, increased citizen participation, and an enhanced global information infrastructure. Referring to knowledge sharing and information, the Canadian International Development Agency's Knowledge-Sharing Plan (2007) explains that it is an effort at "knowledge pooling" amongst diverse participants across all sectors of the economy in order to consolidate extant, generation of fresh, intellectual capital with a view to improving programming in governance and development (CIDA, 2007).

Given the foregoing, the purpose of this article is to address the concept of a connected government as a way of promoting information sharing within government agencies. This article is mainly motivated by the authors' experience with disjoint government agencies in developing countries where getting a full cycle of services is a not easy. This paper therefore makes a modest but significant contribution to the body of knowledge in e-government as the concept of connected government is virtually unknown in a developing country setting, especially in Africa.

The structure of the chapter is as follows: The next section brings to the fore the connected government concept, which in this article is also referred to as a networked government. The next section gives the background information of the country under investigation, which is the Zimbabwe Government. The section that follows explains case study as it is applied to academic research. The cases selected for this study are also described in this section. Findings of the study are presented in the next section which is then followed by the section on analysis of the findings. Recommended solutions to challenges which would have emerged from analysis of findings are presented before a general discussion in the next section. The last section gives a conclusion to the chapter.

The Connected Government Concept

Australia's Management Advisory Committee (2004) report that the distinguishing characteristic of whole of government (connected government) work is that there is an emphasis on objectives shared across organizational boundaries, as opposed to working solely within an organization. It encompasses the design and delivery of a wide variety of policies, programs and services that cross organizational boundaries. The connected government concept realizes the need to position the entire government as one integrated enterprise, underpinned by interoperability among its constituent departments and agencies. Simply put, connected government enables government to work as an integrated body that shows only one face to the public (United Nations, 2006). The concept of connected government can be understood in the context of e-government. As generally accepted, e-government is the use of information technology to support government operations, engage citizens, and provide government services. While this definition captures the essence of e-government, Dawes (2002) unpacks e-government to expose the following four dimensions:

- **E-services:** The electronic delivery of government information, programs, and services often (but not exclusively) over the Internet.
- **E-democracy:** The use of electronic communications to increase citizen participation in the public decision-making process.
- **E-commerce:** The electronic exchange of money for goods and services such as citizens paying taxes and utility bills, renewing vehicle registrations, and paying for recreation programs, or government buying supplies and auctioning surplus equipment.
- **E-management:** The use of information technology to improve the management

of government, from streamlining business processes to maintaining electronic records, to improving the flow and integration of information.

The connected government philosophy's thrust rests upon the last dimension, that is, e-management. According to the UN e-Government Survey Report (2008), the concept of connected government is derived from the e-government-as-a-whole concept which focuses on the provision of services at the front-end, supported by integration, consolidation and innovation in back-end processes and systems to achieve maximum cost savings and improved service delivery. By having related agencies across different levels of governments and also different agencies with different functionality communicate with each other, these consumers, according to Layne and Lee (2001), will then view the government as an integrated information base.

Governments in recent years have primarily embraced e-government to focus on Dawes' first three dimensions rather than re-organizing and re-designing government agency functions and services. Describing initial promoters of web-enabled applications, Davision et al (2005) indicate that typical approaches involved automation of the front-end web presence so as to spark e-commerce activity, but failed to integrate and re-design the business as a whole in order to make it truly web-centric. Ntiro (2000) produced a model which has three focal domains for e-Government initiatives, namely: (i) improving process – e-Administration, (ii) Connecting citizens – e-Citizens and e-Services and (iii) Building external interaction – e-Society, as shown in Figure 1.

While all initiatives are pertinent in Ntiro's e-government model, the e-Administration initiative appeals well to proponents of the connected government theory. Ntiro (2000) describes the following e-government initiatives within the domain that deals exclusively with improving the internal workings of the public sector:

- **Cutting process costs:** improving the input-to-output ratio by cutting financial costs and/or time costs
- **Managing process performance:** planning, monitoring and controlling the performance of process resources (human, financial and other)
- **Making strategic connections in government:** connecting arms, agencies, levels and data stores of government to strengthen capacity to investigate, develop and implement the strategy and policy that guides government processes.
- **Creating empowerment:** transferring power, authority and resources for processes from their existing locus to new locations.

Ntiro (2000) explains that through connecting citizens, relationships between government and citizens are enhanced; and through building external interactions, relationships between government and other institutions are improved. These institutions may be businesses, Non-Governmental Organizations (NGOs) or other countries.

Background Information on the Government of Zimbabwe

Zimbabwe has a parliamentary type of government that is run by the president and prime minister. The government controls senior appointments in the public service, including the military and police, and the independent Public Service Commission is charged with making appointments at lower levels on an equitable basis. Zimbabwe has been beleaguered by economic, social, and political turmoil since 2000 which has had a debilitating effect on its economy. The government's controversial land reform program has reportedly been the cause of significant damage to the commercial farming sector rendering the country a net importer of food after having traditionally been the source of jobs, exports, and foreign exchange (Isaacs, 2007). This

Figure 1. Focal domains for e-government initiatives. Source: Adapted from Ntiro (2000)

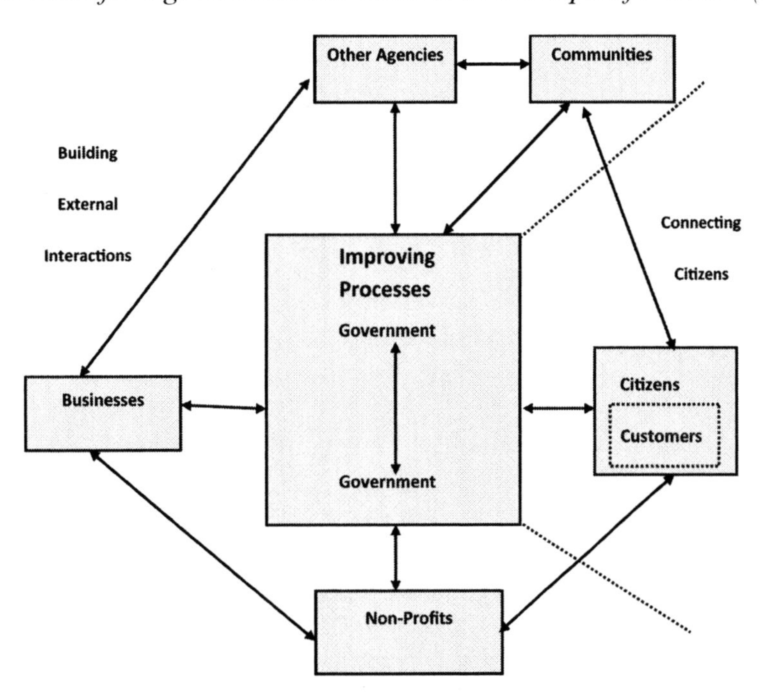

led to the isolation of Zimbabwe from the international community. Zimbabwean economy has since 2000 been beset with crises, characterized by an unsustainable fiscal deficit, an overvalued exchange rate, rampant inflation, above 80 per cent unemployment, decline in manufacturing and agricultural production due to farm invasions, company closures, human capital flight, and many other problems. It is in the light of the crises that the Zimbabwe government was forced to adopt a new structure in February 2009 that accommodates different political parties which were in tension. Some of the aims of the new government are to eliminate internal conflict, restore normal relations with the international community as well as creating a conducive environment for manufacturing and agricultural production.

The new government of Zimbabwe has thirty-one ministries (apart from ministries of state) and seven independent departments that are headed by principal directors. It is however important to point out that this study is carried out at a time the structure of the government is undergoing changes. Some ministries have parastatals or semi-government institutions under them. For example, directors of the following institutions report directly to the minister of Agriculture: Agricultural Bank of Zimbabwe (AGRIBANK), Tobacco Industry and Marketing Board (TIMB), Pig Industry Board (PIB), Agricultural Rural Development Authority (ARDA), Grain Marketing Board (GMB), Tobacco Research Board (TRB) and Cold Storage Commission (CSC). It is expected that these institutions that fall under one ministry should closely interoperate. However, this is not the case in these institutions. Marche and McNiven (2003) report that public administration has a general reputation of functional insularity, that is the tendency to not integrate service provisioning across government departments when responding to citizen's needs. Absence of collaboration and connectedness between departments government agencies leads to duplication of data, information and effort; slow response times; departmental barriers and bottlenecks; poor coordination and ineffective planning; and resource wasting. The next section describes the research method that was employed to carry out the investigation.

CASE STUDY APPROACH

As a research strategy, Yin (2003) assets that the case study is used in many situations to contribute to knowledge of individual, group, organizational, social, political, and related phenomena. Yin (2003:13-14) defines a case study as "an empirical enquiry that investigates a contemporary phenomenon within its real-life context, especially when the boundaries between object of study and context are not really evident". Dul and Hak (2007) add savor to Yin's definition by expressing that scores obtained in case studies are analyzed in a qualitative manner. A remarkable strength of case study research is that it yields in-depth information of the phenomena under investigation. Stake (1995), and Yin (2003) identified at least six sources of evidence in case studies, namely: documents, archival records, interviews, direct observation, participant-observation and physical artifacts. This allows for a rich description of the phenomena under study as it occurs in the natural environment.

According to Feagin et al (1991), a case study is an ideal methodology when a holistic, in-depth investigation is needed. A case study selected in this study is a problem-oriented one, which analyses a real life situation where existing problems need to be solved. As discussed earlier, a single case study is considered sufficient in this study to expose the connected government requirements of organizations.

Selected Cases

The complexity of the area under study rendered it impossible for data to be collected from all ministries and departments. Nonetheless, our experience of the disjointedness of systems in government agencies in Zimbabwe contributed to unraveling of the picture of these entities. Telephonic interviews were conducted with information system administrators of the National Blood Services of Zimbabwe (NBSZ), the National Aids Council (NAC), the Zimbabwe Tourism Authority and the National Social Security Authority (NSSA). Same interview was held with the deputy director at the Ministry of Science and Technology Development. We sought to establish whether the organizations are computerized, if they have local area networks (LANs), wide area networks (WANs), if they electronically share documents internally and if they collaborate externally with other organizations. Table 1 shows the interview questions that were used to gather primary data. It is our view therefore, that the underlying analyses of data thus collected from selected government institutions will be applicable to all ministries and departments as well. The five departments under investigation were randomly selected. The rationale behind random selection is that every government institution does not operate in total isolation but requires horizontal integration with other departments. Information is therefore deemed to flow in and out of each government agency in order to create a networked government. We have considered five departments as a fairly sizeable number to draw meaningful research conclusions from.

National Blood Services Zimbabwe (NBSZ)

NBSZ is the only licensed operator in human blood services and products in Zimbabwe. The organization operates within the ministry of health and child welfare. One of the company's core values says, "The voluntary non-remunerated blood donor drives the organization and the needs of the donor will be central to the operations of the NBSZ". Blood donors are either institutions or individuals. The Service operates according to international standards as set out by the World

Table 1. Interview Questions

QUESTION	*ANSWER*
1. How many users are in your organization	
2. How many PCs (including laptops) does your organization have?	
3. Which tasks do you use your computers for? (Tick all that apply)	Desktop Applications
	Line of business application
	Billing/accounting
	E-mail
	Web research
	Other (Specify)
4. Do you do the following? (Tick all that apply):	Share a printer, fax machine, or other office equipment?
	Share documents, client information and an appointment calendar?
	Use the Internet for work purposes
	Access company e-mail from anywhere
	Retrieve business information when travelling?
	Work with employees in different business locations
5. What kind of computer network does your organization currently have?	Peer-to-peer
	Server-based
6. Do you have a website?	
7. If your answer above is Yes, do you post downloadable information on your website – for use by your customers, suppliers or other organizations?	
8. If your answer above is Yes, what type of downloadable information do you post?	
7. Do you have any need to share information with other ministries/ departments?	
8. If your answer above is Yes, which ministries and or departments do you exchange information with?	
9. Do you have in information sharing policy or Strategy document?	
10. In general (do not include details), what kind of information do you exchange?	
11. How do you currently exchange information?	
12. How does your business connect to the Internet?	Dial-up (e.g 56k modem)
	High Speed (e.g Cable, DSL)
	No internet connection yet
13. What Operating Systems do you run on your PCs?	
14. What Network Operating Systems do you run?	
15. What tools do you use for Data Management & Analysis	Business Intelligence
	Database/Data Warehouse
	Knowledge Management
	Other (Specify)
16. What tools do you use for Enterprise Back Office?	

continued on following page

Table 1. continued

QUESTION	*ANSWER*
17. How do you rate ICT literacy and ICT skills in your organization?	Poor
	Good
	Satisfactory
	Excellent
18. Do you have a stand-alone IT department?	
19. If your answer above is yes, indicate how many of the following you have in the IT department	Technicians/Operators
	Programmers/Database/Network Administrators
	Senior Managers in IT

Health Organization (WHO). Within the context of this study, NBSZ exchanges data and information with other organizations as the WHO, the parent ministry, hospitals, medical school, National Social Security Authority (NSSA) and others. The organization has a wide area network (WAN) covering five cities – Harare, Masvingo, Bulawayo, Gweru and Mutare. Data that is recorded is blood donor (companies, schools, and individuals), stocks of blood, testing results, notification letters and other relevant information. Both the WAN and Local Area Network (LAN) are widely used as employees share network resources like printers and fax machines. Documents are also shared through file sharing facility across the WAN. Emails are extensively used especially in communicating with blood donors externally and with other employees internally. Information sharing with other institutions is done offline with a lot of paper trail.

National Social Security Authority (NSSA)

NSSA is a public institution, a government initiative that culminated in the introduction of social protection to Zimbabwean workers and their families. The Authority is mandated by government to administer Social Security schemes. The contingencies for which members are protected against are retirement, invalidity, death of a member and injury at work. In this way the Authority provides the first line of defense against poverty and destitution. NSSA sought the services of Zimbabwe Post (Zimpost), the government-owned postal operator, to act as a distribution agent for NSSA pensions owing to Zimpost's countrywide postal network. This relationship is designed to relieve pensioners of traveling costs, time and other inconveniences associated with traveling to towns and cities where NSSA offices are located. However, lack of electronic collaboration between the two organizations has negatively affected many pensioners as they still have to travel to NSSA offices. NSSA has a well established information technology department with WANs connecting offices in major towns.

Ministry of Science and Technology Development (MoSTD)

Some of this ministry's functions were taken over by the newly established ministry of ICTs. This study however took place when the MoSTD was still responsible for government computerization including e-government activities. All technology policies and programs were commissioned by the MoSTD. The Zimbabwe's e-readiness survey report of 2005 and the ICT policy framework of 2006 were commissioned by MoSTD. Despite the MoSTD having its own LAN at the head office, there was no online information exchange with the rest of government organizations except with ZArnet, a department in the MoSTD and the government telecommunication agency (GTA) which was also responsible for government wide network implementation and the government web portal.

Zimbabwe Tourism Authority (ZTA)

ZTA is an organization falling under the ministry of tourism and hospitality, whose mission is to market Zimbabwe as a leading tourist destination, set, and monitor standards, provide market research and statistics, and assist in creating and enabling environment, for the benefit of the nation and its visitors. For statistical purposes, ZTA exchanges tourist information mainly with the department of immigration and the central statistical office. ZTA also shares foreign exchange information with the Reserve Bank of Zimbabwe. ZTA has a LAN and hosts an efficient website but information sharing with other organizations is done offline and manually.

National Aids Council (NAC)

NAC is an organization enacted through the Act of Parliament of 1999 to coordinate and facilitate the national multi-sectoral response to HIV and AIDS. NAC's mission is to provide for measures to combat the spread of HIV and management, coordination and implementation of programs that reduce the impact of HIV and AIDS. In addition to interview question in Table 1, we also sought to establish from NAC if they knew at any given time the number of local and international HIV/AIDS workers by region, and also if they had any direct link with hospitals to obtain updates on HIV/AIDS related patients and how often they update their database from information obtained from other organizations that run HIV/AIDS mitigation programs in Zimbabwe like the Zimbabwe Aids Network (ZAN), ZNNP+ and SAfAIDS. Directly integration between the NBSZ and NAC is also expected as the latter collects statistics of HIV/AIDS contaminated blood. NAC has both LANs and a WAN and they run a website that is mainly informative. The website is regularly updated to reflect current information and news bulletins. However, requested information was not immediately available because of lack of collaboration among these institutions.

FINDINGS

The findings were that all five organizations do have computers running on Windows Operating System that are on server-based LANs and WANs. However, only regional offices in major towns and cities have online connections especially in the case of NSSA, NBSZ, ZTA and NAC. All five organizations are on the Internet on DSL connections and all do have online presence through websites. Most of the websites are predominantly informative without any online transactions or provision of downloadable material except for NSSA and ZTA where the former has all pension application forms on the website and downloadable in portable document format (pdf). NSSA also provides downloadable documents regarding occupational health and safety (OHS), that OHS policy, OHS constitution; OHS work plan, strategy document, OHS statistics, and annual reports. The ZTA website has interactive search facility where visitors can perform a search by for example hotel type or location. Electronic file sharing takes place internally in all five organizations.

Information exchange with external entities is done manually in all the five organizations. NAC and NBSZ exchange information with the parent ministry of health, the central statistical office, WHO, medical school, ZAN, ZNNP+, SAfAIDS and other organizations like the donor community in the case of NBSZ. As indicated earlier, ZTA collaborates with the department of immigration on foreign tourists, central statistical office, the central bank for foreign exchange remittances information and other tourism and hospitality associations. ZTA also passes information to the parent ministry of tourism and hospitality. The MoSTD operated with the support of the Research Council of Zimbabwe (RCZ), Scientific and Industrial Research and Development Centre (SIRDC), Zimbabwe Academic Research Network (ZARNet) and Biosafety Board. In order to play its role of unlocking the scientific and technological resources in Zimbabwe, the MoSTD had to do coordination, facilitation, monitoring and evaluation, advocacy, popularization, mainstreaming, harmonizing science and technology policies, seed financing and regularization of technology services throughout government organizations. It means therefore that MosTD had a wider community of organizations to do collaborate with. Horizontal integration with NSSA involves even more organizations that MoSTD. Every employer in Zimbabwe shares information concerning employee details and payments with NSSA. NSSA is supplied with this information and data electronically via companies' payroll systems and also through financial institutions like banks. However NSSA does not have electronic links with Zimpost and the parent ministry of social welfare.

Another pertinent observation is that all the five organizations have small to medium databases for data management and analysis. This makes integration with other organizations even simpler. Each of these organizations has an established information technology department, though a major problem is that of high staff turnover and no staff at all in the case of MoSTD by the time of the interviews for this study. None of the organizations under this study have some information sharing policy or strategy.

Analysis of Findings

It has emerged from the findings that there are some problems and issues which act as inhibitors to inter-departmental integration. A major challenge is that of poor technological infrastructure across the whole country. The underlying factor is believed to be the political and economic despair Zimbabwe is reeling under. All the WANs of organizations in this study connect offices that are in towns and cities while the majority of smaller stations are electronically isolated. NAC officers travel the whole country but can only capture data and information in manual documents for later transmission to computer systems. The same applies for NBSZ whose blood donor community and individuals, schools and other institutions spread throughout the country. Organizations cannot establish communication links in rural and marginalized areas because of lack of electricity. Zimbabwe is currently in a state of economic and political isolation, so it is therefore very difficult to finance any electrification projects due to severe foreign currency shortage. The Rural Electrification Program whose master plan was approved by cabinet in 1997 and funded by the African Development Bank (Mapako & Prasad, 2004) was a noble venture until financial resources ran dry. Zimbabwe is currently a net importer of electricity which is the determinant in rolling out ICTs to the whole country. Internally, Zimbabwe can generate 750megawatts (MW) from Lake Kariba, 920MW from Hwange Power Station and 300MW from other smaller stations at Munyati, Bulawayo and Harare. The domestic power consumption requirement is estimated at more than 2700MW, so the remainder is imported from the Democratic Republic of the Congo (DRC), Zambia, Mozambique and South Africa.

Another major challenge to deploying connected government applications in Zimbabwe concerns human capital flight. Projects to establish connected government architectures cannot be considered without taking cognisance of the human capital needed to drive the efforts. Severe political and economic pressure exerted on Zimbabwean companies and individuals has led to mass exodus of qualified people. Zimbabwe has lost more than four million people to neighboring countries and overseas in less than a decade. Company closures affected all industries including software and hardware businesses where organizations like Microsoft and IBM closed their Zimbabwe operations. Maintaining software from such software houses like IBM's progress database system at NBSZ then became almost impossible. The effects of human capital flight coupled with the realistic consequences of HIV/AIDS scourge, have dealt a major blow to the Zimbabwean economy as a potentially influential pool of talent has been lost. USAID (2007) reported that Zimbabwe's scientific and technological capacity is comparable to regional benchmarks, but integration of new technologies through foreign investment is partly being blocked by flight of skilled workers. The deputy director at MoSTD indicated that the information technology manger position had been vacant for more than three months and it was difficult to find a suitable person owing to low remuneration and generally poor economic environment obtaining in the country.

Low rate of computer literacy and pre-requisite skills to use new technological tools and applications is another notable challenge in Zimbabwe. This problem is closely related to lack of ICT infrastructure in the sense that where ICT facilities are not accessible, there is little or no knowledge of technology. ICT skills in the public sector leave a lot to be desired. This is a particular problem in the developing countries, where the chronic lack of qualified staff and inadequate human resource training has been a problem for years (UNPA & ASPA, 2001). Government employees and their managers have less knowledge and appreciation of ICTs as compared to their counterparts in the private sector where e-commerce has established itself. MoSTD indicated the lowest number of computer literate people followed by NSSA, ZTA, NAC and finally NBSZ. During the time for interviews for this study, lack of ICT skills had even forced MoSDT senior executives to connect computers on dial-up for internet purposes because of problems on DSL connection. There were no skilled people from within the organization at the time who could fix any computer-related problem.

Another observed phenomenon is that some systems used in organizations are closed systems. A closed information system is one that is usually used within an organisation where data can be exchanged or accessed by users with computers in an organisation. Such systems use only the supplier's communication standards or protocols without allowing other external systems to communicate with them. The progress database at NBSZ is a closed system because it is not open for external connections. It is therefore impossible to establish a networked environment when an underlying system is not flexible. Dissimilar information systems within various departments in the Government of Zimbabwe make interoperability difficult to implement.

Absence of information sharing policy is another inhibitor to implementing connected government applications across departments. Organizations that are committed to working in partnership with other agencies involved in providing services to the public require an inter-organization policy on information sharing. All the five organizations in this study indicated they never have such policy. Information sharing policy or protocol sets out a framework for information sharing across the respective organizations. The policy commits the partner organizations to draw up information sharing agreements to meet specific business needs in an agreed format. Such policy would make it easy for organizations such as NAC, ZAN or Research institutions to electronically extract HIV/AIDS data from the NBSZ database. The same goes for all other public organizations like the ZTA which is currently sharing information with the immigration department. Information is often an organization's most valuable resource, which means it becomes a high priority that information be secured with a high degree of confidentiality, integrity, and availability. An information sharing policy is a scaleable security measure because it can exist at any level of an organization; from global corporate mandate down to workgroup issue-specific policy (Gilbert, 2003).

In the following section, we present recommended solutions to the problems and issues identified during the analysis of the findings.

Recommended Solutions

During the analysis of the findings, it has emerged that many challenges are faced when efforts are made towards diffusing connected government applications in public enterprises in Zimbabwe. It is our view that the proposed solutions we present in this section will assist to form a government with coherent internal workings while providing seamless products and services to consumers.

The problem of poor technological infrastructure in Zimbabwean public organizations can be overcome by implementing information architectures that take care of the agency's or department's information needs as well as facilitating interoperability across departmental boundaries. The information and network architecture at one government agency should allow for flexibility and interoperability with other departments. Standard data definitions must be defined across networks to facilitate cross-transfer of information. A standardization plan should be formulated for system-wide network, data exchange, system standards, etc. A connected government's full potential, at least in the consumers' view, can be achieved by lateral integration of government electronic services across different enterprises' functional walls. The roadway to reaching this level is the establishment of a government portal, which by definition, is a website that constitutes the central access or entry point to all available government information and electronic services. The portal should make available an array of online services including government information publicity, government and citizen interaction, government service for individuals, government service for business, culture services, etc. Basic services that many e-government portals throughout the world are offering, but not yet available in Zimbabwe are: voter registration, ordering birth, death, and marriage certificates, filing state taxes, hunting and fishing licenses, accessing to medical information, etc.

The problem of closed systems such as the inflexible progress database at NBSZ can be solved by implementing a collaborative system environment. As the need for a connected government emerges, then communication and integration-oriented technologies become more imperative. Replacing closed systems with open enterprise resource planning (ERPs) software, customer relationship management (CRMs) systems and integrated databases will facilitate collaboration of the whole public sector, resulting in improvement of user-focused services as well as internal and external delivery effectiveness. These large modeling systems will facilitate the processing and sharing of diverse management information such as human resources information, marketing, procurement, financial reporting, etc. The command-based progress software at NBSZ does not permit any communication with other software such as Pastel which is used for the financial processing within the same organization. This is a similar case with other public organizations in Zimbabwe where islands of information are processed and stored in legacy systems that do not connect to other software. A study on ERPs by Seddon (2005), suggests that ERP systems provide benefits such as improved information visibility, personnel and inventory reduction, productivity improvement and new improved processes. The main source of different and unrelated software is that most systems are imported into the country. There are also no guiding principles at government level regarding acquisition and deployment of these systems. The Zimbabwe government can be spurred by the example of developing countries, such as India, that seem to be "riding the ICT wave" successfully. India and indeed many other countries, has established technology park concept, commonly known as technoparks to promote research and development (R&D) especially in software development. Many developing countries such as Turkey, China, Malaysia, Singapore, Taiwan, Thailand, have sought to emulate these successes and reap the benefits of ICT-led growth for their own economies. Applications that are locally developed in these technoparks are then deployed to various private and public organisations, fostering interoperability within and across these entities. The Turkish government passed a technology development zones law in 2001 whose thrust was to promote the establishment of technoparks under the guidance and lead of universities. According to this law, companies are encouraged to invest more into R&D and software development through tax incentives and according to this law, any kind of software development activity is considered an R&D activity. Phenomenal expansion of ICT applications (e.g., e-commerce, e- learning and tele-medicine etc.) especially in rural and remote areas in Turkey has been witnessed since then.

The challenge of low ICT literacy rate can be overcome if the government would make a deliberate policy to implement e-literacy programs. The impact of human capital flight would also not have as catastrophic as it is today had the Zimbabwean population been e-literate.

Another recommendation is the formulation of a government information sharing policy. This policy is concerned with the planning, introduction and use of IT resources for the benefit of all public enterprises. The new paradigm of value creation in the information-based economy requires information system strategies that create the platform to integrate and optimize the value chains in the extended institutions. We concur with Dawes' (2007) regarding her proposed contents of such an information system strategy. Dawes (2007) posits that the strategy should emphasize the principles, standards, and infrastructure that make it possible for all agencies to work in consistent ways. This might include legal and policy infrastructure, telecommunications infrastructure, standards for data and technologies, rules and mechanisms for information use and sharing, and a host of other elements.

Discussion

In developing countries there are significant demographic differences in both the access to, and use of the Internet, for personal, professional, or political purposes, either at home, at work, or in the community (Riley, 2004). This difference makes it impossible to effectively implement and deploy connected government applications as the majority of consumers of government services do not have access to the e-services. Explaining the challenges of digital divide, the United Nations World Public Sector Report (2003) reported that the potential of e-government as a development tool hinges upon three prerequisites – a minimum threshold level of technological infrastructure, human capital and e-connectivity – for all. E-government readiness strategies and programmes will be able to be effective and "include all" people only if, at the very minimum, all have functional literacy and education, which includes knowledge of computer and Internet use; all are connected to a computer; and all have access to the Internet. The primary challenge of e-government for development therefore, is how to accomplish this. According to Riley (2004), the concept of the digital divide is based on the hypothesis that there are both "information-haves" and "information-have-nots" in the Internet Age. This is very true in Zimbabwe as all the five organisations in this study only reach people in urban areas more often than rural folks because of the wide digital divide. Low rates of access to ICTs in rural areas of developing countries are a concern to proponents of connected government concept. It was indicated that e-mail is the mode of communication in all the organisations surveyed, but this facility is only a preserve of those who are not only in urban centres of Zimbabwe, but also those that are computer literate.

Zimbabwe's capacity to deliver on the sustainable development front, including e-government initiatives, has been seriously hampered by unprecedented political and economic turmoil the country is reeling under. Social services have crumpled and so is education, industry, agriculture and all other sectors of the economy. Exodus of professionals en masse is a result of this political and economic decay. Institutional reforms can therefore take place only if there are serious positive changes on the macro economic and political front.

SUMMARY AND CONCLUSION

The conclusion to the study is presented in three subsections as follows: summary, limitations and related further study.

Summary

An information society begins with a connected government and ICTs are the bedrock and founding pillars of such information societies. In this paper, we introduced the connected government concept which is a new philosophy in e-government implementation. The focus of this concept is that governments should manage their back-end processes which are the backbone of information integration and collaboration. Whereas the traditional e-government focused on e-services, connected government emphasizes the value of those services. We then presented an investigation, a case study involving five government organizations, namely the NBSZ, NAC, ZTA, MoSTD and NSSA. The investigation was aimed at determining the extent of connectedness within and across government agencies, with the idea of stimulating some thinking within and among public administrators, around the possibility that a connected government can indeed be established in a developing country setting. Naturally, service delivery inhibitors were exposed as e-government initiatives have not yet been fully accepted and diffused in the developing world. We documented these shortcomings and suggested their possible causes as we believed that the identification of such problems is a small step on the road towards an integrated and networked government. A set of proposed solution to problems was presented in the paper. The establishment of digital infrastructures as network connections, subsequent implementation of modeling technologies and e-literacy program roll-out were among the recommended solutions. Another important recommendation was the formulation of a sound information sharing policy to guide sharing of information within and across organizations.

The study presented in this article is deemed by the authors to be applicable not only to the Zimbabwe environment, but to all other developing countries in similar setting especially in Africa.

Limitations

As with all case study research, there are limitations to the interpretation of the results in attempting to generalize these analyses to a broader community of organizations. In this study, although government institutions in Zimbabwe face similar challenges, the choice of five organizations to investigate the extent of inter-departmental integration may have introduced a bias. Inhibiting factors affecting connected government implementation at the five organizations may not necessarily be the same across all government agencies in Zimbabwe. Another limitation of this study may be a scarcity of hard quantitative information with regard to the actual contribution a connected government can have in the context of the Zimbabwean economy.

Related Further Studies

We have exposed the inhibitors to the deployment of connected government applications within government institutions in Zimbabwe. We have also presented our recommended solutions to overcome the hindrances. Further research is however needed on how the Zimbabwe government can employ ICTs to extend electronic service delivery to citizens particularly under the current economic environment. On the same note, further research is required that can spell out the actual quantifiable benefits of investing in connected government applications in the developing economies such as Zimbabwe

REFERENCES

Akther, M. S., Onishi, T., & Kidokoro, T. (2007). E-Government in a developing: citizen-centric approach for success. *International Journal of Electronic Governance*, *1*(1). doi:10.1504/IJEG.2007.014342

Anderson, G. L., Herr, K., Nihlen, A. S., & Noffke, S. E. (2007). *Studying Your Own School: An Educator's Guide to Practitioner Action Research*. Thousand Oaks, CA: Corwin Press.

Artkinson, D. R. (2003). *Network Government for the Digital Age*. Washington, DC: Progressive Police Institute.

Canadian International Development Agency. (2007). *Knowledge-Sharing Plan*. Retrieved February 19, 2008 from http://www.acdi-cida.gc.ca/CIDAWEB/acdicida.nsf/En/EMA-218122154-PR4

Chan, O. J. (2005). Enterprise Information Systems Strategy and Planning. *The Journal of American Academy of Business*, *2*(3).

Davision, M. R., Wagner, C., & Ma, C. K. L. (2005). From government to e-government: a transition model. *Information Technology & People*, *18*(3), 280–299. doi:10.1108/09593840510615888

Dawes, S. S. (2002). *The Future of E-Government*. Retrieved July 3, 2008, from www.vinnova.se/upload/EPiStorePDF/vr-06-11.pdf

Dul, J., & Hak, T. (2007). *Case Study Methodology in Business Research*. Oxford, UK: Butterworth-Heinemann.

European Commission. (2003). *Linking-up Europe: the importance of interoperability for e-Government services* [European Commission working document]. Retrieved July 20, 2007 from http://europa.eu.int/ISPO/ida/

Gichoya, D. (2005). Factors Affecting the Successful Implementation of ICT Projects in Government. *The Electronic . Journal of E-Government*, *3*(4), 175–184.

Haricharan, S. (2005). *Knowledge Management in the South African Public Sector*. Retrieved July 5, 2008 from http://www.ksp.org.za/holonl03.htm

Heeks, R. (2002). eGovernment in Africa: Promise and Practice. *iGovernment Working Paper Series*, Paper 13.

Heeks, R. (2003). *e-Government Special – Does it Exist in Africa and what can it do?* Retrieved September 30, 2007 from http://www.balancingact-africa.com/news/back/balancing-act93.html#headline

Ifinedo, P. (2005). Measuring Africa's e-readiness in the global networked economy: A nine-country data analysis. *International Journal of Education and development using ICT, 1*(1).

Isaacs, S. (2007). *Survey of ICT and Education in Africa: Zimbabwe Country Report*.

Kaaya, J. (2004). Implementing e-Government Services in East Africa: Assessing Status Through Content Analysis of Government Websites. *The Electronic . Journal of E-Government, 1*(2).

Kumar, V., Mukerji, B., Butt, I., & Persaud, A. (2007). Factors for Successful e-Government Adoption: a Conceptual Framework. *The Electronic . Journal of E-Government, 5*(1), 63–76.

Layne, K., & Lee, J. (2001). Developing fully functional E-government: A four stage model. *Government Information Quarterly, 18*(2), 122–136. doi:10.1016/S0740-624X(01)00066-1

Marche, S., & McNiven, J. D. (2003). E-Government and E-Governance: The Future isn't what it used to be. *Canadian Journal of Administrative Sciences, 20*(1), 74–86.

Parajuli, J. (2007). A Content Analysis of Selected Government Web Sites: A Case Study of Nepal. *The Electronic . Journal of E-Government, 5*(1).

Ruhode, E., Owei, V., & Maumbe, B. (2008). Arguing for the Enhancement of Public Service Efficiency and Effectiveness Through e-Government: The Case of Zimbabwe.' In *Proceedings of IST-Africa 2008 Conference*, Windhoek, Namibia.

Seddon, P. B. (2005). Are ERP Systems a Source of Competitive Advantage? *Strategic Change*, John Wiley & Sons, Ltd.

Stake, R. E. (1995). *The Art of Case Study Research*. Thousand Oaks, CA: Sage Publications.

Tapscott, D. (1995). Leadership Needed in Age of Networked Intelligence. *Boston Business Journal, 11*(24).

Tellis, W. (1997). Introduction to Case Study. *Qualitative Report, 3*(2).

United Nations, E-Government Survey Report (2008). Retrieved June 10, 2008, from unpan1.un.org/intradoc/groups/public/documents/UN/UNPAN028607.pdf

Uzoka, F. E., Shemi, A. P., & Seleka, G. G. (2007). Behavioural Influences on E-Commerce Adoption in a Developing Country Context. *The Electronic Journal of Information Systems in Developing Countries, 31*(4), 1–15.

Yin, R. K. (2003). *Case Study Research Design and Methods* (Vol. 5, in Applied Social Research Methods Series).

Chapter 9
Zambia and e-Government:
An Assessment and Recommendations

Neal Coates
Abilene Christian University, USA

Lisa Nikolaus
Abilene Christian University, USA

ABSTRACT

E-government can benefit developing countries by enhancing the economy, increasing access to health care, improving bureaucracy, and consolidating democracy. Sub-Saharan countries have lagged behind the world in adopting this system of communications. A variety of reasons explain the lag, namely lack of national resources and an illiterate population. Zambia serves an example of democracy on a continent where freedom and peace are lacking, but also as a country where e-government is only beginning. This evaluation is the first to examine e-government there, and is carried out at five distinct levels: Current communication systems; Zambia's ICT policy; key central e-government websites; e-government at the provincial/municipal level; and at the individual level. As a result, this case study will evaluate how a developing country is struggling to provide government access and enhance the economy and suggests improvements needed if Zambia's e-government will become adequate and sustainable.

INTRODUCTION

Many developed states now maintain an extensive electronic government structure, providing invaluable access to economic and quality of life information and greatly enhancing government services. Likewise, e-government is often heralded as the way forward in developing countries to increase their rate of development and allow for greater democracy (e.g., Bhatnagar, 2002; Krishna & Walsham, 2005). But developing states, desperate to enhance the economy, increase health care access, and consolidate democracy, have only recently taken steps to implement this internet-based tool, and these initiatives often lack the needed financing and proper infrastructure and face constraints such as an illiterate and poor citizenry (Basu, 2004; Ndou, 2004).

E-government refers to the use by government agencies of information technologies to provide

DOI: 10.4018/978-1-60566-820-8.ch009

a wide range of services and programs via the internet, wide area networks, and mobile computing (About E-Government, 2005). It is sometimes referred to as "online government" or "internet-based government." These initiatives have the ability to transform relations among citizens, businesses, and the various arms of government. At its most basic level, citizens have increased access to information and government services, such as acquisition of drivers' licenses, voting registration, and renewal of passports. Another important aspect of e-government is the potential to boost the economy—businesses find their costs are diminished and they can operate with less paperwork. E-government also allows citizens to hold government officials more accountable by increasing transparency. One example is the ability of offices to publish their policies on the internet, helping set a uniform standard of regulations. Health clinics across a country can also be linked to a centrally-located physician for diagnoses, saving lives and much money from lost productivity. Finally, countries are in favor of e-government because of the significant cost savings. Use of this service by citizens and businesses saves large amounts of face-to-face and telephone contact by local, provincial, and national bureaucrats, making government more efficient.

This chapter evaluates the current status of e-government in a peaceful though poor African state, Zambia, asking what problems need to be addressed to better provide e-government services there. This is a vital question as many citizens and businesses in Zambia already understand the possible benefits of e-government and want it to succeed (Kasumbalesa, 2005; Mupuchi, 2003). For example, Zambia's banks are modernizing by offering e-commerce services to their customers. Several thousand customers of Finance Bank and the Zambia National Commercial Bank are now accessing their accounts from anywhere in the country or the world, verifying their bank balance, learning exchange rates, confirming the status of checks, and receiving statements. Computer-based and cell phone-based banking are reducing time for customers and enabling the banking sector to cut costs by reducing the amount of labor involved in processing transactions (Malakata, 2007).

It has further been established that Information and Communication Technologies (ICT) can help transform the lives of rural Zambians. For example, the women of Kalomo District are, with support from the International Institute of Communication and Development of the Netherlands and Step Out, a private firm, using the internet to market their goods. In 2004, they formed the Kalomo Bwacha Women ICT club, generating annually 1.5 to 2 million Kwacha. They distribute to 32 area women's' clubs five KGs of seeds such as maize, groundnuts, beans, and sunflowers. After harvest, 50 KG bags are returned in payment. The group of 11 women in Kalomo has an office assistant and trains others how their operations would be enhanced with the internet. Mushinge (2005) reports that the club now markets dry food stuffs, vegetables, baskets, clothes, and crafts, and the women sell the produce via email and a digital camera to clients in the area and even in some neighboring countries.

It is estimated, though, that only 231,000 Zambians use the internet, out of a population of 11,500,000. That two percent is very low, although how it compares to many of the world's other developing states is unclear due to the lack of statistics from many countries. In neighboring and stable Botswana, the rate is three percent, but in wealthy South Africa the usage rate is 12 percent. By contrast, the United States rate is 68 percent (Background Notes—Zambia, 2007; World Factbook—Zambia, 2007). In short, the internet is not an integral part of daily life in Zambia, nor is it tied to the delivery of government services. This is significant because many scholars explain that utilizing the internet to improve living standards is crucial for lifting persons out of poverty. In fact, other than education, Adesisa (2001) claims that ICT use is the fastest method of moving Africa toward the 21st century. Although cell phone usage is widespread in Zambia, it appears to be the only ICT used widely in the country.

Thus, this chapter serves the purpose of evaluating Zambia's e-government, and does so at five levels: Current communication systems; Zambia's ICT policy; key central e-government websites; e-government at the provincial/municipal level; and at the individual level. This review is by nature qualitative due to the nascent status of Zambia's efforts and because this study is the first to examine Zambia's ICT policy and websites. This chapter also relies on interviews with Zambians, officials, and expatriates. For example, Zambians attending college in foreign countries use the internet to renew government documents and keep updated with family.

In essence, this chapter considers whether Zambia's e-government can currently be classified as successful or as a failure, based on whether the standards of knowledge, service, and governance are being provided, and whether Zambia's own ICT goals are being met. It is hoped that the various government ministries in Zambia will consider the information and constructive criticisms within this chapter and take action. This chapter also provides a template for countries still further behind Zambia in development of their internet-based government services. Other case studies have just been completed such as Parajuli (2007), who examined the ministry websites of Nepal, finding that 17 of 20 had dedicated sites, but that web features critical in fostering government openness, citizen-government communication, and citizen participation and satisfaction are infrequent or completely absent.

Zambia's checkered history, its current position as a relative beacon of freedom in Africa, and its potential for economic growth provide a justification for the adoption of a workable e-government system. At the same time the government has immense hurdles to overcome in running such an initiative. For three decades after independence, the country functioned as a one-party state overseen by a strong president. Although it has since adopted multiparty democracy, corruption has seeped into many political offices and powerful persons have exercised control of the media. It is a tribute to Zambians that despite drought, poor leadership, and inefficient use of resources during their country's 45-year history that they have not moved toward acceptance of brute force or civil war as their neighboring countries have. E-government, if properly implemented, can further move Zambia on the road toward development and consolidation of democracy.

E-GOVERNMENT IN DEVELOPING COUNTRIES

E-government in any country can be evaluated by examining the strength of its basic elements, e-knowledge, e-service, and e-governance (Heeks, 2006). This has already been seen during the past several years in evaluations of developing countries such as China (i.e., Coates & Rojas, 2006; Lollar, 2006). As Zambia's individual ministry websites mature and a national portal site is added, a more quantitative examination than this study of first impression can later be conducted.

Lollar examined four content features in 29 of China's provincial and municipal websites in China—information available, services provided, transparency, and citizen outreach. All sites examined by Lollar had an index, 96 percent had documents and agency links, 93 percent provided laws and regulations, 72 percent had governor's or mayor's email boxes, 57 percent had links to media, 41 percent had bidding information, 27 percent posted approved public goods prices, and 10 percent had civil service postings. With these results, Lollar believed that overall e-government in China has played a large role in improving transparency and citizen involvement but did not yet provide unbiased information or efficiently deliver services. Zambia, with no provincial sites and only one municipal website to date, still has much work to do at the regional and local levels of e-government. Its ministry websites and the executive and legislative branches' sites, however, can be examined.

Regarding e-knowledge, several types of information should, ideally, be available via government websites: 1) governmental processes and projects; 2) officials' duties; 3) documents and information about file centers; 4) routine governmental affairs; 5) rules, regulations, and policies about specific fields; and 6) information about trading and transaction markets. E-knowledge is the most basic of e-government services and the easiest to implement and maintain.

E-service involves providing government services online, and includes interactive and non-interactive aspects. This online information helps citizens or businesses efficiently carry out activities. Subjects include weather, air quality, local maps, job information, and instructions on various procedures such as passport application, vehicle registration, and business operation guides. Interactive services include those in which users have more control over content and includes e-filing for license renewal and incorporation, e-payment for bills and fines, and e-mail service.

Finally, a primary goal of e-government should be to facilitate e-governance by creating government-to-citizen connections through features such as online forums and polls, surveys, chat rooms, and complaint forms. Additionally, links to websites and emails of officials and policymaking bodies are in this category. Civic engagement in governance is an important way to promote citizen empowerment and connect people to their country's leaders, increase accountability by officials, and raise trust in government. Effective e-governance is critical in order to improve governmental processes and enhance economic development (Heeks, 2001).

Regardless, although considered a cure for many of the ailments of developing countries, the stark reality is that most e-government projects fail. Several authors (Avgerou & Walsham, 2000; Dada, 2006) have warned of this so that officials could anticipate potential problems. According to Heeks (2003), 35 percent of implementations are classified as total failures (e-government was not begun or was immediately abandoned), and 50 percent have been partial failures (major goals were not attained and/or there were undesirable outcomes). Besides inadequate funding, the reason so many e-government projects fail in developing countries is that gaps exist between the design and the reality of the system.

One reason that less developed countries participate in e-government projects, though, is that such systems are thought to equate to "good governance" and increase the amount of foreign aid received. Ciborra and Navarra (2005) criticize this motive because implementing electronic government is often not successful. Perhaps only the wealthy will have access to services, corruption may continue, or levels of democracy may not change. In addition, e-government projects in developing countries are usually dependant almost totally on aid from donors. Once this financing ceases, there may be an "all or nothing" approach to systems development, rather than incremental improvements over time. This is the case with Lusaka's City Council's system. Thus, further political and social changes are required in countries such as Zambia when electronic mediums are implemented.

In developing countries there is often a large gap in the physical, cultural, and economic contexts between the location of the software designers and the place where the system is installed. Heeks (2002 and 2003) has identified three situations where design-reality gaps are common. First, "Hard-Soft Gaps" are the difference between the technology and the reality of the social context in which the system operates. The design and implementation of an information system cannot ignore resources, skill-levels, beliefs, and motivations. In addition, Jaeger and Thompson (2003) assert that e-government can fail if citizens are not educated about its value. This is aggravated in developing countries such as Zambia where literacy rates are low and educational institutions are lacking, and where there are little computer skills and training.

Second, Heeks explains that "Private-Public Gaps" occur when systems designed for the private sector do not work within the reality of the public sector. Governments have to change their view of the recipients of these e-government projects from citizens to customers. Ciborra and Navarra (2005) identify numerous problems with seeing a citizen as a customer, because the later needs the ability to choose among options. This is not possible for e-government if it operates as a monopoly, as it mostly does in Zambia where few ministry websites have been privatized.

Last, "Country Context Gaps" exist when trying to use the e-government system prepared for a developed country in a developing country. This is due to differences in access to technology and infrastructure (Ebrahim & Irani, 2005). Zambia has a poor IT infrastructure—there is not consistent and reliable electricity, telecommunications, and internet access. Power is rationed almost daily because the electricity infrastructure lacks greatly. Only 20 percent of households are wired for power in cities and only three percent in rural areas, and the Zambia Electricity Supply Company, Zesco, only enrolls 10,000 new customers a year. Blackouts are so common for the 300,000 current customers that service in Lusaka is completely unreliable. Zambia claims it will build its way out of the power crunch, wanting to add $1.2 billion in upgrades and capacity financed by China and India (Wines, 2007).

POLITICS AND POVERTY IN ZAMBIA

History and Government

To further understand the context of Zambia's e-government efforts, it is helpful to know some of the country's background, economics, and culture. Zambia is a landlocked country in the heart of sub-Saharan Africa. Formally part of the colony of Rhodesia, the British left their footprint through language, Christianity, and exploitation of resources. After a peaceful transition to independence in 1964, multiple challenges arose as Zambia had to trade through the neighboring countries of Tanzania, Malawi, Mozambique, Zimbabwe, Botswana, Namibia, Angola, and Democratic Republic of Congo. Most of these surrounding nations have been politically or economically unstable (Background Notes, 2007; World Factbook, 2007).

Despite its considerable mineral wealth, it was unclear how Zambia would modernize. There were few trained and educated Zambians who could run the government, and the economy was largely dependent on foreign expertise. President Kenneth Kaunda and the United National Independence Party controlled the legislature and pushed through the 1973 constitution which provided for a strong executive and a unicameral National Assembly. During the next decade, the central government supported revolutionary movements such as the Union for the Total Liberation of Angola, the Zimbabwe African People's Union, and the African National Congress in South Africa. South Africa responded by raiding ANC targets in Zambia, and conflicts with Rhodesia resulted in its closing the border with Zambia, causing severe problems with international transport and power supply.

In 1990, after riots throughout Lusaka and a coup attempt, President Kaunda signed legislation allowing a multi-party election in 1991. Growing opposition to UNIP's monopoly had resulted in the rise of the Movement for Multiparty Democracy. In response to the growing demand for real democracy, Zambia then enacted a new constitution also in 1991. It enlarged the National Assembly, established an electoral commission, and allowed for more than one presidential candidate. In October 1991, MMD candidate Frederick Chiluba resoundingly defeated Kaunda with 81 percent of the vote. To add to the landslide, in parliament the MMD won 125 of 150 seats.

Table 1. Country statistics

Population	11,477,447
Area	752,614 sq. km. (290,585 sq. mi.); slightly larger than Texas
Major cities	Lusaka (capital), Kitwe, Livingstone, Ndola
Languages	English (official), about 70 local languages and dialects, including Bemba, Kaonde, Lozi, Lunda, Luvale, Nyanja, and Tonga
Environment	Mostly plateau savanna, generally dry and temperate; suffers periodic drought and tropical storms (November to April)
Literacy	Men 81%, Women 60%
Education	No compulsory education; 7 years free education; primary school enrollment 67%
Life expectancy	33 years
Infant mortality rate	95/1,000
High rate of disease	Diarrhea, hepatitis A, typhoid fever, malaria, schistosomiasis
HIV rate	17% (2003)

The 1990's were marked by the reforms of selling some state owned services, designed to revitalize the economy, but the country suffered increasing poverty, devastating droughts, and the onset of AIDS. Compounding the difficulty was that although in principle Zambia was a multiparty democracy, Chiluba ran the country with a tight fist. Corruption also became rampant, and the hope of Zambia becoming a strong democratic champion seemed lost. By the end of Chiluba's first term in 1996, the MMD's commitment to political reform had faded. Relying on the party's overwhelming majority in parliament, however, Chiluba orchestrated constitutional amendments which eliminated Kaunda and other prominent opposition leaders from the elections. In the elections held in November, Chiluba was re-elected and the MMD won 131 National Assembly seats (see Table 1).

By the turn of the century, Zambians were ready for a change in leadership. This was cemented when President Chiluba tried to amend the constitution to seek a third term. Presidential, parliamentary, and local government elections were held but encountered serious irregularities. MMD candidate Levy Mwanawasa (who had served as Vice President under Chiluba) won a plurality of the vote and took office in 2002. During Mwanawasa's first term, he pushed a special task force to prosecute corrupt officials. He was re-elected in 2006, and the ruling MMD party won a slim majority. In his second term, the President boosted the economy by about five percent annually, and foreign donors were impressed enough to forgive all Zambian debt. Troubling reports, though, continued to surface regarding suppression of the press and corruption. In 2007, the President announced plans to hold another constitutional convention, but he died of a stroke in 2008. The road ahead for true democracy in Zambia is still long.

The Economy and Poverty

Zambia began its slide into poverty in 1973 when copper prices drastically declined. The Kaunda socialist government had nationalized the mines, and then tried to make up for falling revenue by increasing borrowing and by printing greater amounts of Kwacha, the national currency. After the 1991 multi-party elections, President Chiluba privatized many government-owned corporations, endorsed free market principles, and the eliminated subsidized food prices. In fact, since 1993 more

Table 2. Economic statistics

Average income	U.S. $627
GDP	$5.795 billion
GDP growth rate	5.8%
Work force	Agriculture—75%, Services—19%, Mining and Manufacturing—6%
Oil pipelines	771 km. (480 mi.)
Paved roadways	20,117 km. (12,500 mi.)
Airports	10 (with paved runways)
Exports	Copper, cobalt, electricity, tobacco, flowers, cotton
Imports	Machinery, transportation equipment, petroleum products, electricity, fertilizer, foodstuffs, clothing

than 200 of Zambia's 326 state enterprises have been privatized. Now the country is the world's fourth largest copper producer and the largest of cobalt. The copper industry has grown to produce over 80 percent of the country's export earnings, resulting in about 15 percent of GDP. In addition, the maize harvest has been good since 2005 as a result of adequate rainfall (see Table 2).

With regard to foreign debt and inflation, Zambia has recently had good news. In 2006, inflation dropped into single digits for the first time in several years. Following privatization of the copper industry, production rebounded, reaching almost 400,000 metric tons in 2004 and 440,000 metric tons in 2005. But the country's total foreign debt stood at about $7 billion when Zambia reached the Highly Indebted Poor Countries Initiative completion point in 2005. In December of that year, the U.S. and Zambian governments signed an agreement for cancellation of $280 million in bilateral debt. The IMF will forgive all debt incurred by Zambia before 2005, amounting to $577 million (International Monetary Fund, 2005; Zambia: Govt Cautions, 2006). In addition, in July 2007 the government gave a positive outlook regarding the country's trade surplus, predicting it would reach $1.2 billion dollars. The growth was fueled by the increase in the price of copper to record levels, and production was expected to reach more than 600,000 tons, compared to 500,000 in 2006 (Sinyangwe, 2007). The central government must decide how it will use any additional revenues generated by the better economy, including e-government at the national and regional levels.

Poverty, however, continues to present a major obstacle for Zambia. Of the population of 11.5 million, an estimated 70 percent live in poverty. Per capita annual incomes of $627 place the country among the world's poorest. Social indicators continue to decline, particularly in life expectancy at birth and maternal mortality. The country's economic growth rate cannot support the increase in population or the strain which HIV/AIDS places on government resources. The disease is the nation's greatest challenge, with an official 17 percent prevalence. Nearly one million Zambians are HIV positive or have AIDS, an estimated 100,000 adults died in 2004, and over 750,000 Zambian children have been orphaned. In addition, measles, malnutrition, and diarrhea contribute to a very high infant mortality rate. Malaria, pneumonia, yellow fever, and cholera are major health problems.

It goes without saying, but poverty creates problems such as lack of education, poor healthcare and living conditions, decreased purchasing power, and a digital divide with the Western world due to the cost of computers and the internet. Much of the continuing poverty is blamed on the government's inability to plan and make basic decisions. For example, during June 2007 rumors spread throughout the Southern province that the most recent major shortage of diesel in Zambia was because a bureaucrat "ordered the wrong crude oil." (E. Hamby, personal communication, August 5, 2007). Similarly, citizens were told at health clinics that the lack of medicine was because someone in Lusaka "forgot to place the six month order." One could conclude that if delivery of basic goods cannot be carried out on a regular basis that it is questionable whether e-government can be implemented fully. In response to Zambia's extensive poverty, millions of dollars are given each year for a wide array of programs. The donors include the World Bank along with the IMF, EU, UN, and the African Development Bank. The U.S. is Zambia's largest country donor— USAID administered more than $141 million in 2006 to fight corruption, make customs more efficient, increase trade, improve basic education, and to fight HIV/AIDS. No funds, however, were designated to improve Zambia's e-government system.

Culture

The various aspects of Zambian culture influence the use of the internet and the effectiveness of the country's e-government efforts. Rural life revolves around small-scale farming—the large amount of time required and low income produced from this lifestyle results in few persons having the funds or time to use computers, let alone e-government. Each year Zambia's Human Development Index value is in the bottom quartile of 175 nations. Zambians have few opportunities to pursue personal goals, with women lagging far behind men in income and literacy.

Although education is free, the requirement that pupils provide their own school materials and uniforms precludes many prospective students. About 80 percent of Zambian children attend primary school, but insufficient facilities and funding result in only 20 percent moving on to secondary schools, and only two percent to institutions of higher education. There are just two universities, 14 teacher-training colleges, and 14 vocational and technical institutions in the entire country. Learning is not yet valued highly. Add onto this that Zambians, especially in rural areas, take life as it comes and are not concerned that they do not use the World Wide Web.

Several aspects of the culture indicate that acceptance of computers and e-government could take hold. With half of its people living in cities, Zambia is very urbanized. Just under half of the population is younger than age 15, and younger persons are open to new ideas and technologies. In addition, English is the official and administrative language, and the internet uses mainly English. Western-style clothing is common, and the ideas and technology from developed countries are generally accepted. The existence of an educated labor force, at least in the larger towns, and the huge number of young Zambians indicates that computers could be used to improve the economic situation for many, especially with adequate commitment from the government.

ZAMBIA'S COMMUNICATION SYSTEM

Whether Zambia's e-government can succeed rests partly on the state of its communications equipment and how the country has gone about creating its e-government system. The general assessment of Zambia's telephone system is that its facilities are aging but are still among the best in sub-Saharan Africa. High-capacity microwave radio relay connects most large towns and cities, and several cellular telephone services are in operation. Internet service is widely available. Zambia is a member of Intelsat and has an Earth Station in Lusaka, which provides world-wide

Table 3. Communications statistics

Phones—land lines	**94,700 (2005)**
Phones—mobile	946,600 (2005)
Radio stations	AM 19, FM 5, shortwave 4 (2001)
Television stations	9 (2001)
Internet country code	.zm
Internet hosts	3,227
Internet users	231,000 (2005)

telephone, television, telefax, and e-mail links (see Table 3).

Zambia has three main Internet Service Providers. Zamnet was the first ISP, begun in 1994, and has about 80 percent of the market. It is now privately owned and has the contract for several ministries' homepages. Zamnet (www.zamnet.zm) was one of the first ISPs on the continent and was developed by the University of Zambia's Computing Department. Unfortunately, and ironically, the University does not currently have an internet presence.

Zamtel is owned and managed by the government. It operates a fixed line network, post paid services, and a prepaid service under the brand name Zamtel Prepaid, a cellular network under the brand name Cell Z, and an internet service provider under the brand name Zamtel Online. In 1994, the Zambia Telecommunications Company Limited and the Zambia Postal Services were created through the Telecommunication Act. According to the future privatizations planned for the Development Agency, Zamtel is being considered for sale in some manner; the ministry's homepage states that project has not yet been considered. Because the development of the infrastructure of the new media economy frequently involves national telephone companies controlled by politicians, corruption is still an issue and it is doubtful at this point if Zamtel will be privatized anytime soon.

Coppernet (www.coppernet.zm) is also a major ISP competitor. There are also smaller companies including Microlink, UUNet, Mweb, and Africa Online. Currently AfriConnect is the only internet service providing wireless high broadband speeds—iConnect is only available in Lusaka and is delivered by a single base station (Syed, 2007).

As of June 30, 2007, Zamnet charged for internet service per month for one household with one user K376,000, which is about U.S. $100. For a small to medium business the charge is K1,864,000, $497. When compared to the average income of a Zambian of $627, there is no possibility that many can have internet access, let alone a connection to local or national e-government services. Coppernet rates including tax are U.S. $30. Zamnet is $23 including tax for 80 hours, and the other companies charge about the same.

The number of ISPs is still inadequate, reflecting the overwhelming need of the country to modernize. In a similar vein, there were not more than 5-10 web designers in Zambia as of 2000. Today there are many more, but much design work is still contracted to professionals in other countries. Regarding telecommunications and the internet, inadequate capital is the main barrier to wider access. For example, web access is accomplished via U.S./European satellites, meaning an email sent between Zambian users may be routed through Florida, and each document request takes about 0.25 seconds to reach the destination server. Lack of financial resources and political will implies that the connection to one of the undersea telecommunications cables remains far off (About Zambia, 2007).

One Zambian interviewed for this chapter and who lives in Southern Province along the major highway in the country earned an interdisciplinary college degree in computers in the United States and then returned to his home country. He operates the only internet cafe in his town and reports this about e-government in Zambia:

Yes, I do use the e-government services. These include one under the Ministry of Education, Ministry of Home Affairs, and also the Zambia Revenue Authority. For the Ministry of Education I was getting information on ICT (Information and Communication Technologies) in Learning Institutions and also for the high school teachers' postings. From the Ministry of Home Affairs I was getting information on registration for our children with the birth certificates. From the Zambia Revenue Authority which is under Customs and Excise I have been getting information and filling out forms for tax registration for the business I run in Zambia called El Pantano Complex. I must say all the government sites I had visited so far I found the information though in some cases it was not readily available and seemed incomplete. But with several tries I got around and found what I was needing. My experience with the Zambian e-government worked though slow. It worked because the relevant information is available on these pages and the services are adequate considering that Zambia is not a developed nation. The Internet Service Providers in Zambia, and Zamnet being one of them, together with other IT specialists have worked very hard to post useful information on most websites both on government sites and private sites. ... I use the Internet to keep updated about Zambia. I mainly use Zamnet for news updates and then I could log on to the local radio, TV and newspapers (S. Mbumwae, personal communication, August 8, 2007).

In contrast, a person interviewed in the Northwestern Province said he has used Coppernet for two years but has never received a bill. Further, he responded "Zero" to the question of how many people he knows who are using any Zambian e-government services (B. Davis, personal communication, August 9, 2007). Clearly the experiences vary greatly across the country and according to whether persons have enough resources to access the internet.

ZAMBIA'S ICT POLICY

This case study now turns to explore the foundations of the government's commitment to e-government by focusing on the goals laid out in the National Information and Communication Technology Policy, drafted by the Zambian Ministry of Communications and Transport. The NICTP was prepared with assistance from experts from other countries, the World Bank, and USAID and was approved just on March 28, 2007. It is available online at the Zambia Ministry of Information and Broadcasting homepage at www.mibs.gov.zm.

The ICT Policy designates the Communications Department within the Communications and Transport and Ministry as the entity to coordinate and oversee the policy's implementation and to set standards to be used for infrastructure such as computers, telephones, satellites, and television and radio transmission frequencies. The NICTP does not name any one ministry, however, to oversee and implement all aspects of e-government in Zambia. Although it tasks the Communications and Transport Ministry with setting the policy direction for the industry and country at large, a yet to be created Independent Broadcasting Authority will oversee the development of the legal framework for the ICT industry and its broadcasting. As a result, currently there is no one Zambia e-government web page, such as in the United States (www.usa.gov) or South Africa (www.gov.za). Nor is there a single ministry tasked with running a central homepage or for regulating other ministries' sites. It is left to each branch of government, ministry, and bureaucracy to run its own homepage.

To add to the confusion, the Communications and Transport Ministry is required to work with an additional ministry to develop e-government. Within the Finance and National Planning Ministry, the Computer Services Department has been charged as the government's central computing agency. In this role it is supposed to provide the various ministries and departments with information on system services, consultancy services, advice on interoperability, and to provide a secure communications network. Finance and National Planning is also charged with crafting a policy defining how the government should allocate its resources to enhance its IT service delivery.

Beyond this convoluted policy-making structure, the NICTP fails to emphasize private sector involvement. It will, for example, set up initiatives to spread agricultural technologies and relay weather information to farmers, but only on the government's terms. Nor does the ICT policy address any involvement in the East African Submarine Cable project that will connect Africa to the rest of the world through a high-speed fiber-optic cable. The only telecommunications company so far signed up for the project, Zamtel, is state-run and enjoys a near monopoly over the project in Zambia (Zambian ICT Policy, 2007).

Although the NICTP is direct about the obstacles facing Zambia's ICT system, the policy document appears too ambitious at times and too general at times, and does not identify a source for the large amount of funding needed for proper implementation. For example, the policy correctly states there are not enough trained personnel to teach and educate how to use ICTs, that Zambia faces a shortfall in critical ICT skills at managerial, professional, and technician levels, and that only a very small percentage of pupils are ICT literate when they leave secondary school. Regardless, the reason for implementing the policy is laudable, to work toward the achievement of the Millennium Development Goals. These include the eradication of extreme poverty, achievement of universal primary education, reduction of child mortality, and combating HIV/AIDS, malaria, and other diseases.

NATIONAL E-GOVERNMENT SITES

Because there is no Zambia portal to evaluate for e-knowledge, service, and governance, this chapter turns to examine the currently-existing individual portions of e-government that have been assembled to date. What follows is a description of the executive and legislative branch homepages (the judiciary has none), followed by many of the ministries and agencies within the Appendix. In the process the positive aspects and those parts needing improvement are noted. It is hoped that bringing these deficiencies to light will lead to their remediation.

Executive Branch

The Republic of Zambia executive branch's official internet site is known as "Statehouse" and is located at www.statehouse.gov.zm. Touted by other websites as the main page for Zambia's e-government, Zamnet designed and hosts the Statehouse site; it is maintained by the Press and Public Relations Office. The homepage for the president and vice president, however, while mildly informative, is on the whole inadequate. As explained below, the site provides the barest of e-knowledge functions with basically no e-service or e-governance.

The site does contain a picture of the president, a biography, and is replete with links to many of his speeches. It also contains a short one-page policy statement for "The New Deal Administration Vision." A list of cabinet ministers is provided, but there is no link to their ministries' homepages or the other agencies of government. There is a "Contact Us" selection, but an email sent to the website was not returned. There is no selection for current news, press conferences, executive orders, or nominations. These problems could be rectified, but without those additional steps the office of the presidency maintains a firm grip on the dissemination of information. As a result, according to the website itself at its "Who's Online" function, only one or two persons are usually listed as visiting at a time, indicating that few Zambians or other persons even bother to read Statehouse.

It is not beyond expectation that a person in Zambia would want to visit the President's homepage to learn about important issues such as the topic of HIV/AIDS. This is the most devastating disease afflicting the country. The lack of e-knowledge in the current homepage, though, is painfully obvious. For example, the search engine on the President's homepage only searches for speeches that he has made, and the very first "hit" in a search for "AIDS" provides the result "Levy Greets Italy." This nonsensical response is linked to a short speech in which President Mwanawasa thanks the country of Italy for its donated funds for education, agriculture, and fighting HIV/AIDS. A citizen who may have sought help in determining how the disease is spread or how it can be combated finds nothing—knowledge, service, or governance—at the President's e-government site, not even a link to the Ministry of Health.

Another glaring oversight of the Statehouse site is the Vice President's portion dealing with national emergencies. Besides representing Zambia at some formal events, one of the main responsibilities of the Vice President is to respond to disasters and emergencies, handled by the Disaster Management and Mitigation Unit. Yet if a citizen or local official in Zambia accesses this web page at a time of distress, they find that the emergency contact directory is empty. Perhaps this is not a problem, though, because the site surprisingly lists no disasters in the country since 1994 on its Disasters Report page. This is woefully deficient, and needs to be corrected.

Legislative Branch

The National Assembly homepage (www.parliament.gov.zm) is much more informative and functional than the executive branch's website. The duties of Speaker and Deputy Speaker are prominent, along with a presentation of the legislative process. Bills and committee membership are listed, along with a link about the Constitution. These features generally indicate a commitment to providing e-knowledge.

In addition, the site has an easy search method for citizens to find and identify their Member of Parliament by name, party, or by selecting a region on a map of the country. Persons cannot, however, contact their MP by phone, address, or even email because the "Contact Details" by each members' name is blank. This is a basic requirement of e-governance, and can be easily corrected. In addition, there are no links to ministry web pages other than the Electoral Commission, another problem easily altered. There is also no homepage apparently for the House of Chiefs. Although in the past considered under the quasi-authority of the Local Government Ministry, the Chiefs are attempting to be located under the auspices of the President or Vice President. In short, the National Assembly's homepage is failing to provide e-service and e-governance.

Perhaps the most serious e-government problem that Parliament must address is its inability to maintain control of its site. As of mid-July 2007, Parliament homepage requests were suddenly forwarded to the homepage of Zamnet. Officials including the First Secretary of Political Affairs at the Zambian Embassy in Washington, D.C. had no explanation for this (J. Mulutula, personal communication, July 20, 2007). It is this type of refusal or inability to provide communication service that the central government and members of Parliament must decide they will correct if Zambia's government internet system is to develop into a program that can be trusted for reliability and content.

Bureaucracies

A list of the major ministries, agencies, and commissions in Zambia, along with their homepage address, if any, is found in the Appendix. This listing is first to be compiled for Zambia—surprisingly there is no such listing available from the government. Only one ministry does not have a homepage, which out of 35 is commendable and indicates actions are being taken to try to implement the national ICT policy. Several homepages, such as for the Central Statistics Office, however, only work intermittently. Unfortunately, 13 of the ministries' homepages could not be accessed, as noted in the Appendix. For example, a connection could not be made after multiple attempts with the Commerce, Trade and Industry homepage, a website critical to economic growth. Users must depend on a search engine to find ministry addresses because the Zambian government has yet to create a centralized homepage.

The Appendix details the short but direct examination of the individual ministry web pages. These homepages were evaluated during June and July 2007 based on whether and how their content provides e-knowledge, e-service, and e-governance. For example, the Education, Agriculture, and Health ministries' homepages, while providing some basic information, do not provide the information or service they should. Of those ministries to which a connection could be made, the Appendix records what was observed regarding whether the elements of e-knowledge or e-service were present along with a brief description. For most of these bureaucracies, even if one of the elements was noted as being present, it was normally not robust. Only a few of the web pages provide adequate basic information, and it is rare that e-service and especially e-governance were present. The web page company and designer are also listed, if known. Strikingly, only one of these homepages exhibited any e-governance—the Ministry of Communications and Transport.

It is unfortunate that citizens and businesses do not have adequate e-government interaction with the government of Zambia. For example, when international aid volunteers try to access various agencies to learn the details of regulations or to determine how to properly register visiting medical personnel they are unsuccessful or are often unable to make contact. So NGOs still find themselves in the situation that they must to travel to the country and meet face-to-face with officials to learn about Zambia's education, health, and agriculture systems before they can plan to give aid (C. Hurst, personal communication, July 13, 2007; K.B. Massingill, personal communication, July 25, 2007; C. Varner, personal communication, July 13, 2007). This is no better than before the implementation of e-government.

E-GOVERNMENT IN PROVINCES AND CITIES

Zambia is divided into nine provinces, each administered by an appointed deputy minister. In addition, there are 72 local authorities or councils consisting of four City Councils, 14 Municipal Councils, and 54 District Councils. The Ministry of Local Government and Housing has a website that lists each of these. With such a large number of appointed and elected officials, the opportunities for providing e-government services have tremendous potential in Zambia, especially considering that almost half of the population lives in cities (Butler, 2003). Surprisingly, though, the only entity with a homepage is the City Council of Lusaka. It is a major failure of regional and local government in Zambia, and the national government, that over 99 percent of regional and local governments (one of 153) do not yet have e-government as of 2007. For a list of major cities, see Table 4.

For example, the second largest city, Kabwe, and home of Zambia's first political rally leading to independence, does not have an internet homepage. Considering the many environmental

Table 4. Major cities

Major City	Province	Population
Chingola	Copperbelt	150,500
Chipata	Eastern	75,000
Kabwe	Central	213,800
Kasama	Northern	91,056
Kitwe	Copperbelt	305,000
Livingstone	Southern	108,100
Luanshya	Copperbelt	124,800
Lusaka	Lusaka	1,218,200
Mansa	Luapula	42,277
Mongu	Western	52,534
Mufulira	Copperbelt	131,000
Ndola	Copperbelt	347,900
Solwezi	North Western	30,000

problems that arose from extensive mining there, it may be that local will is not yet present to make available more e-knowledge or e-governance. Kabwe is Africa's most polluted city and has gained the dubious distinction of being ranked as the world's fourth most polluted site. The lead mine there was run without pollution controls by the government-owned Zambia Consolidated Copper Mines (UN: Kabwe, 2006). Likewise, Ndola and Kitwe have very strong mining interests, but the officials there apparently have not seen the imperative nature of setting up e-government.

Even Livingstone, with a great number of tourists each year due to the nearby Victoria Falls, has yet to implement e-government. This is surprising based on the potential profits from advertising for the city's website and due to the increase in tourism that would result from better communication. This is not limited to tourists learning about opportunities while planning their vacations—once in a game park, they could benefit from facilities such as cell towers, which in turn would spur further ecotourism. There are numerous private tourist homepages for the Livingstone area because they already recognize the opportunity and currently serve as the best source for local information. Provincial capitals and major cities do not have to wait for authorization from the national government to start a website, and the incentive to take such action is seen in the example of Livingstone.

The Lusaka City Council by default serves as the local government internet example for the entire country, because it is the only one. The site, at www.lcc.gov.zm, has a number of features that signify the type of success that e-government can achieve, such as allowing selections describing Lusaka, Services, Council and Committees, Departments, Projects, and Recreation. The homepage, though, also lacks certain aspects. For example, it lists City services but has no contact phone number listed or active email link. It lists the City Council minutes but does not provide emails for the council members. After a greeting from the Chief Executive and a history of the capital city, an examination of the site reveals it has a number of e-knowledge components: (1) governmental processes and projects; (2) officials' duties; (3) documents and information about file centers; (4) routine governmental affairs; (5) rules, regulations, and policies about specific fields; and (6) information about trading and transaction markets. But there are few interactive e-services such as filing for license renewal, e-payment for bills and fines, and e-mail service. The homepage is deficient when examined for e-governance.

Lusaka is not alone in its difficulties. African municipalities have cumbersome bureaucracies with unclear procedures for handling requests from city residents. It is common for citizens to have to physically visit City Hall to access information and pay utility bills. To address this, in 2001 UNESCO and the Danish International Development Agency operated a three-year project for the capitals of Mali, Mozambique, Niger, Tanzania, and Zambia. Mambwe (2005) reports that this jump-started the Lusaka City Council project, and now supposedly as many as half of City employees there are now using ICTs. But these cities usually fail to discuss the practices of other national or regional municipalities, and their systems are all underfunded and are underperforming. Whether the Lusaka experiment is successful will greatly impact whether any other local e-government can operate well in Zambia. As for now, true transparency and good governance continue to elude even the people of Lusaka, where e-government's potential is not being fully utilized.

LACK OF COMPUTERS IN ZAMBIA

The lack of desktop or laptop computers in Zambia is a real detriment to trying to empower persons to access e-government. Only one in 1,000 Zambians owns a computer. Many high schools do not have a computer lab, and those that do often have hardware that is a decade old. This out-of-date equipment usually is not adequate for accessing what e-education is available from the Ministry of Education, let alone the internet (R. Riddle, personal communication, July 7, 2007). However, personal computers may soon be available to children and their parents under the nonprofit project called the One Laptop Per Child. Five countries—Argentina, Brazil, Libya, Nigeria, and Thailand—have made tentative commitments to purchase the computers for millions of students, with production expected to begin by mid-2007. Zambia is considering joining the program.

This particular computer has features favorable for rural areas. It uses only two watts of power because of the lack of a hard drive and because the microprocessor shuts down when the computer is not processing information. It uses the freely available Linux operating system, a simple Web browser and word processor, and a number of learning programs. It has a carrying handle, built-in speakers, wireless data connection, hand crank to generate power, and a screen visible in bright sunlight. The project will use several methods to connect to the internet—satellite downlinks, existing cellular data network, or specially designed long-range WiFi antennas. Last, the cost is about $150. Other possibilities for inexpensive computer ownership are also being prepared such as Intel's development of a $400 laptop and an education program for teachers. Microsoft is offering an inexpensive cell phone which can be configured into a computer by connecting it to a TV and a keyboard.

There have also been setbacks. The Indian Education Ministry rejected a proposal to order a million computers, saying the money would be better spent on primary and secondary education. Critics have raised questions about the laptop, pointing to the price of internet connectivity which can cost $24 to $50 a month in developing nations. But the One Laptop project responds that the laptops would connect in a mesh network of up to 1,000 computers wirelessly sharing one or two land-based internet connections (Markoff, 2006; One Laptop Per Child, 2007).

Having such a computer for students and/or their parents would provide a missing link in the ability of Zambia to provide more effective remote access to e-government. Already similar success has been seen in rural places in Uganda. Near Fort Portal, four villages are using a system designed by the San Francisco nonprofit Inveneo to harness solar-powered computers, wireless networks, and telephones. Villagers now surf the internet, use email, build databases, and place phone calls. This has resulted in farmers obtaining market prices for produce before deciding whether or which marketplace to travel to. Computer literacy among the 3,200 residents is increasing. City government is utilizing the technology to organize and computerize records files, and hand-held computers are transmitting information about patients from field offices to the regional ministry of health (Villano, 2006).

Regardless, laptops are not the solution for all development ills. For example, many Western school districts have distributed laptops to close the digital divide between students who have computers at home and those who do not. After several years, however, there has been no evidence the programs impacted student achievement as measured by grades and standardized testing. This, along with escalating maintenance costs and problems when students used laptops to exchange test answers, hack into businesses, and download inappropriate material has led multiple U.S. school districts to terminate the programs and save millions of dollars. Many teachers in these districts reported that laptops were actually a distraction to the educational process, and instead preferred pencil and paper for subjects such as math, along with the use of calculators. In short, technology is often embraced as a solution, but traditional low-cost solutions often are better-suited for success (Hu, 2007). Schools in Zambia, not to mention businesses and the government, must learn from these mistakes and be careful about their use policy for donated or purchased laptops.

POSSIBLE SOLUTION FOR ZAMBIA E-GOVERNMENT WOES

Up to this point, this chapter has provided a description of the positive and negative aspects of Zambia's e-government. Prescription is now in order, and it is suggested that one need not look further than successful e-government programs in other countries and even in Texas to find a sustainable model. Texas' official portal, TexasOnline.com (also at texas.gov), provides quick access to almost 800 state and local government services and is currently recognized as the best site in the United States by the Taubman Center for Public Policy at Brown University.

E-services in Texas are divided into 15 categories and organized alphabetically. They include searchable license records databases, commonly used government forms, access to vital records, renewable vehicle registrations and driver licenses, payment of utility bills and parking tickets, and the filing of court documents. An emergency preparedness portal gives information for hurricane season and other natural disasters and provides weather updates and evacuation routes. A business portal breaks down the process of starting a business in Texas to four easy steps. In addition, a majority of the websites feature audio and video clips, and there is a Spanish version of nearly every page. Twenty-four hour customer assistance is available. Two online services generating a significant amount of online revenue collection for the State are the WebFile program, which enables businesses to file and pay sales tax returns via the internet, and a service in which Texas A&M University students pay their tuition and fees.

A company called BearingPoint, located in McLean, Virginia, has operated the TexasOnline website since 2000 with close involvement and oversight by the Texas Department of Information Resources. It is a contractual public/private partnership—in other words, the State's homepage is partially privatized. BearingPoint provides a variety of managed services such as Web application development, payment processing, call center management, and intrusion detection monitoring. Persons access TexasOnline at no cost. Every month the homepage receives over 2 million visitors and in the process BearingPoint conducts over a million financial transactions associated with the website.

All e-government project expenses are paid by BearingPoint and it is reimbursed via project revenue. The company assumed all financial risks and no State tax dollars went into the financing of the project. Revenues consist mainly of convenience and subscription fees paid as part of the financial transactions of the project. From 2001 through most of 2006 the project operated with significant investment and debt on BearingPoint's part. In 2006, BearingPoint began to make a profit. By contract, 20 percent of the gross income from the project is paid to the State of Texas General Revenue and also offsets costs that other state agencies incur to provide assistance and services to TexasOnline.

An example of how those fees are charged can be demonstrated by the Texas vehicle registration system. The base part of the fee is for the vehicle registration to support the Department of Transportation and local counties in the processing of registration and the vehicle plates or stickers. It is the same amount whether a person visits their local county tax office, renews by mail, or processes it online through TexasOnline. If a person processes by mail or online, there is a $1 fee to cover the additional costs the local county tax office must incur to document, package, and put postage on the plates or stickers they provide via mail. When a citizen processes by mail they can only pay by check. For processing transactions online, BearingPoint offers the additional service of using a credit card. When a person does so, the credit card companies charge TexasOnline a merchant fee for processing the credit card just like they do any retail merchant. This averages about 2.25 percent of the total dollars charged plus about 25 cents for other fees TexasOnline incurs. The average price in Texas for a vehicle license renewal is $70, and the cost to TexasOnline for processing a credit card or electronic check is $1.83. This convenience fee is rounded up to $2. The other 17 cents compensates BearingPoint for running the homepage. After expenses, net income from the project is shared equally between BearingPoint and the State of Texas until the contract ends August 31, 2009. The contract will then be bid again as is normal with government contracts, and it is expected BearingPoint will be one of the bidders (A. Martin, personal communication, July 11, 2007).

Besides providing Zambia with a model of partial privatization, it should be noted that Texas' e-government did not come about simply because of healthy revenues from other sources or because the political structure provided an easy path to success. With regard to finances, in 2004 Texas was spending roughly $2 billion each year on information technology but was in a deficit situation. Governor Rick Perry took the step to hire the chief information in the State of Pennsylvania who had helped it save $270 million by consolidating data centers, putting telecommunications services under one contract, and building a common e-mail platform for state government. But as director of the Texas Department of Information Resources, Larry Olson faced a difficult government structure for making change. Texas has a governor with no cabinet, a strong legislature, and independent agencies overseen by governing boards that govern many of Texas roughly 250 agencies. In other words, Texas has dispersed authority compared to other states, and Olson and the DIR faced enormous hurdles convincing the Legislature and multiple independent bureaucracies to work together to accomplish cost savings on an enterprise-wide basis (Welsh, 2004).

One of Olson's first moves was to develop a strategic IT plan. Zambia has such a policy. And Zambia has the advantage that it has a more centrally strong government structure than Texas, so it could make the same types of changes that Texas has implemented. As a part of this, it would be of great aid if Zambia would adopt one, single homepage for national government.

As a result of these changes, the Texas web page, operated by DIR and BearingPoint, was named in August 2006 as the best e-government site in the United States in the seventh annual study of state and federal e-government conducted by the Taubman Center for Public Policy at Brown University. Researchers evaluated various features including the number of online services, online publications, language translation, disability access, privacy policies, and security (Texas Department of Information Resources, 2006). Texas' DIR set a goal of providing citizens with greater access to government services while reducing service delivery costs. They have accomplished that and are now working to maintain and improve service. In short, the public/private partnership has allowed Texas to build almost 800 services online without any direct government investment (Texas Web Site Surpasses $1 Billion, 2004).

CONCLUSION

Has Zambia's e-government been successful? Has it been relatively successful in its short lifetime? Has it instead been a failure, or is it heading that way? The answer is mixed. Currently it is accurate to describe e-government in Zambia as immature, based on this study's evaluation. Aspects of e-government have been created, and several ministries have reasonably effective websites, but the simple action of emailing elected officials is not among e-government activities that Zambians are currently able to accomplish because email addresses are not given for Members of Parliament on its site. The widespread type of electronic government interfaces available that one would expect for businesses are also not available—e-commerce in Zambia is still in an infancy stage with problems such as poor connectivity and lack of secure electronic transactions. Will the central government dedicate enough resources to have a quality e-government program? It is crucial that e-knowledge and e-service continue to expand to meet the needs of Zambia's citizens.

Thus, the main suggestions in this study are that the Zambian government must construct a national portal, partially privatize that site, ensure all ministry websites work and have elements of e-knowledge, service, and governance, make needed improvements in the ministry sites as noted in the Appendix, provide links to pertinent addresses, and create provincial and city websites. For the most part these are inexpensive improvements, but they will take political courage and commitment. These will provide the base from which other additions can be made, and will demonstrate that e-government can work for Zambians and foreigners who need to use the system. Of highest priority is to allocate resources to allow citizens to access the internet. This is not through the acquisition of grants to purchase computers—instead, it is by taking steps to make the internet available for those who want to spend their own funds to acquire an ICT device which gives them access. To use an analogy, like building a highway, those who wish to purchase their own car will travel often and as they see fit, but the government helps by constructing the roadway. That freedom will improve education, commerce, and democracy in Zambia.

A number of difficulties exist and demand many resources. For example, illiteracy and the multiplicity of languages in Zambia are barriers to network building. In addition, Zambia has not been able to give much attention to national needs for communications and information. Resources are swallowed up by concerns over health care, education, and transportation, but the correlation between poverty and access to information needs to be communicated to political leaders. The country's highways are derelict—there are only 64 kilometers (40 miles) of four-lane, divided and paved highway in the country, running from Ndola to Kitwe. It may be that the government is just too strapped for cash to properly implement e-government in the next several years. It has limited resources and a population with very limited education and it has to choose among providing which vital government services.

Likewise, Zambia's ICT policy is adequate, if not ambitious, but the central government lacks the funds to implement it. Until more resources are put into e-government, it faces an uncertain future unless it is partially privatized, something not authorized at present.

Few Zambians own computers, and only two percent use the internet. It goes without saying, but hardly any purchased an iPhone, introduced in 2007. It is cost-prohibitive for Zambians, costing in the U.S. $499 for the 4 GB model and $599 for the 8 GB model. But as the cost comes down, perhaps Zambians will acquire these just as they did cell phones, which have provided real improvement in the lives of persons in developing countries. About one million Zambians now have a cell phone. The iPhones have internet, email, and other features on the touch screen, easy enough for illiterate persons to use. Zambians could access e-government from anywhere in the country with this type of tool.

The last hurdle that exists to building an effective e-government in Zambia is overcoming the lack of trust that exists between citizens and their government. Much of the media in the country is still state-owned, and how the news is presented is markedly different when comparing the *Times of Zambia*, controlled by Lusaka, and the *Post of Zambia*. The Media Institute of Southern Africa website shows the Zambian government is still involved in information control—the continued existence of Radio One and Radio Two as propaganda machines this many years after liberation suggests that any centrally-controlled site will be viewed skeptically by many. This exercise of control was also seen in the removal of the Parliament homepage in July 2007. Until the national government releases control of most of the media, it will be difficult for Zambians to be willing to believe what they read on any state-run electronic government website.

Regardless of these problems, a prescription is provided in the model of Texas to make e-government sustainable, and other models for advancement also exist. The government of Zambia has empowered the City of Lusaka to develop a homepage and related services. But can the City maintain an effective system, which can grow to meet already existing needs? Can it be made to operate more like the homepage of Texas, a place not that different in size and topography from Zambia? The next few years will be critical in this regard. An important step will be to create a national portal and to partially privatize the website, just as Texas has done.

Beyond the recommendations to mirror the efforts of other successful e-government projects, it is suggested that a survey be administered of citizens, businesses, and bureaucrats now and in five years to measure any improvement. An assessment similar to this study can then be done, providing another evaluation of progress. Last, to bolster efforts to pull the country out of poverty, it is recommended that Zambia make internet kiosks available in post offices and towns throughout the country because women are especially affected by increased access to political and economic services. This would be consistent with NICTP Goals 29 and 31 to use postal facilities and public libraries as primary points for e-government and e-commerce. Women in Zambia, like in the rest of Africa, have suffered discrimination in education, health, business, and government.

Hopefully this chapter will encourage others to explore associated questions and further make Zambia's e-government perform better. What ranges of e-government services will become available in the future, and what are the implications for the economy and the spread of democracy in Zambia? Can e-government help spur main industries such as copper and agriculture? Could e-government help small businesses acquire micro-loans and allow business owners to trade their products and services online? As Zambians become more computer literate, will their demands for public services and e-government grow?

It is too early to determine if Zambia's e-government will be successful or will be a failure.

There are signs of both. But with adequate funding and true dedication to improving what has already been created, it can be a success at some point in the future. The central government can take this information from the first examination of Zambia's program and, combined with its own evaluation, build on it to make a truly successful and sustainable e-government system.

REFERENCES

About, E. -Government. (2005). *World Bank*. Retrieved April 27, 2007, from http://go.worldbank.org/6WT 3UPVG80

About Zambia. (2007). *OpenSource Zambia*. Retrieved July 21, 2007, from www.opensource.org.zm/

Adesisa, O. (2001). Governance in Africa: The role for information and communication technologies. *Economic Research Papers, 65*, 5–30.

Avgerou, C., & Walsham, G. (Eds.). (2000). *Information technology in context: Implementing systems in the developing world*. Brookfield, VT: Ashgate.

Background notes—Zambia. (2007). United States State Department. Retrieved July 29, 2007, from www.state.gov/r/pa/ei/bgn/2359.htm

Basu, S. (2004). E-government and developing countries: An overview. *International Review of Law Computers & Technology, 18*(1), 109–132. doi:10.1080/13600860410001674779

Bhatnagar, S. (2002). E-government: Lessons from implementation in developing countries. *Regional Development Dialogue, 24*, 164–174.

Ciborra, C., & Navarra, D. (2005). Good governance, development theory, and aid policy: Risks and challenges of e-government in Jordan. *Information Technology for Development, 11*(2), 141–159. doi:10.1002/itdj.20008

Coates, N., & Rojas, M. (2006). China and e-government. In D. Remenyi (Ed.), *Proceedings of the 2nd International Conference on E-Government* (pp. 34-41). Pittsburgh: University of Pittsburgh Press.

Dada, D. (2006). The failure of e-government in developing countries: A literature review. *Electronic Journal on Information Systems in Developing Countries, 26*(7), 1–10.

Ebrahim, Z., & Irani, Z. (2005). E-government adoption: Architecture and barriers. *Business Process Management Journal, 11*(5), 589–611. doi:10.1108/14637150510619902

Heeks, R. (2001). *Understanding e-governance for development*. Manchester, UK: Institute for Development Policy and Management. Retrieved July 15, 2007, from http://idpm.man.ac.uk/ publications/wp/igov/igov_wp11.pdf

Heeks, R. (2002). Information systems and developing countries: Failure, success, and local improvisations. *The Information Society, 18*(2), 101–112. doi:10.1080/01972240290075039

Heeks, R. (2003). Most e-government-for-development projects fail: How can risks be reduced? *iGovernment Working Paper Series*, No. 14. Retrieved July 21, 2007, from http://unpan1.un.org/intradoc/groups/public/documents/CAFRAD/UNPAN011226.pdf

Heeks, R. (2006). Benchmarking e-government: Improving the national and international measurement, evaluation and comparison of e-government. *iGovernment Working Paper Series*, No. 18. Retrieved July 21, 2007, from www.sed.manchester.ac.uk/idpm/research/publications/wp/igovernment/documents/ iGWkPpr18.pdf

Hu, W. (2007, May 4). Seeing no progress, some schools drop laptops. *New York Times*. Retrieved May 4, 2007, from www.nytimes.com/2007/05/04/education/04laptop.html?ex=1180670400&en=fae7a1a9906 70 e83&ei=5070

International Monetary Fund. (2005, December 23). *IMF to extend 100 percent debt relief to Zambia under the multilateral debt relief initiative*. Press Release. Retrieved June 20, 2007, from http://imf.org/ external/np/sec/pr/2005/pr05306.htm

Jaeger, P. T., & Thompson, K. M. (2003). E-government around the world: Lessons, challenges, and future directions. *Government Information Quarterly*, *20*(4), 389–394. doi:10.1016/j.giq.2003.08.001

Kasumbalesa, F. (2005). The untapped potential of e-governance. *Times of Zambia*. Retrieved April 24, 2007, from www.times.co.zm/news/viewnews.cgi?category=8&id=11309

Krishna, S., & Walsham, G. (2005). Implementing public information systems in developing countries: Learning from a success story. *Information Technology for Development*, *11*(2), 123–140. doi:10.1002/itdj.20007

Lollar, X. (2006). Assessing China's e-government: Information, service, transparency and citizen outreach of government websites. *Journal of Contemporary China*, *15*(46), 31–41. doi:10.1080/10670560500331682

Malakata, M. (2005). Zambia's leap into e-banking. *iConnectAfrica: ICTs at Work*, 2(2). Retrieved April 24, 2007, from www.uneca.org/aisi/iconnectafrica/v2n2.htm#5

Mambwe, K. (2005). Boosting local authorities: E-governance in Zambia. *IConnectOnline*. Retrieved June 29, 2007, from www.ftpiicd.org/iconnect/ICT4D_Governance/EN_Governance_ZM.pdf

Markoff, J. (2006, January 30). Microsoft would put poor online by cellphone. *New York Times*. Retrieved May 23, 2007, from www.nytimes.com/2006/01/30/technology/30gates.html?ex=1178683200 &en=f7fded26ce9f0dc9&ei=507

Markoff, J. (2006, November 30). For $150, third-world laptop stirs big debate. *New York Times*. Retrieved May 23, 2007, from www.nytimes.com/2006/11/30/technology/30laptop.html?ei=5070&en=040c21fc7ec4bec4&ex=1178683200

Mupuchi, S. (2003, August 23). Government should consider e-governance. *Post of Zambia*. Retrieved April 2, 2007, from http://unpan1.un.org/intradoc/groups/public/documents/un/unpan014273.htm

Mushinge, G. (2005). ICTs help transform lives of rural women in Zambia. *ConnectAfrica: ICTs at Work, 2*(2). Retrieved July 24, 2007, from www.uneca.org/aisi/iconnectafrica/v2n2.htm#5

Ndou, V. D. (2004). E-government for developing countries: Opportunities and challenges. *The Electronic Journal of Information Systems in Developing Countries*, *18*(1), 1–24.

One laptop per child. (2007). *One Laptop Per Child*. Retrieved May 23, 2007, from http://laptop.org/en

Parajuli, J. (2007). A content analysis of selected government web sites: A case study of Nepal. *Electronic Journal of e-Government, 5*(1), 87. Retrieved September 9, 2007, from www.ejeg.com/volume-5/vol5-iss1/v5-i1-art9.htm

Sinyangwe, C. (2007). Zambia's trade surplus to reach $1.2 billion. *Times of Zambia*. Retrieved July 6, 2007, from www.zamnet.zm/newsys/news/viewnews.cgi?category=6&id=1183627066

Syed, M. (2007). Zambia's first wireless ISP offers high speed broadband solution. *Balancing Act News Update, 250*. Retrieved September 9, 2007, from www.balancingact-africa.com/news/back/ balancing-act_250.html

Texas Department of Information Resources. (2006, August 28). *TexasOnline.com named best state e-government site in the nation*. Press Release. Retrieved July 4, 2007, from www.dir.state.tx.us/dir_over view/ pressreleases/20060825txo/index.htm

Texas web site surpasses $1 billion in state revenue collection. (2004, April 27). *Government Technology*. Retrieved July 4, 2007, from www.govtech.com/gt/articles/90060

UN. Kabwe, Africa's most toxic city. (2006, November 10). *IRIN*. Retrieved July 8, 2007, from www.irinnews.org/reporttest.aspx?ReportId=61010

Villano, M. (2006, November 13). Wireless technology to bind an African village. *New York Times*. Retrieved November 13, 2006, from www.nytimes.com/2006/11/13/us/13tech.html?ei=5070&en=93831ac1f7064429&ex=1178683200

Welsh, W. (2004). Texas-size opportunity: Hard-charging CTO Larry Olson aims to make the lone star state a national leader in IT innovation. *Washington Technology, 19*(18). Retrieved July 4, 2007, from www.washingtontechnology.com/print/19_18/25090-1.html?topic=cover-stories

Wines, M. (2007, July 29). Toiling in the dark: Africa's power crisis. *New York Times*. Retrieved July 29, 2007, from www.nytimes.com/2007/07/29/world/africa/29power.html

World factbook—Zambia. (2007). *CIA*. Retrieved June 22, 2007, from www.cia.gov/library/publications/ the-world-factbook/geos/za.html

Zambia: Govt cautions about spending debt savings. (2006, March 21). *IRIN*. Retrieved June 1, 2007, from www.irinnews.org/PrintReport.aspx?ReportID=58494

Zambian ICT policy fails to address key issues. (2007, April 4). *SciDev.Net*. Retrieved June 15, 2007, from www.scidev.net/News/index.cfm?fuseaction=readNews&itemid=3537&language=1

APPENDIX

Table 5.

<table>
<tr><th>Ministry</th><th>Website</th><th>E-Government Level</th></tr>
<tr><td>Agriculture, Food and Fisheries</td><td>www.agriculture.gov.zm</td><td>Could not connect</td></tr>
<tr><td>Anti-Corruption Commission</td><td>www.acc.gov.zm</td><td>e-knowledge</td></tr>
<tr><td colspan="3">The ACC is an independent government organization designed to fight corruption. The website provides downloadable newsletters, the ACC strategic plan, and workshop and poster materials. Annual reports can be accessed, but the latest is from 2001. The names and pictures of the Commission's officers are available, but no contact information for them individually is present. Under "Contact Us", the site does list contact information for ACC offices in various cities; included in this information is email, telephone, and address. Under "Latest News", the report is outdated as it reads July 2006. It is unfortunate there is no form of e-governance with this particular homepage because corruption is widespread in Zambia. One of the President's goals is to fight corruption, and an efficient way to do this is by including discussion boards, polls, and complaint forms in this site. This would encourage citizen interaction with the government and increase transparency. The website was designed and is hosted by Zamnet.</td></tr>
<tr><td>Bank of Zambia</td><td>www.boz.zm</td><td>e-knowledge</td></tr>
<tr><td colspan="3">At first blush, this site appears helpful and even impressive. It advertises functions such as Latest News, Publications, and Financial Markets, Economics, Banking and Currency, and Payment Systems. There are a number of good features such as treasury bill tender invitations and payment systems information. There is also an active "Enquiry" email service. Unfortunately, though, several of its selections return an error message of "The page cannot be found".</td></tr>
<tr><td>Central Statistics Office</td><td>www.zamstats.gov.zm</td><td>e-knowledge</td></tr>
<tr><td colspan="3">Designed and hosted by the CSO Web Unit, this site offers statistical information about Zambia in many areas such as census, consumer price index, the economic census, a demographic heath survey, and a sexual behavior survey. Citizens can download press releases and statistical reports. Contact information is listed for headquarters and provincial offices.</td></tr>
<tr><td>Commerce, Trade and Industry</td><td>www.mcti.gov.zm</td><td>Could not connect</td></tr>
<tr><td>Communications and Transport</td><td>www.mct.gov.zm</td><td>e-knowledge, e-service, e-governance</td></tr>
<tr><td colspan="3">This ministry is more advanced than many of the others because it has a basic system of e-governance. Citizens can register and sign in to the ministry and participate in online discussions. There are links to other government bureaucracies, but these links are empty and selecting them only brings the reader back to the MCT website. Several of the web page's internal links are also empty. For example, by selecting "Regulations", "Ministry structure", or "Licensing" the page turns up blank. Citizens can download the ICT policy, but it is riddled with misspellings and typographical errors. Under the "Contact Us" button it lists the ministry's address, but lacking are individual member's names and emails. There is an online feedback and inquiries form. The site was designed by RKC and is hosted by ZM.COM, a United Kingdom web development company.</td></tr>
<tr><td>Community Development and Social Services</td><td>No website</td><td></td></tr>
<tr><td>Defence</td><td>www.defence.gov.zm</td><td>Could not connect</td></tr>
<tr><td>Development Agency</td><td>www.zpa.org.zm</td><td>e-knowledge</td></tr>
<tr><td colspan="3">This site designed by ZPA lists state-owned enterprises for sale in an attempt to privatize government services and assets. Descriptions, costs, and contact information for these entities is included. There are descriptions of the major investors, but no link is available. Contact information for employees is up-to-date including email addresses. Visitors can download annual reports and press releases. A more effective way to manage this homepage would include allowing for investors to bid online for enterprises, and e-governance could be better implemented by allowing for online discussions regarding privatization. Links to relevant sites are empty.</td></tr>
<tr><td>Drug Enforcement Commission</td><td>www.deczambia.gov.zm</td><td>e-knowledge</td></tr>
<tr><td colspan="3">This site is highly inefficient. There are links regarding different forms of street drugs, but when selecting a link it does not allow the user to click on the information. Also, all links from the homepage are empty. The site could improve by including a search engine to allow persons to find information about drugs and addictions; another would be to include laws and regulations about drug use and trafficking. The site is designed and maintained by the DEC.</td></tr>
<tr><td>Education</td><td>www.education.gov.zm</td><td>Could not connect</td></tr>
<tr><td>Electoral Commission</td><td>www.elections.org.zm</td><td>e-knowledge</td></tr>
</table>

Ministry	Website	E-Government Level
The Electoral Commission's homepage lists presidential and parliamentary candidates but has no link to their party web page or candidate contact information. There is a graph depicting percentage of registered voters. E-knowledge information includes the Electoral Code of Conduct, press releases, the constitution, and the mission statement of the organization. This site would be more helpful if it allowed citizens to register to vote online. In addition, it should allow for conversation between voters and officials through discussion boards, chat rooms, and forums. Although this site was important enough to farm out to a company to design and maintain, the company, HiyaCON, needs to make a number of improvements.		
Energy and Water Development	www.energy.gov.zm	Could not connect
Environmental Council	www.necz.org.zm	e-knowledge
The Council provides on its website information regarding pollution, regulations, projects, and reports. The homepage also has a feedback form. One fault is that the news section was last updated in 2006. Contact information including email is present, however, only for the Council itself and not for individual members or employees.		
Finance and National Planning	www.mofnp.gov.zm	e-knowledge
Developed by the Ministry's Centralised Computer Services Department, much of this site is still under development. The information that was available describes the purpose of the ministry and several reports, and provides press releases. However, the "News" was last updated in 2004, three years earlier.		
Foreign Affairs	www.foreign.gov.zm	Could not connect
Gender and Development	www.gender.gov.zm/index.php	E-knowledge, e-service
Designed by WebCon, this homepage provides access to policies, reports, documents, and publications. The homepage has an extensive contact list with email addresses. There is also an online feedback form and a list of working links to pertinent websites. One interesting item is the disclaimer regarding reliability or accuracy of the information on the site.		
Health	www.moh.gov.zm	e-knowledge
The Ministry of Health provides only the most rudimentary information about services. There are no links to their publications or statistics. It is possible to download an Acrobat document that lists detailed information about the individual health facilities in the country, however, they do not provide addresses or contact information. They have a news section with important topics listed such as HIV, breastfeeding, and malaria, but it is set up as link and is not accessible. Zamnet designed and hosts the site.		
Home Affairs	www.home.gov.zm	Could not connect
Information and Broadcasting	www.mibs.gov.zm	e-knowledge
Designed and hosted by CopperNET Solutions, this site provides some access to information through policies, news, and press statements. One thing that stands out about this website is that the news is actually updated, but this is one of the purposes of the ministry. Under "Contact Information" there is an address and telephone number but no email.		
Investment Centre	www.zic.org.zm	e-knowledge
This homepage provides information regarding investing, such as what is needed to acquire a license, a general list of investment opportunities, and services offered. There is no direct access to any of these services, though. If a client is interested they must make an appointment for an in-person visit. This homepage would be much more beneficial if it provided e-service because this ministry is very important to the economic growth of Zambia. Foreign investors would be inclined to invest if they could have access to the Investment Centre through the site because it would help them do business more efficiently. This page was designed and is hosted by Zamnet.		
Labour and Social Security	www.labour.gov.zm	Could not connect
Lands	www.ministryoflands.gov.zm/mol/default.aspx	e-knowledge
The site, powered by WebAssemblet.NET, appears well designed and professional, but the large amount of information available only qualifies this homepage as being in the e-knowledge category. Regardless, the breath of information available on the Ministry of Lands' site is outstanding.		
Legal Affairs	www.legal.gov.zm	Could not connect
Local Government and Housing	www.mlgh.gov.zm	e-knowledge
Designed and hosted by LGH, this site gives adequate notice of the purpose of the ministry and a brief description of a project underway. E-service features an e-mail system and online feedback form. However, the feedback form is not accessible due to construction of the website. Some links are empty such as "News", "Speeches", and "Events". There is a management page with pictures and names of the ministry employees but no contact information. There are links to other pertinent organizations.		
Mines and Minerals Development	www.mines.gov.zm	Could not connect
National Tender Board	www.tenderboard.gov.zm	e-knowledge

<table>
<tr><th>Ministry</th><th>Website</th><th>E-Government Level</th></tr>
<tr><td colspan="3">Designed by WebCon, this site provides information regarding tenders including barred bidders, archives, invitations to tender, and procurement guidelines. It also includes an online feedback form. Contact information is present, but no email address is available. There is a link called "Advertise on this Website" but it is a blank page.</td></tr>
<tr><td>National Tourist Board</td><td>www.zambiatourism.com</td><td>e-knowledge, e-service</td></tr>
<tr><td colspan="3">This is the most efficient and aesthetically appealing website. It offers much information regarding hotels, places to visit, things to do, and culture. The site is well organized with links to businesses, airlines, activities, and hotels. There are also links to weather, news, photo galleries, and maps, and Google advertisements on the web page. Two things that are missing, though, are a search engine and information regarding places to eat. The site is not run by a ministry, instead it was designed by Africa Insites, a private firm that designs other websites in Zambia and elsewhere. It should not be surprising that this is one of the most accessible and enjoyable sites among all the ministry websites—this is the second most important business in the country behind copper. The Tourist Board web page gives hope that e-government in Zambia can be successful.</td></tr>
<tr><td>Revenue Authority</td><td>www.zra.org.zm</td><td>e-knowledge, e-service</td></tr>
<tr><td colspan="3">The website provides some tax information such as incentives, statistics, and budget tax changes. Forms are also downloadable, but cannot be submitted online—this lags very much behind what other countries offer. There is a human resources page listing jobs and qualifications required, however an applicant cannot apply online. There is an online inquiry form under "Contacts" that can be submitted via the web. The only contact information available is a telephone number.</td></tr>
<tr><td>Science, Technology, and Vocational Training</td><td>www.mstvt.gov.zm</td><td>e-knowledge</td></tr>
<tr><td colspan="3">This homepage provides information about the STVT Ministry's services, policies, programs, and policies. There are several PDF files that require Adobe, and the website allows for persons to download Adobe reader. The ministry also has an internal e-mail system. Contact information is up to date with email and addresses, however, there is no contact information for individual employees.</td></tr>
<tr><td>Securities and Exchange Commission</td><td>www.sec.gov.zm</td><td>e-knowledge</td></tr>
<tr><td colspan="3">The Zambia SEC has a quality e-government website with a wide menu, containing much information about the Exchange, listed and quoted companies, and latest company news. Commissioners are listed by name, but the email for the SEC is inactive. The Securities Act and other statutes are available. Market reports and media statements constitute two of the many internal pages.</td></tr>
<tr><td>Sport, Youth and Child Development</td><td>www.sports.gov.zm</td><td>Could not connect</td></tr>
<tr><td>Welfare</td><td>www.welfare.gov.zm</td><td>Could not connect</td></tr>
<tr><td>Works and Supply</td><td>www.supply.gov.zm</td><td>Could not connect</td></tr>
<tr><td>Zamtel</td><td>www.zamtel.zm</td><td>e-knowledge, e-service</td></tr>
<tr><td colspan="3">Perhaps due to the important business nature of this government-owned utility, adequate resources have been dedicated to this site. It has information and services regarding dial-up internet, leased lines, email services, domain name services, and web designing and hosting. Interesting, Christian sites are also featured along with tourist and business homepages.</td></tr>
</table>

Chapter 10
Globalized Workforce Development and Responsible Global Citizenship through e-Literacy Capacity Building Programs for Low Income African Countries

Benjamin A. Ogwo
State University of New York, USA

Vincent E. Onweh
University of Uyo, Nigeria

Stella C. Nwizu
University of Nigeria, Nigeria

ABSTRACT

The skewed global workforce interactions during the agricultural and industrial revolutions which still bother the antagonists of globalization could be straightened by progressive workforce development policies that will mutually benefit high and low income countries. In addition, the e- literacy and information technology boom have further narrowed spatial perception of geographic distance thus providing low-income countries insights on policy dynamics of high income countries and its impact on the rest of the world. Thus in order to attain equity and balanced global workforce development, this chapter explores the rational and different paradigms for capacity building on e-literacy in low income African countries so that their workforce would contribute to the globalized economy and civic responsibility. The chapter contends that e-literacy empowerment should be regarded as a human right issue and that through other ethical globalization efforts every person on earth should form part of the workforce for sustaining the global village.

DOI: 10.4018/978-1-60566-820-8.ch010

INTRODUCTION

The exponential global impact on workforce development as well as the politico-social changes occasioned by the agricultural/industrial revolution era makes global interaction a familiar issue, with perhaps, different connotations for the two global economic divides (haves and have-nots). At that time there was explicit coercive muscle drain from the colonized territories (have-nots) to the imperial lands (haves). Several cruel methods and policies were employed by the rich and powerful (haves) to balkanize and exploit the colonized peoples of the world. Whereas previous global interactions had questionable imperialists' intentions for the colonized territories, its poor human right fallouts created appalling feelings on the moral spectrum of thinking global citizens. It is against this backdrop that some civil rights activists resent globalization. Be that as it may, various forms of economic thoughts have been acclaimed to guide the actions of governments of different independent nations namely, capitalism, communism, socialism, and the mix economies. These economic thoughts have faced serious litmus tests in the recent past, collapse of Russian socialism and, questionable state of China's communism and the bedridden condition of capitalism in United States of America, such that economists are in doubt on the practical relevance of these countries' claims to be abiding by any of these economic principles. Hence this chapter would refrain from discussing the needed reengineering of the world economic order from the framework of any of the aforementioned economic thoughts rather the focus would be to institute an ethical basis to develop the workforce potentials of low income countries in a manner that would optimize their human capitals and guarantee responsible global citizenship for all.

The workforce development experiences of countries in Asia occasioned by outsourcing and recent trends on globalization in which cost reduction informs contracting foreign supply of service/parts/products, have proved profitable for low income countries. Incidental or not, outsourcing as fallout of globalization by high income countries has proved a veritable source of direct foreign investment, technological learning, reverse engineering as well as capacity building for skilled workforce in these countries. Specifically, Japan, and lately India and China have gained immensely from outsourcing in information communication technology. The developments of human capitals in information communication technology (ICT) in these countries have been phenomenal. It is noteworthy that governments of these countries pursued diligently policies that engendered sustainable workforce development of ICT professionals. The result among others is increased income per capital and relatively more responsible participation in global affairs by citizens from that region.

On the other end of the spectrum, the low income countries in Africa have not significantly profited from globalization. The desperate efforts for illegal migration to Europe and North America by Africans are some of the practical showcase of the checkered workforce imbalance/perception in that region. Whereas there was need for muscle drain during the agricultural and industrial revolutions in Europe and North America from healthy Africans in the 18th and 19th centuries, same is not true today. At best what is preferred is the brain drain of proficient Africans professionals (with ICT competency) who migrate to the West for greener pasture. Without sufficient e-literacy even with all their muscles Africans migrating to Europe and North America would not make good slaves now. Worse still, in these African countries, outsourcing cannot be feasible because the human capital therein is relatively undeveloped and not e-complaint hence they would remain "non-outsourcing" locations. The irresponsible acts by Somalia pirates, restive youths in Nigerian Niger Delta and the debilitating poverty in the region are acclaims that global efforts are needed to enhance capacity building on e-literacy in Africa.

The obvious danger in focusing developmental programs on poverty alleviation in Africa at the expense of e-literacy is that the region is further stunted as e-literacy becomes more and more the Lingua Franca in global workforce developments. It is apparent that governments in Africa cannot muster the political will to design and sustainably implement these e-literacy programs therefore the international agencies interested in human rights, poverty, democratization and workforce development should consider this as 21st century emergency for global citizenship. Thus the objectives of this chapter is to explore the rational for capacity building on e-literacy in low income African countries so that their workforce would contribute effectively to the globalized economy and civic responsibility thereby mitigating the pangs of poverty, brain drain as well as sustaining the local economies.

Background

Defining the concept of literacy has remained very difficult even among scholars. Being able to read and write were once considered basic literacy however these abilities do not in themselves enable an individual to function effectively within the society hence the concept of functional literacy. According to (Martin, 2005) literacy is used to describe those baseline skills, shown over a range, by an individual in reading, writing, communication and comprehension in situations entailed by every day activities. Reading or writing as mere skills do not suffice as functional literacy unless they facilitate one to make better decisions that would have been impossible without these skills. As cited by Martin, UNESCO (2002) defines literacy as operating within a continuum and which is more than the ability to read, write and do arithmetic but comprises other skills needed for an individual's full autonomy and capacity to function effectively in a given society. It follows by UNESCO's position that a literate person while exercising the literacy skills should show proper understanding of the society in which he/she is operating and in the context of a globalized world show appreciation of larger global understanding of whatever is read or written. When literacy is given this wider spectrum, it connotes other abilities such as requiring the individual to recognize when information is needed and having the ability to locate, evaluate and use effectively the read or written information. (American Library Association, 1989 referred to as information literacy (cited in (Joint, 2003)). The description of functional literacy helps to conventionalize the lately developed concept of e-literacy.

The concept of e-literacy like basic literacy has been facing consensus issues and fortunately it is in recent times among scholars of different disciplines acquiring a more common meaning. UNESCO, which is the foremost global organization tackling the issue of literacy, has moved beyond the simple notion of a set of technical skills of reading, writing and calculating to one that encompasses multiple dimensions of these competencies. These competencies are incorporated in the advancement of information and communication technologies (ICTs) and used in conjunction with these basic literacy skills for the advancements related to the recent economic, political and social transformations- including globalization (UNESCO, 2008). In its most simplistic description, e-literacy is whatever literacy means plus the use of electronic medium. Literature is replete with such synonymous terms to e-literacy such as digital literacy (Martin, 2005), ICT literacy, silicone literacy and technological literacy (Martin, 2006). For example Martin, (2005) gave a detailed definition of digital literacy that can suffice for e-literacy:

Digital Literacy is the awareness, attitude and ability of individuals to appropriately use digital tools and facilities to identify, access, manage, integrate, evaluate, analyze and synthesize digital resources, construct new knowledge, create media, expressions, and communicate with others, in the context of specific life situations, in order to enable constructive social action; and to reflect upon this process (pg 135 – 136).

The use of electronic media has very broad spectrum to include mobile phones, computers, hand held and heavy duty electronic devices applied in personal, civic and occupational situations. Certain skills are identifiable that are common to the use of these electronic equipment and it is on that basis that e-literacy could be considered a generic term. Doczi (2005) described it as the skills, knowledge and attitudes to use information communication technology by a person in order to have maximum advantage and to keep up skilling for personal well being, civic and work situations. However, the foremost thing in using these electronic devices is to deride the technophobia associated with attitude of avoiding their use. In terms of national application of the concept of e-literacy, four primary dimensions have been identified:

- **Attitudinal:** Perceiving ICT skills and access to the Internet as value-adding and important to future well being; perceiving the Internet as having meaningful and relevant content; being confident and motivated to go online.
- **Financial:** Being able to connect to the Internet from home, work or a community location in a way which enables a person to learn ICT skills and carry out required activities, through affordable access to a PC with connectivity to the Internet.
- **Skills:** Having sufficient skills, or access to tuition to develop and increase skills, to use ICT to an optimal level for personal and economic gain.
- **Infrastructure:** Having sufficient levels of bandwidth available to carry out e-commerce, e-government or educational interactions. Having hardware and software which meets specific needs, such as people with physical or learning disabilities, or language difficulties might have. (Doczi, 2005, p. 3).

Just like basic literacy was regarded as imperative for surviving in the civilized world, e-literacy is indispensable for living in the e-world. Thus governments should provide the infrastructure and financial resources while the individual should show positive attitude for both basic e-literacy skill acquisition and up skilling. The possession and use of e-literacy skills are increasingly becoming the sine qua non for living in the 21st century. In formal and non-formal school settings, e-literacy skills are required for synchronous and asynchronous learning settings. In today's e-world, commerce, governance, medicine, industry, and daily living have been barged by e-literacy issues. What started initially as a status symbol, supplementary skills, e-literacy, has now become mandatory if you are to live and work in today's e-world. Lack of e-literacy skills in today's e-world is as challenging as lack of literacy and numeracy skills in the civilized world. The use of electronics associated with e-literacy has reduced the geographical perception of locations, enhanced communication, facilitated government activities, learning and medical care. At work the place, most occupations are now entailing the possession and use of e-literacy for securing and retaining employment. Wish it away, resent it or embrace it, e-literacy has become a fact of modern life. Agreeably, it is an additional challenge to pursue the goal of capacity building on e-literacy for low income African country when proportionately the population is illiterate. However, the rest of the world has moved on into the e-world and the low income countries are facing increased pressures for closer global interactions/citizenship and they cannot afford to pay the price of e-literacy isolation in a world undergoing globalization.

Globalization is the in thing within the diplomatic circles. Regional governments in Europe, Africa and the Americas, Asia (for example, European Union, African Union, etc) have formed alliances that afford them more competitive advantage in a globalized world where there is little room for the small countries and increased difficulties for the economically challenged. The concept of globalization has been used to connote internationalization, liberalization, universalization, westernization and deterritorialization (Scholte, 2000 cited in Smith, 2002). Some of these various ways of viewing globalization tend to evoke apprehension among some peoples of the world. For example, the notion of globalization as liberalization of trade places low income countries at a disadvantage since they only export raw materials that are always of less value while westernization denotes neo-colonization and many former colonies are scared of it. The notion of globalization as adopted in this chapter is that of deterritorialization which entails a 'reconfiguration of geography', so that social setting and space are no longer wholly mapped in terms of territorial places, territorial distances and territorial borders (Smith, 2002)). In this respect, policies and actions at local levels are influenced by their transnational impact such that you think globally and act locally. It is in this context of globalization that e-literacy and its use in intensifying and extensifying personal, communal, national, international, regional and interregional socio-economic interactions and transactions are considered vital for enhancing global workforce development as well as global citizenship.

Overview of Global Economic Distribution: Income, Education, and Production Indicators

Most of the transactions as well as classification of countries are based on the World Bank ratings of economies based on their gross national income (GNI) per capita. The classification is labeled as low, middle (subdivided into lower middle and upper middle) and high income countries. Both the low and middle income countries are collectively referred to as developing countries while the high income countries are referred to as developed countries. According to the World Bank (2007, pg 19), Low-income economies are those with a GNI per capita of $875 or less in 2005. Middle-income economies are those with a GNI per capita of more than $875 but less than $10,726. Lower-middle-income and upper-middle-income economies are separated at a GNI per capita of $3,465. High-income economies are those with a GNI per capita of $10,726 or more. These classifications are used for the convenience of taking economic decisions in terms of grants, other incentives offered by World Bank and International Monetary Fund. In addition, GDP is defined by the World Bank as gross domestic product and measures the total output of goods and services for final use occurring within the domestic territory of a given country, regardless of the allocation to domestic and foreign claims. GDP could be used to infer the level of internal economic activities within a country and in turn extrapolate the nature of the workforce within the economy. Table 1 below gives the graphic illustration of the economic nature of the globalizing world. It could be readily observed that the GDP of low income countries is far lower than the total world output as well as those of the middle and high income countries. It follows that relatively less economic activities are going on in these countries even as they allocate disproportionate portion of their land on agriculture and still live on starvation diets.

The table above shows disturbing statistics on the imbalance on the global economies. If globalization is to be done on fair distribution of world GDP, then the low income countries should be aided to shovel up their GDP which would entail developing their workforce to provide their basic items rather than over dependence on importation. The table shows the vicious circle of these low income countries investing less than the high income countries on education. Singularly, the under investment on education would perpetuate the underdevelopment of these countries because

Table 1. Global economic distribution according to population, income and production indices

Population and Economic Activities	World	Income level Indices		
		Low	Middle	High
GNI per capita, *World Bank Atlas* method ($)	7,011	585	2,647	35,264
Population (millions)	6, 438	2, 352	3,074	1, 011
Land Area (1000 sq. km)	129, 606	28, 185	68, 518	32, 904
GDP ($ billions)	44, 645	1, 416	8, 554	34, 687
Urban population (% of total)	48.8	30.0	53.9	77.6
Urban population growth (average annual %, 1990–2005)	2.2	3.1	2.4	1.1
Population growth (average annual %, 1990–2005)	1.4	2.0	1.1	0.7
Agriculture	38	45	35	39
Agricultural land (% of land area)	18.4	24.3	18.5	11.8
Irrigated land (% of cropland)	1,043	668	1,179	1,202
Fertilizer consumption (100 grams/ha arable land)	543	589	593	325
Population density, rural (people/sq. km of arable land)				
Energy	4.7	4.4	4.2	5.2
GDP per unit of energy use (2000 PPP $/kg oil equivalent)	1,793	513	1,451	5,511
Energy use per capita (kg oil equivalent)	10.3	47.8	10.5	3.1
Energy from biomass products and waste (% of total)	2,607	375	1,840	9,609
Electric power consumption per capita (kWh)	65.9	74.0	70.8	61.9
Electricity generated using fossil fuel (% of total)	16.0	23.4	21.5	11.9
Electricity generated by hydropower (% of total)				
National accounting aggregates	20.8	28.1	30.0	18.7
Gross savings (% of GNI)	12.6	9.1	11.0	13.1
Consumption of fixed capital (% of GNI)	4.4	3.3	3.6	4.6
Education expenditure (% of GNI)	4.1	9.8	12.1	2.0
Energy depletion (% of GNI)	0.3	0.7	0.9	0.1
Mineral depletion (% of GNI)	0.0	0.6	0.0	0.0
Net forest depletion (% of GNI)	0.4	1.1	1.0	0.3
CO2 damage (% of GNI)	0.4	0.7	0.7	0.3
Particulate emission damage (% of GNI)	7.4	9.5	7.8	7.7
Adjusted net savings (% of GNI)				

Source: World little data book, 2007

the developed countries will be developing further and further while the developing countries would be experiencing further underdevelopment. If the low income countries could not invest enough for basic education, it is highly inconceivable to convince them to commit funds to e-literacy capacity programs. The countries in Asia that served as outsourcing location for the developed countries had skilled or semi-skilled workforce that were trainable. Their workforce situation justified the direct foreign investments made in manufacturing sectors of their economy. In time these countries are being reclassified from low income to lower/upper middle income countries. The same cannot be said of many low income African countries thus resulting in the incongruity of the developed countries getting more developed and the developing ones getting more underdeveloped.

Granting that most of these low income countries operate agrarian economies, e-agriculture should be introduced to them in order to enhance their productivity. On the other hand, introducing e-government has its own promises such as cost effectiveness in government and public operations, significant savings in areas such as public procurement, tax collection and customs operations, with better and continuous contacts with citizens, especially those living in remote or less densely populated areas (Lanvin, 2002). However, many a new agricultural practices could be gradually introduced to farmers using electronic medium and followed subsequently by other elements of e-government. Where individual access and skills are limited, community centers could still be used to afford more people the opportunity to update their knowledge and upgrade their skills. Even the use of mobile phone could be used as part of e-literacy tool to facilitate e-agriculture without necessarily waiting for sophisticated gadgets. Fortunately, there are increasingly cheaper laptops produced to serve the needs of developing countries. If this development is coupled with efforts to provide cheaper and more bandwidths to these countries the battle to win e-illiteracy would have been half won. By up skilling, the workforce would graduate from unskilled to semi-skilled at the least. Being that raw materials are under priced in the world market, if the workforce is enabled to produce semi finished products, this would add value to their productive efforts as well as the economy.

Rational for Globalized Workforce Development and Responsible Global Citizenship

The rational for globalizing workforce and human capital development finds relevance on broadening the globalization efforts beyond economic gains of privileged countries at the expense of others and for creating more decent employment for all. As a result of the increasingly widespread conviction that decent employment is the only route out of poverty, full and productive employment and decent work for all has been introduced as a new target under Millennium Development Goal (MDG), to halve the share of people living in extreme poverty by 2015 (ILO, 2008). If globalization is made systemic whereby every subsystem of its workforce is identified and strengthen then every country would most likely benefit from the interactions. In this regard, where the human capitals of low income African countries are deemed to be deficient especially regarding e-literacy, the countries should be assisting with capacity development programs that are country sensitive. Basic to this rational is the fact that the weakest subsystem determines how strong the entire system of globalization will be in relation to its acceptability and sustainability. Ironically, how strong these low income countries are would eventually determine the sustainability of the globalization process. If these countries are left to wallow in e-illiteracy to the detriments of their GDP then poverty would increase as well as associated vices of irresponsible global citizenship. It is rather obvious that those countries and regions that have adaptable workforces with a rich mixture of skills and a supportive economic environment are able to add value to the free flow of capital, information and technology that characterizes the global economy as well as being able to benefit from integration into the global economy; while those countries, regions and people that lack necessary skills are destined to fall further behind (Porter, 2002).

In contemporary economic thought, the importance of human capital to production has received better appreciation and understanding such that the workforce competence of any organization/nation determines how well it would utilize other resources. Human capital is the stock of knowledge and skill inside the organization or nation which can only improve the nation's prospects of engaging in innovation and Research & Development (R&D) activities, but the skills of the citizens do also increase the country's ability to absorb new knowledge and technologies Marotta, Mark, Blom, and Thorn (2007). Thus a competent workforce, which is the classification of human capital according to areas of focus in the production chain, could turn waste to wealth, manage scarce resources, and think through difficult situations. With highly skilled and e-literate workforce, Japan endowed with few natural resources is able to develop a robust economy compared to Nigeria with over 34 commercially viable natural resources that is unable to sustain its own economy with its fundamentally e-illiterate workforce. Development of the workforce do not just happen like discovering

Table 2. Vulnerable employment as share of total employment (%), world & regions, 1997 to 2007

	1997	1998	1999	2000	2001	2002	2003	2004	2005	2006	2007
World	52.8	52.1	52.5	52.2	51.9	51.8	51.7	51.2	50.9	50.5	49.9
Developed Economies and European Union	11.0	10.8	10.6	10.3	10.1	9.8	9.7	9.7	9.5	9.4	9.2
Central and South-Eastern Europe (non-EU) & CIS	20.1	20.0	22.6	22.1	21.6	20.4	20.6	20.9	19.7	19.8	19.3
East Asia	63.2	61.5	60.8	60.4	59.6	59.3	58.7	58.1	57.2	56.5	55.7
South-East Asia and the Pacific	63.4	58.2	63.0	63.5	62.0	62.4	61.7	60.5	60.7	60.1	59.4
South Asia	80.0	80.1	80.0	80.5	79.6	79.1	79.6	79.2	79.1	78.3	77.2
Latin America and the Caribbean	31.4	31.2	31.3	32.0	32.8	32.8	32.8	32.9	32.6	33.1	33.2
Middle East	39.7	43.3	39.6	37.2	38.2	37.0	36.8	35.9	34.5	33.4	32.2
North Africa	36.9	37.7	35.6	32.9	33.3	33.3	33.2	35.1	33.5	32.0	30.7
Sub-Saharan Africa	77.2	78.1	77.7	76.0	76.8	76.7	76.1	74.7	74.9	73.9	72.9

Source: ILO, Global Employment Trends Model, November 2007

informal sector economy is hardly accounted for in most developing countries and it follows that the capacity building for these countries should include developing models to re-engineer this sector without changing its fundamental nature. When the toll of the global recession is taken, more hardship awaits those in vulnerable employment situations and they would certainly need assistance in order to cope with their adverse economic circumstances.

The table above depicts a daring situation of vulnerable employment situations for South Asia and Sub-Sahara Africa. The authors dare state that so far as the globalization process continues and no globalized efforts made at up skilling these workers, they would sink deeper into abject poverty. According to ILO, if the proportion of vulnerable workers is sizeable, it may be an indication of widespread poverty and the poverty connection arises because workers in the vulnerable statuses lack the social protection and safety nets to guard against times of low economic demand and often are incapable of generating sufficient savings for themselves and their families to offset these times. It is not surprising as Table 2 shows that the developed countries have low percentage of vulnerable employment situations since they have less of the informal economy and every citizen is accounted for economically.

With the widening use of ICT and international media, many citizens of the low income African countries are aware of the economic situations of the developed nations and would bid anything including their lives to migrate to richer nations. In presenting these points for solidarity in globalized capacity building, the authors are in no way oblivious of the near lack of government in these low income countries. It is envisaged that the e-literacy skills would equally empower the citizens to demand as well as work for more effective government. The longing to migrate to the developed countries accounts for the stemming tide of dangerous and dehumanizing illegal migration. It would be fool hardy to expect that tighter immigration laws and its implementation would calm the nerves of these economic refuges. Conversely, if their home countries are made better by conscientious building of capacity for e-literacy that would turn around their economy, then and only then would they most likely refrain latent natural resources in commercial quantity but it evolves through careful policy provisions and implementation. According to Alford, Chasteen & DeRosear (2008), the business of workforce development is to serve as the bridge connecting education and economic development efforts. It is a policy area that uses both education and economic indicators in order to measure its relevancy in today's globalized economy and its impact on business retention, expansion and recruitment. The role of workforce development is to identify business and industry talent needs by understanding the current and emerging occupational and job-specific skills and knowledge, and inventorying the "quantity and quality" of people willing and able to seek employment in critical occupations (Alford, Chasteen & DeRosear). The quality of the workforce could be classified as unskilled, semi skilled and skilled. A highly skilled workforce would possess the human capital needed to transform the economy for globalization. If these low income African countries are not assisted to develop their workforce, their economy would continue to under develop under the weight of globalization and their citizens would continue to wallow in poverty that weakens the systemic structure of the globalized economy. In the spirit of globalization, suffering in its variety of forms requires empathy and solidarity by all and transcends a politically correct ideology (Kössler & Melber, 2007).

The prevalent global recession is affording the world a graphic picture of the intricate national interrelatedness of globalization. For example, some faulty government regulatory lapses in the United States as well as vaulted greed by speculators/investors have occasioned a landslide economic crisis across the world. In the same vein, the increasing number of working poor is creating serious moral pressures on the minds of thinking global citizens. With the prevailing economic order, the low income African countries can hardly be expected to escape poverty on their own efforts especially in reducing the number of working poor and those in vulnerable employment situations. It is not just a perceptual problem but real nagging issues of empowering people who if left alone will remain debilitated by physiological needs of hunger and want of shelter. The International Labor Organization (ILO) decries this situation thus:

With five out of ten people in the world in vulnerable employment situations and four out of ten living with their families in poverty, despite working, the challenges ahead remain daunting. Economic progress does not automatically lead to progress in the world of work. Active engagement and the proactive decision to put labor market policies at the centre of growth and macroeconomic policies are needed to ensure that economic progress is inclusive and does not lead to increasing inequality. And, only if countries use their labor markets to make growth inclusive, will their progress have a real chance of being sustained (ILO, 2008, p.5).

Vulnerable employment indicator represents the mainly agrarian and subsistent entrepreneurial ventures located within the informal sector of the economy which are predominant in the low income countries. The employment situations are characterized by little or no direct relationship with the formal economy and are very fluid and hardly responsive to the larger economic structure of the country (Ogwo, 2007). These workers are mainl illiterate (mostly stark e-illiterate), reactionary i terms of contributing to macro-economic polici and lack access to funds. Vulnerable employme is a newly defined indicator that calculates t sum of self-employed workers and contribut family workers as a share of total employm within a country's workforce. These contribu family workers and self-employed workers are likely to have formal work arrangements, w allows for the usage of the indicator on vulne employment to confirm or refute claims increasing informalization of labor market 2008). In spite of its contributions to the G from these dangerous immigration escapades. Dzvimbo (2003) distinguished between factors that "push" people out of their home countries, and factors that "pull" them to a new or "receiving" country. Among the former, we can mention adverse domestic conditions such as inadequate educational capacity, lower living standards, technology limitations, inadequate coordination between education and labor market, and uncertainty about the future, and lack of realistic workforce development policies. The situation is so desperate that a heavy dose of global intervention is required in order to stem the tide. This is especially so if the e-literacy capacity building and skill empowerment is considered among the human rights to be enforced by the United Nations. Outsourcing succeeded in Asia because the governments prepared the workforce for sustaining the direct foreign investments that were orchestrated by globalization. The same cannot be said of most low income African countries that are short sighted about the impact of local actions on the global economic landscape or are too politically naive to bother. By providing for capacity building of the workforce, the African countries would become good outsourcing destination even for the Asian countries.

Paradigms for Functional Workforce Development on e-Literacy for Low Income Countries

The focus of e-literacy for workforce development in low income countries is not about rhetoric but a serious research and practical issues that directly affect lives of more than four billion of the world's population. In fact from systemic perspective, since globalization affects every nation thus increasing the potentials of the weak would positively affect the strong. UNESCO as the lead agency for the United Nations Literacy Decade (UNLD), 2003-2012, launched by the UN General Assembly has pursued remarkable programs in realizing its objectives. The UNLD emphasizes the goal of literacy for all people of all ages, in all regions of the world, in rural and urban areas, in school and out-of-school, within families, communities and the work place (UNESCO, 2008). However, as laudable as the UNESCO initiatives are for specifically basic literacy, there should be explicit program focusing on e-literacy because every advancement on the e-world makes those citizens already basically literate more handicapped by the needed skills in e-literacy. The global community would not stand still developmentally for every citizen in the low income country acquires basic literacy so the argument of succeeding in basic literacy before e-literacy is hardly tenable. In fact many e-literacy programs promise to facilitate instructional process of acquiring basic literacy. In addition, many a people in these remote communities are already using the global system for mobile (GSM) communication technology to communicate within and outside their countries, and could be said to possess the basic artificial intelligence required for e-literacy programs. What is required in such a case is to build upon their use of the cell phone and move from known to unknown (cf Geddes, 2006). Furthermore, granting that these countries depend heavily on subsistence agriculture, it stands to reason to have the building capacity programs based essentially on e-agriculture. Recent and best practices would be easily communicated to them at their farm locations since many do not even have good road network.

The major paradigm of building capacity on e-literacy is to regard such empowerment program as human right. It is a fundamental right to equip everyone with the world's lingua Franca in order to have everybody relatively on the same page. In doing this, there should be global solidarity similar to the ones exercised in clamoring for democracy across the world. In such solidarity, e-government would be facilitated as well because the citizens would be availed easy access to information regarding their rights and how to legitimately go for them. The basis for the global solidarity is based essentially on two premises namely affinity and restitution (Kössler & Melber, 2007)). The affinity objective is about the common global good for every citizen in having decent job that would result when production process is enhanced by acquisition of e-literacy skills for workforce empowerment. Global solidarity based on the notion of restitution is about the recognition of the historical fact of the injustices perpetrated by developed economies pursuing beggar-my-neighbor policies and directly contributing to the underdevelopment of the low income countries. In making these suggestions, the authors risk the outrage of some scholars who would disagree with any form of restitution for the low income countries. However, they would find it difficult to contest the fact that if globalization is not based on good morals and positive perception of each other's actions then it might be unnatural to expect responsible global citizenship from the poor whose disadvantage fuels the success of the rich. e-literacy capacity building should be regarded more a process than a state of mind or an achievement of once and for all goal thus it has to be debated and even fought over, in an ongoing effort to investigate and define common ground and goals or to hammer out possible forms of joint action which may imply widely diverse priorities and goals, depending not only on different perceptions of the problems but also more fundamentally on different concerns (Kössler & Melber, 2007).

The authors subscribe to the UNESCO modalities of capacity development for literacy which should also be adopted for e-literacy to include training, study visits, peer reviews, South-South and North-South exchanges, networking and partnership-building. The main areas of capacity-building in the fields of literacy and non-formal education used by UNESCO, which the authors suggest should be pursued in inter-organizational level to include such agencies like ILO, UNDP that have cognate experiences at workforce program to include: policy formulation and implementation, institution building, planning and management, curriculum development and materials design, teaching and learning strategies and methodologies, training of trainers, as well as facilitators, developing support structures and mechanisms, as well as learner performance assessment, monitoring and evaluation. While commending UNESCO's UNESS (United Nations Education Support Strategy) country-based approach, it should be recognized that many of these governments do not really appreciate the long term essence of planning in a globalized economy hence UNESCO should have ideals but optional programs on sustainable e-literacy programs that low income African countries that participate are uniquely rewarded. There are many cases of failing nation states and political juntas occupying the seats of government that may require serious external jolt in order to fulfill their roles in globalization, after all, globalization has removed the issues of human rights violation and social injustice outside the control of the nation states (Ouma, 2007).

For specific implementation purposes, the authors suggest the involvement of multinational corporations like Microsoft, CISCO, etc in developing e-literacy programs aimed at skills empowerment for low-income African countries for maximizing their workforce potentials. They would eventually have return on their investments and those who invest on cheap and affordable computers/bandwidth should be given global incentives. Curriculum designers should personalize e-literacy capacity building programs so that it would be more end user driven. Furthermore, the e-literacy programs should be integrated into existing country programs and make them compulsory components of the curriculum (basic and post basic education programs, adult and non formal education). Software developers should develop programs that should be socio-culturally acceptable to the countries.

CONCLUSION

Globalization has come to stay and would not be leaving any time soon. Being a systemic concept, its success would rely on reconciling effectively the functioning of the various subsystems. For this world order to be sustainable, every member of the global village should be empowered to be proficient in the use of the Lingua Franca which is e-literacy. The workforce needs capacity building in e-agriculture among others in other to be competitive and secure decent jobs that would make them responsible global citizens. Country based immigration laws of the developed countries cannot possibly stem the tide of illegal migration nor brute force eliminating terrorism fuelled by historic injustices. The notion of nation-state as previously held is being challenged by globalization as regions are forming larger units to pursue common goals which by extrapolation should be made global for the good of all. When capacities of the workforce of the low income African countries are built, it would enhance their productivity and equally enable them benefit from the gains of e-government. International solidarity is needed to effect these suggestions and the global system cannot withstand, for a long time, the perils of impoverished low income countries in the new world order.

REFERENCES

Alford, T., Chasteen, T., & DeRosear, K. (2008). *It's the educonomy, stupid!: Redifining America's workforce sysetm: A practitioner's guide.* Kingston, Tennessee: Worldwide Interactive Network, Inc.

Aluko-Olokun, I. (1997). Institutional building for effective capacity development. In ITF (Ed.). *Realizing National Vision by the Year 2010: The Human Resource Development Imperatives* (pp. 69 – 76). Jos, Nigeria: Industrial Training Fund (ITF).

Anheier, H. K. (2007). Bringing civility back in – Reflections on global civil society. *Development Dialogue*, (49): 41–49.

Ayogu, M., & Ogbu, O. (2002). *Globalization and technology: Africa's participation and perspectives.* Nairobi, Kenya: African Technology Policy Studies Network (ATPS).

Bond, P. (2007). Linking below, across and against – World social forum weaknesses, global governance gaps and the global justice movement's strategic dilemmas. *Development Dialogue*, (49): 81–95.

Carr-Hill, R., & Leach, F. (1995). *Education and training for the informal Sector.* London: Overseas Development Administration.

Catts, R., & Lau, J. (2008). *Towards Information Literacy Indicators.* Retrieved December 20, 2008 from http://unesdoc.unesco.org/images/0015/001587/158723e.pdf

Chen, M. A., Jhabrala, R., & Lund, F. (2002). *Supporting workers in the informal economy: A policy framework* [Working Paper on the Informal Sector No. 2]. Geneva: Employment Sector, International Labour Office.

Dean, G. J. (1994). *Designing instruction for adult learners.* Malabar, Australia: Krieger Publishing Company.

Doczi, M. (2005, December 13). *What "access" and "e-literacy" mean.* Retrieved December 20, 2008 from Information and Communication Technologyies and Social Economic Inclusion http://www.med.govt.nz/templates/Multipage-DocumentPage_9666.aspx#P27_7816

Dzvimbo, K. P. (n.d.). *The international migration of skilled human capital from developing countries.* Retrieved November 1, 2008, from The World Bank http://siteresources.worldbank.org/INTAFRREGTOPTEIA/Resources/peter_dzvimbo.pdf

Fanon, F. (1963). *The wretched of the earth.* New York: Grove Press.

Garsten, C., & Jacobsson, K. (2007). Corporate globalization, civil society and post-political regulation – Whither democracy? *Development Dialogue*, (49): 143–157.

Geddes, G. (2006). Old dogs and new tricks: Teaching computer skills to adults. *eLIT, 5* (4). Retrieved December 29, 2008, from http://www.ics.heacademy.ac.uk/italics/vol5iss4/geddes.pdf

Haan, C. H. (2002). *Training for work in the informal sector: New evidence from eastern and southern Africa.* Retrieved March 20th 2003 from www.itcilo.it/english/bureau/turin/whatisnew/flyers/Training%20for%20work.pdf

International Labour Organization. (2008). *Global employment trends: January 2008.* Geneva: International Labour Organization.

Joint, N. (2003, June 11). eLiteracy and social exclusion:A global perspective. In *eLit2003:2 International Conference on Information and IT Literacy*. Retrieved December 20, 2008, from http://www.strathprints.strath.ac.uk/2357/1/strathprints002357.ppt

Kössler, R., & Melber, H. (2007, November). International civil society and the challenge of global solidarity. *Development Dialogue*, (49): 29–39.

Lanvin, B. (2002). *The e-government handbook for developing countries.* Retrieved December 20, 2008, from Center for Democracy & Technology, http://www.infodev.org

Liimatainen, M. (2003. *Training and skills acquisition in the informal sector: A literature review.* Geneva: International Labour Office.

Mani, S. (2001). *Globalization, markets for technology and the relevance of innovation policies in developing economies* (ATPS Special Paper No. 2). Nairobi, Kenya: ATPS.

Marotta, D., Mark, M., Bloom, A., & Thorn, K. (n.d.). *Human capital and university-industry linkages' role in forstering firm innovation: An empirical study of Chile and Colombia.* Retrieved November 1, 2008, from The World Bank http://www-wds.worldbank.org/external/default/WDSContentServer/IW3P/IB/2007/12/13/000158349_20071213130640/Rendered/PDF/wps4443.pdf

Martin, A. (2005). DigEuLit - a European framework for digital literacy: A progress report. *Journal of eLiteracy, 5*(4), 130 – 136. Retrieved December 29, 2008, from http://www.jelit.org/65/01/JeLit_Paper_31.pdf.

Martin, L. (2006). Enabling eliteracy: Providing non-technical support for online learners. *eLIT, 5* (4), 97 – 108. Retrieved December 20, 2008, from http://www.ics.heacademy.ac.uk/italics/vol5iss4/martin.pdf.

McKenzie, D., & Sasin, M. J. (2007). *Migration, remittances, poverty, and human capital: Conceptual and empirical challenges*, World Bank Working Paper No. 4272. Retrieved on October 15, 2008 from http://www.worldbank.org/migration/wps4272.pdf.

Miranda, M. (2007). Global civil society and democracy – A difficult but unavoidable task: Visions from the south. *Development Dialogue*, (49): 97–107.

Mytelka, L. K., & Tesfachew, T. (1999). *The role of policy in promoting enterprise learning during early industrialization: Lessons for African countries.* Geneva: United Nations Conference on Trade and Development (UNCTAD).

Ogwo, B. A. (2004). *Technological developments in the maintenance operations of imported used automobiles and their policy implications for the automobile industry in Nigeria.* Nairobi, Kenya: African Technology Policy Studies Network (ATPS). Rodney, W. (1973). *How Europe Underdeveloped Africa.* London: Bogle-L'Ouverture Publications.

Ogwo, B. A. (2007). Informal sector technical skills development experiences in the maintenance of modern automobiles in Nigeria. *Network for Policy Research Review and Advice on Education and Training (NORRAG) News*, (pp. 25 – 27).

Ogwo, B. A., Oranu, G. C., & Oranu, R. N. (2008). Application of ICT-based open learning principles for market/mechanic village schools in south eastern Nigeria. In *IST-Africa 2008 Conference proceedings held in Windhoek, Namibia*. http://www.ist-africa.org/Conference2008/default.asp?page=my-page

Ouma, S. (2007, November). Civil society and the nation-state - The case of Kenya. *Development Dialogue*, (49): 109–118.

Porter, M. E. (n.d.). *Workforce Development in the global economy.* Retrieved December 30, 2008, from http://www.gwit.us/global.asp#main.

Sander, C., & Mainbo, S. M. (2003). *Migrant Labor Remittances in Africa: Reducing Obstacles to Developmental Contributions.* World Bank Working Paper No. 64. Retrieved on November 6, 2008 from http://www.worldbank.org/afr/wps/wp64.pdf Scholte, J. A. (2007). Global civil society – Opportunity or obstacle for democracy? *Development Dialogue*, (49), 15 – 27.

Smith, M. K. (2002). Globalization and the incorporation of education.In *The encyclopedia of informal education*. Retrieved December 24, 2008, from http://www.infed.org/biblio/globalization.htm

Stiglitz, J. E. (2002). *Globalization and its discontents.* New York: W. W. Norton & Company, Inc. The World Bank Africa Region (AFR), (September, 2007). *African diaspora initiative concept note.* The World Bank: Washington DC

The World Bank. (2007). *The Little Green Data Book 2007.* Washington DC: The International Bank for Reconstruction and Development/THE WORLD BANK.

The World Bank. (November, 2007). *Mobilizing the African Diaspora for Development.* Working document presented at The African Diaspora Open House held at the World Bank Headquarters in Washington DC on November 29, 2007.

Town, J. S. (2004). *Modelling eLiteracy: Politics, truth and beauty.* Retrieved December 20, 2008 from http://www.sconul.ac.uk/groups/information_literacy/publications/papers/st_elit_2004.ppt%20

UNESCO. (n.d.). *United Nations literacy decade (2003 - 2012).* Retrieved December 29, 2008, from UNESCO Education Plans and Policies: http://portal.unesco.org/education/en/ev.php-URL_ID=53553&URL_DO=DO_TOPIC&URL_SECTION=201.html

UNESCO-UIS/OECD. (2005). *Education Trends in Perspective – Analysis of the World Education Indicators.* Retrieved October 6, 2008 from http://www.uis.unesco.org/TEMPLATE/pdf/wei/WEI2005.pdf

Chapter 11
Building Capacity for Electronic Governance in Developing Countries:
Critical Success Factors

Rogers W'O Okot-Uma
Studies Forum International, UK

J.K. Ssewanyana
Makerere University, Uganda

ABSTRACT

This chapter presents the essence of critical success factors with focus on building capacity for electronic governance (eGovernance) in a developing country jurisdiction. The results are borne of the authors' years of experience with regard to national eGovernance implementations in developing member countries of the Commonwealth. Critical success factors (CSFs) denote those aspects of, or associated with, the new information and communication technologies (ICTs), which may be perceived as comprising core, key or critical factors against which the level of capability of National Capacity for ICT or eGovernance may be assessed, measured and/or interpreted. CSFs, perceived to be critical for the success of any eGovernance initiative is best modelled as a three-tier minimalist framework, comprising CSFs at levels described as macro-, meso-, and micro- levels. The nature of any given ICT initiative which is appropriate nationally in central government, locally in local government, or in the public service, in the civil service, or in some selected sector or jurisdiction of the national economy, whether existing or planned, and whether implicit or explicit, must take cognisance of the need for the identification of CSFs at the inception stage of the initiative.

INTRODUCTION

The rapid development, deployment and proliferation of the new and emerging information and communication technologies (ICTs) herald new opportunities for growth and development in countries around the world. Governments worldwide are seeking to harness the potential offered by these new technologies to create new dimensions of economic and social progress. Immediate challenges relate

DOI: 10.4018/978-1-60566-820-8.ch011

to the need for requisite efforts by governments to transcend the digital divide by narrowing the digital gap through, incrementally or otherwise, the following factors:

- Putting in place the necessary national information infrastructure;
- Developing and nurturing the necessary human resource and other 'soft-wired' structures and systems required to operate the national information infrastructure; and
- Providing adequate financial resources to implement both the infrastructural and human resource requirements.

The three factors outlined above may be considered to comprise part of a broader scope of factors – critical success factors. For developing countries deemed to be at the initial stage of making the entry into, or at an early stage on, the roadmap to the information society, or the information economy, a 'checklist' or 'quasi-template' of critical success factors becomes a useful practical 'toolkit', that need to be put into consideration as a necessary preamble. Critical success factors (CSFs) are used to denote those aspects of, or associated with, the new information and communication technologies (ICTs), which may be perceived as comprising core, key or critical factors against which a measure of level of capability of National Capacity for ICT, in general, or *e*Governance, in particular, may be assessed, measured and/or interpreted.

This chapter presents, from an empirical standpoint, an articulation of critical success factors (CSFs), as they pertain to building capacity for information and communication technologies in general, and electronic governance (*e*Governance), in particular. Over the years, during the mid 1990s, one of the authors above implemented over 50 ICT initiatives, the majority being specialist training programmes, in which the participants were routinely required to identify, state, indicate and/or discuss with other participants the issues or factors that, in their respective individual or collective opinion, they regarded as constituting critical success factors with regard to national initiatives aimed at building capacity for the new ICTs in their respective national jurisdictions. The result was a long list of issues or factors which, on analysis, was reduced to a 17-entity listing of 'critical success factors'. CSFs, perceived or deemed to be critical for the success of any *e*Governance initiative (or any other national ICT initiative), is best modelled as a three-tier minimalist framework, comprising CSFs at levels described as *macro-*, *meso-*, and *micro-* levels. Individual jurisdictions can broadly 'scope' the individual CSFs or the three levels, to take account of their unique circumstances. During the early 2000s, it became apparent that it would be useful to investigate the CSFs further, particularly with respect to each other in perceived relative terms. It was therefore arranged to carry out a survey exercise amongst alumni of the various previous ICT initiatives. The purpose was to obtain a general relative ranking of the critical success factors, with the objective of articulating a broad overview of priority levels, temporal or administrative, with regard to the critical success factors. The objective of this chapter is simply to establish, at least from an empirical standpoint, how the critical success factors rank across the macro-, meso-, and micro- levels.

Background

For the benefits of governments and businesses, citizens and the public at large, human or organisational, the need arises for evolving or building national capacity for ICT to levels that are consistent with a nation's aspirations in development. The roadmap to *e*Maturity is, inevitably, long, tortuous, resource-intensive and can be, more often than not, unpredictable. It becomes necessary for Governments to put in place National Information and Communication Policies (NICPs) or, more specifically, National Information Technology Policies (NITPs), which may be referred to, simply, as *e*Policies. A good *e*Policy must be visionary and widely owned by society for the benefits which the policy is formulated. A good *e*Policy must, in addition, be capable of, or be amenable to, implementation. To be amenable to implementation, a good *e*Policy must have strategies for implementation. Strategies for implementation of *e*Policies comprise National Information Technology Strategies or, simply, *e*Strategies. It is good practice to advocate for flexible, simple policies that allow implementation to adapt to concrete situations in which decisions are made.

Electronic Governance (*e*Governance) will be taken to be contextually inclusive of *e*Government, *e*Democracy and Government *e*Business (Okot-Uma, 2000a). Building capacity for ICT, in general, and for *e*Governance, in particular, will necessitate consideration with respect to a diverse number of issues, including infrastructure, human resource, legal framework, regulations, national information infrastructure and many more. Focus on CSFs has, as one of its objectives, the task to streamline what issues, in broad terms, are necessary and which, therefore, need to be taken into account in any initiative aimed at building capacity for ICT or *e*Governance in any national jurisdiction. While the issues relevant or critical to building capacity for ICT or *e*Governance have variously been discussed elsewhere (Okot-Uma, 2005), the concept of bringing together these issues under the auspices of a single umbrella described by the CSFs is new and unique. The framework presented by the CSFs consideration is in no way to be construed as complete in itself. It is, however, presented as a practical checklist that countries on the roadmap to making some initial ingress into ICT or *e*Governance planning and implementation may find useful. Elsewhere in the literature, CSFs have been discussed, somewhat implicitly, in terms of 'basic components' in 'toolkits' for *e*Readiness or *e*Preparedness of national jurisdictions (IRMT methodology, 2004; Havard methodology, 2002; WEF methodology, 2001; McConnel approach, 2002 ; BMi-T approach, 2002) of selected developing countries in the African region

METHODOLOGY

The creation of this chapter was initially motivated by the authors' motivation that a set of issues in the implementation of *e*Strategy initiatives, including *e*Governance, could be characterised as critical success factors, namely, a set of issues that were perceived to be critical pre-conditions for the success of any *e*Governance initiative (or any other national ICT initiative). This was empirically put to the test over the period 2002/2003. Specifically, survey questionnaire (see APPENDIX I) was constructed, based on 17 parameters and administered in June 2003. A 4-Point Scale assignment procedure was chosen for use in the assessment of the parameters with regard to their level of importance to building capacity for *e*Governance. The choice of a 4-Point Scale, as contrasted against a Likert Point Scale was arbitrary. A sample of 63 potential respondents, who had previously participated in public sector informatics initiatives implemented under the auspices of one of the authors, over the period from 1999 through 2002, was chosen from a list of alumni during the said period. The respondents were required to assess the 17 parameters using a nominal 4-point scale assignment procedure. They were also required to rank the 17 parameters in the order of importance, as individually perceived by them. A response rate of 44% was achieved. The respondents were drawn from a set of developing countries made up of the set of all the Commonwealth countries of the Africa region, from Botswana to Zimbabwe.

FINDINGS

Table 1 depicts the data recorded in the survey questionnaire, and analysis of the data in terms

Table 1. A recording and analysis of questionnaire data for a 4-point scale

	A Listing of Critical Success Factors (CSFs)	A 4-point scale				Computed Aggregate 'Rank'	Derived Nominal Ranking of CSFs
		1 x1	2 x2	3 x3	4 x4	<x>= ∑xifi/∑fi,	X (1 to 17)
		Number of respondents choosing 1,2,3,or 4				i=1,2.3,4	
		f1	f2	f3	f4		
1	Top Level Commitment / (Prime) Ministerial Responsibility	0	0	7	21	3.75	1
2	Vision, Mission, Mission and Strategies	0	4	7	17	3.46	2
3	Organisational Responsibility	0	1	16	11	3.36	3
4	Policy Orientation	1	2	12	13	3.32	4
5	Policy Scope	1	2	16	9	3.18	5
6	Telecommunications Infrastructure	0	4	16	8	3.14	6
7	Internet Infrastructure	3	4	8	13	3.11	7
8	Security	2	3	14	9	3.07	8
9	Lifelong Learning	2	5	13	8	2.96	9
10	Legal and Regulatory Framework	2	8	8	10	2.93	10
11	Change Management	3	4	14	7	2.89	11
12	Marketplace	3	7	8	9	2.86	12
13	Knowledge Management	2	5	17	4	2.82	13
14	Partnerships	3	6	15	4	2.71	14
15	Acculturation	3	13	4	8	2.61	15
16	Availability, Access (including open access, participatory access and universal access) and Affordability	5	9	12	2	2.39	16
17	Human Resource	4	16	7	1	2.18	17

of the 4-Point Scale; Table 2 depicts respondent nominal (direct) rankings of CSF parameters for analysis in terms of computed aggregate ranks and derived nominal ranking of the parameters. The two computed nominal rankings, one indirectly from the 4-Point Scale and the other directly from nominal rankings, are compared for rank correlation of the difference in the rankings (Table 3). Computed aggregate 'rank', <x>, is defined as the mean of the variables $\mathbf{x}_1$, $\mathbf{x}_2$, $\mathbf{x}_3$ and $\mathbf{x}_4$' occurring $\mathbf{f}_1$, $\mathbf{f}_2$, $\mathbf{f}_3$ and $\mathbf{f}_4$ number of times, respectively, where $\mathbf{x}_1$, $\mathbf{x}_2$, $\mathbf{x}_3$ and $\mathbf{x}_4$ represent the 4-point scale assignments 1, 2, 3, and 4, respectively. The ranking of these computed aggregate ranks, from 1 (for highest value or ranking) to 17 (for least value or ranking), gives the derived nominal ranking of the study's parameters, denoted by the variable **X**. This is compared against a variable denoted by **Y**, representing the respondent's direct ranking of the 17 CSFs, in accordance with the questionnaire.

To measure the correlation between the two variables X and Y required a deployment of non-parametric methods. (Blanche and Durrheim, 1999). The variables X and Y are ranked in the order of importance, from 1 to 17. In general, when the variables can be ranked from 1 to N in the order of some characteristic attribute (size,

Table 2. Respondent nominal ranking of the CSFs, computed aggregate ranks and derived nominal ranking

Nominal Ranking	**17**	**16**	**15**	**14**	**13**	**12**	**11**	**10**	**9**	**8**	**7**	**6**	**5**	**4**	**3**	**2**	**1**	**Computed Aggregate Rank**	**Derived Nominal Ranking**
CSF Listing	Number of Respondents Per Nominal Ranking for a Given CSFs																	y>= $\sum y_i f_i / \sum f_i$, i=1, .. 17	Y
	f17	f16	f15	f14	f13	f12	f11	f10	f9	f8	f7	f6	f5	f4	f3	f2	f1		
Vision, Mission, and Strategies									2		1		2		2	13	7	2.74	1
Lifelong Learning				1			1	4	1	4	2	5	3	3		2	1	5.56	2
Organisational Responsibility						2		1	5	5	3	3	1	2	5			5.70	3
Policy Scope						1	3	1	2		5	3	5	3	1	1	2	6.19	4
Top Level Commitment		1	2			1	2			3	3	1		4	3	4	3	6.19	4
Policy Orientation		1		2	1	4		3	2	2	3	3	4'	1			1	6.63	6
Telecommunications Infrastructure		1				1	2	4	3	3	3	1	2	1	3	3		6.81	7
Internet Infrastructure				1		4	1	3		2	1	3	1	3	7	1		6.81	7
Security				2	2	2					6	2	4	2	3		3	6.96	9
Partnerships	2		2			1	2			3	4	1	1	4	2	4	1	7.11	10
Legal and Regulatory Framework					1	1	5	2	5		1	2	3	3	1	2	1	7.30	11
Acculturation		2	2	4	3	2	1				1					1	11	7.52	12
Change Management	1				2	1	3	1	2	4	2	1	3	3	3		1	7.59	13
Availability, Access (including open access, participatory access and universal access) and Affordability	2			3		2		3	3	4	6			2	2			9.07	14
Marketplace			4	4	2	3	2	1	4	1	1	1	1	2	1			9.37	15
Knowledge Management		1		1	3	5	4	5	1	2		3			2			10.07	16
Human Resource	2		8	10	7													14.26	17

volume, importance, priority, etc), it is logical to invoke the Spearman's Rank Correlation (Blanche and Durrheim, 1999). In this study, the variables X and Y are ranked in the order of importance or priority. The Coefficient of Rank Correlation (r) between any two variables X and Y, computed using Spearman's Formula for Rank Correlation, yields the following value: **r** = 0.8554.

The assumption is that **r** has a Student's t-distribution (BLANCHE and DURRHEIM 1999); the t-score or t-statistic for the correlation coefficient **r** = 0.855 is computed to give the fol-

Table 3. Comparing the nominal rankings from the respondent 4-point scale assignments of CSFs with the nominal rankings from respondent direct rankings of CSFs

A Listing of Critical Success Factors (CSFs)		Nominal ranking derived from respondent 4-point scale assignments (X)	Nominal ranking derived from respondent direct ranking of factors of utility (Y)
		(1 to 17)	(1 to 17)
1	Security	8	9
2	Lifelong Learning	9	2
3	Change Management	11	13
4	Legal and Regulatory Framework	10	11
5	Availability, Access (including open access, participatory access and universal access) and Affordability, Regulation, Competition, Private Investment, and Access	16	14
6	Marketplace	12	15
7	Vision, Mission, and Strategies	2	1
8	Knowledge Management	13	16
9	Human Resource	17	17
10	Acculturation	15	12
11	Telecommunications Infrastructure	6	7
12	Internet Infrastructure	7	7
13	Policy Scope	5	4
14	Organisational Responsibility	3	3
15	Top Level Commitment / (Prime) Ministerial Responsibility	1	4
16	Policy Orientation	4	6
17	Partnerships	14	10

lowing value: t = 6.38, which demonstrates that the sample correlation coefficient **r** is significant at the 5% level of significance. That is, the rank correlation between the two different modes of ranking of the CSFs, namely, (i) nominal ranking as derived from respondent 4-point scale assignments, and (ii) nominal ranking as derived from respondent direct ranking of the parameters of CSFs, is not random.

Notably, the above result does not prove a cause-effect relationship in any way between the two different modes of ranking of CSF parameters deployed in the research study. Also, this test has a further restriction as it assumes that the sample of (X,Y) pairs was drawn from a normal distribution.

Given that the rank correlation between the two different modes of ranking of CSFs is significant (5% level of significance), paves the way for a discussion of the observations which can be made with respect CSFs. An interpretation of the results and observations on the assessment may be made more explicit and comprehensive by examining the rankings of the 17 CSFs. The findings demonstrate, specifically, that the five top rankings in the CSF listings are: top level commitment; vision, mission, strategies; organizational responsibility; and policy orientation, and policy scope.

The next set of five CSFs in the CSF rankings are telecommunications infrastructure; Internet infrastructure; security; life-long learning; and legal and regulatory framework; The rest of the

CSFs are ranked as change management; marketplace; knowledge management; partnerships; acculturation; availability, access (including open access, participatory access and universal access) and affordability; and human resource.

Discussions

The CSFs can be presented as a three-tier framework as indicated in figure 1:

- CSFs at the *macro-level*, defined as representing CSFs at the initial level, or top level, and may be represented by the following parameters in the planning and/or implementation of an *e*Governance initiative: top level commitment, a national vision, ministerial/prime ministerial responsibility, organizational responsibility, policy 'scope', and policy 'orientation'.
- CSFs at the *meso-level*, defined as representing CSFs at the next level, midstream, and may be represented by, first, the **'parameters of infrastructure'** in the planning and/or implementation of an *e*Governance initiative, namely, the telecommunications infrastructure; the Internet infrastructure; *e*Security; and availability, access (including open access, participatory access and universal access), and affordability (AAA) with respect to the infrastructure; and, second, the **'parameters of infostructure'**, namely, developing a competitive IT human resource; providing an IT-enabled environment for life-long learning; building an IT culture; a developing a legal regulatory framework for IT/ICT; **marketplace** for national ICT capacity building; **partnership** for national ICT capacity building; and building **knowledge management** for ICT capacity building.
- CSFs at the *micro-level*, defined as

Figure 1. Critical success factor framework for eGovernance

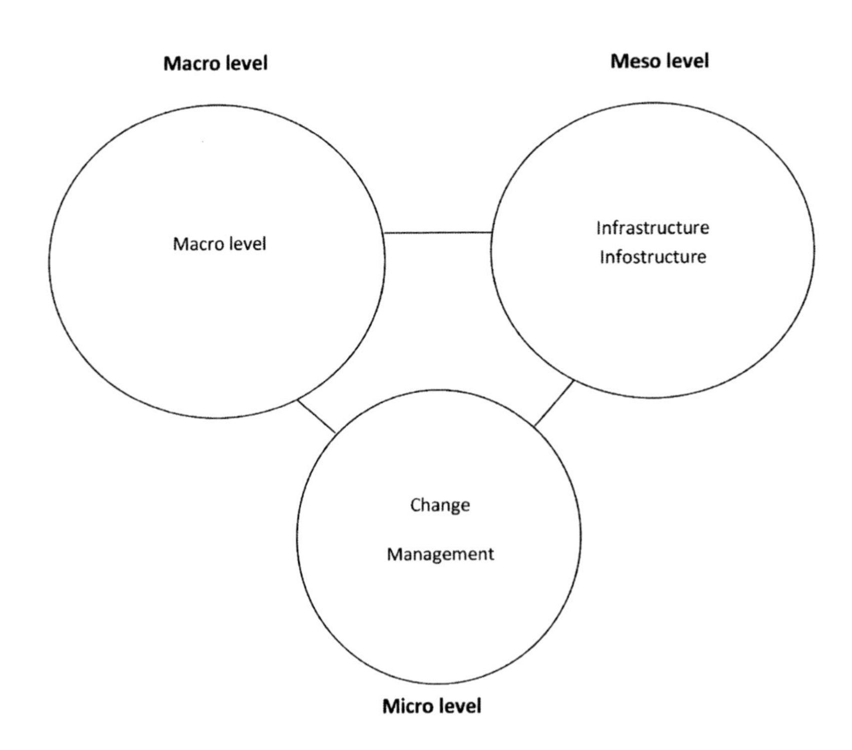

representing CSFs at yet the next detailed layer, downstream, and may be represented by **change management** for national ICT capacity building.

On the basis of the foregoing classification, and the survey questionnaire findings, CSFs at the macro-level would appear to hold a preponderance of importance over the CSFs at the meso-level or micro-level.

This chapter will focus on the macro level component of the CSF framework. CSFs at the macro-level are the most critical at the start of any ICT initiative – they represent immediate preconditions for any capacity building initiative in ICT or *e*Governance. Selected CSFs at the micro-level, specifically change management, although ranked in the lower 50% of the rankings, will require special consideration in our opinion. This is because change management for national ICT capacity building (with focus on *e*Governance), which must take account of process enablement; people enablement; infrastructure enablement; and systems enablement; among others [Okot-Uma, 2003], is very critical. Let us examine the macro level CSFs.

Top Level Commitment

The existence of top-level commitment by government to the recognition that there is a need to have in place a Comprehensive National Information Technology Strategy can be perceived to be a critical success factor (CSF) for most countries. Such commitment can variously be presidential, prime ministerial, or ministerial, with specific responsibility for invoking and driving action on the delivery and implementation of a National Information Technology Strategy, either directly through the government or through a separately constituted body. At the initial stage such role un-ambiguity must be perceived to provide commitment to a sense of urgency and to a planning process to deliver such a strategy, in the immediate term, the short to medium term, or at some future date.

Validation of top-level commitment is only realizable when such commitment transcends mere 'espousal' of intentions. Commitment must be backed up by an *adequate budgetary resource allocation* to provide for the implementation of the strategy.

Vision, Mission and Strategies

A second CSF relates to the need for a **National Vision, Mission** and **Strategies.** The lessons of experience from a number of developing Commonwealth member countries reveal that national vision has been variously enunciated to incorporate, implicitly or explicitly, an embodiment of information technology at the highest national level, as a prime mover, in the national drive to any one or more of the "wannabe" states of excellence or positioning in the international world or global arena in the *context of the national economic development,* as, for example:

- An excellence in the domestic IT product and/or IT services market industry (e.g., India's domestic software industry in the early years) (Okot-Uma 2003);
- A world leader in the export IT product and/or IT services market industry (e.g., India's 'export' of programmers; Malaysia's export of IT services generated under the auspices of Malaysia's multimedia super-corridor) (Okot-Uma 2003);
- A world leader in the creation of the Information Society, as exemplified in examples such as The Intelligent Island (Singapore) (Intelligent Island, 1991, 1997a, 1997b), The Cyber Island (Mauritius) (Cyber Island, 2006, 2008), The Multimedia Super-Corridor (Malaysia) (MSc Malaysia, 2009);
- A new paradigm in the knowledge economy (k-Economy) (Malaysia k-Economy, 2000,

2006a, 2006b, 2007a, 2007b, 2007c);

- A champion of distributed Governance (incorporating *e*Government, *e*Business and *e*Democracy) (UK *e*Strategy, 2000).

Organizational Responsibility

The next CSF relates to the **IT/ICT Organizational Responsibility**. This, almost invariably, takes form of an ICT organizational structure, with a lead agency for ICT. Such organizational structure would be empowered with the functional authority to drive the national ICT initiative, together with a number of *flagship* or various *awareness programs* to direct or drive action, as well as stimulate awareness, within Government or the public sector at large, for the purpose of *e*Governance delivery, nationally, infra-nationally, or sectorally. Such organizational responsibility may manifest as any one, or combinations, of the following:

- An organizational definition of the role, responsibility and scope of the national IT function ;
- An organizational structure for the strategic *national IT policy formulation;*
- An organizational structure for the *national IT co-ordination and/or promotion function;*
- An organizational structure for the *national IT implementation function;* and
- An industry standard taxonomy of the national generic *IT job/skill set* and competencies for IT in the country.

It is imperative that for the functional sustainability of the stated organizational responsibility, there must be determined an assessment of the following resources, which should be procured or made available in a progressive manner:

- The level of *human resource skills capacity and competence* required for the full implementation of a National *e*Governance Strategy;
- The level of *financial resource capacity* required for the full implementation of a National *e*Governance Strategy

Moreover, an existing organizational arrangement *may* have to be backed up by the implementation of a *"national-level corporate governance strategy"* with the following features:

- role defined or planned for a *Chief Knowledge Officer* or equivalent, aimed at executing a national "corporate" function of the Chief Information Officer in the context of *e*Governance (or eStrategy) planning and implementation (e.g., UK's eEnvoy, Canada's CIO, US's "Technology Czar" in the early 2000 and later).
- role defined or planned for an *IT Steering Committee* or equivalent, aimed at guiding and overseeing *e*Governance (or eStrategy) planning and implementation.
- role planned for an *IT Consultative Group* or equivalent, aimed at facilitating stakeholder awareness and intervention for *e*Governance (or eStrategy) planning and implementation.

Policy Scope - Jurisdictional

It is conceivable that a National Information Technology (or *e*Governance) Strategy, existing or planned, would have a defined "policy scope". Such scope can be perceived to be "jurisdictional" in character and to take the form or one, or combinations, of the following:

- **National Approach**, with inputs typified by one or more of the following "functional" components:
    - As part of a sector-independent informatics strategy, applicable to or focusing on components, or parts thereof, of the development 'triad', namely:

- government civil service only;
- public sector only;
- public and private sector only; and
- the development 'triad' comprising government, civil society and the private sector.

- As part of a sector-specific informatics strategy, applicable to various national sectors of the economy, such as the following:
  - telecommunications utility services sector
  - electricity generation and distribution utilities sector
  - water generation and distribution utilities sector
  - fuel suppliers and distribution utilities sector
  - national emergency services (fire, ambulance, police, national security, etc) sector
  - health services sector
  - air transport (airports, air traffic control and customs & excise) services sector;
  - sea transport sector
  - road transport, traffic control systems and customs and excise
  - food processing and distribution sector
  - meteorological and other warning systems sector.
- **Local Government Approach**, with inputs typified by one or more of the following "functional" components: sector-independent as well as sector-specific strategies mirroring the national approach described in the foregoing section.

Policy Orientations – 'Outlook Perspectives'

It is also conceivable that a National Information Technology (or *e*Governance) Strategy, existing or planned, would have a defined **policy orientation**. Such orientation can be perceived to comprise a series of outlook perspectives. A three-tier' approach provides leeway for the following alternate forms:

- 'Inward Outlook' policy orientation, which strategically focuses on a domestic IT market strategy and an import substitution strategy. The strategy aims for the promotion of local patent systems, the creation of national IT venture centers of excellence, and the creation of an IT business venture park. The *raison d'etre* for this policy orientation is to create and promote local enterprise and entrepreneurship at individual, group as well as national levels.
- 'Outward Outlook' policy orientation, which strategically focuses on an export IT market strategy and the various concomitants of licensing agreements, establishment of local enterprises as subcontracting entities, as well as the creation of national IT venture centers of excellence, and the creation of an IT business venture park. The *raison d'etre* for this policy orientation is to create and promote local enterprise and entrepreneurship for positioning at the international level or global market.
- 'Indigenous IT Business Outlook' policy orientation, which strategically focuses, among others, on an eBusiness incubation, ethos for competitive advantage, local technological R&D and innovation, all geared to developing an indigenous business 'infrastructure'.

On the Critical Success Factors Again

The old debate, about choosing between ICT and other development imperatives, has shifted from one of trade-offs to one of complementarity. In recent years, developing countries have pursued diverse strategies: some have focused on developing ICT as an economic sector – either to boost exports (Costa Rica and Taiwan), or to build domestic capacity (Brazil, India and Korea), while others are pursuing strategies which seek to use ICT as an enabler of a wider socio-economic development process. Countries which use ICT as an enabler, may further be subdivided into those which have focused primarily on repositioning the country's economy to secure competitive advantage in the global economy (Malaysia, Trinidad & Tobago) and those which explicitly focus on ICT in pursuit of development goals such as those set forth in the UN millennium Summit (Estonia and South Africa). These varied experiences have revealed important lessons about the role of ICT in development. In particular, from the stand point of critical success factors, as discussed in this chapter, these varied experiences are found to correlate closely with the early state of success associated with capacity building for ICT / *e*Governance associated with a number of developing countries worldwide, notably, India, Singapore, Malaysia, and South Africa {Okot-Uma, 2003b). The *Indian ICT domestic market strategy* has demonstrated that, among others, the importance of the Bhabha Committee, set up by the Cabinet, which mapped India's IT policy development scenario, and set up the Technological Development Council, which was instrumental in the funding of electronics projects on the basis of specific criteria. Although initially aimed for the domestic market, the Indian model has to date excelled in producing for the international market, development of selected applications software such as that for the railway system, and the production of millions of software programmers for the international market. The *Singapore ICT international export-oriented market strategy* owes its success, in part, to the putting in place, early in its national strategy, of a clear vision of purpose, which gradually transformed into the vision "to make Singapore an intelligent island", heralded by prime ministerial authority; an organisational and institutional framework represented by the Committee of National Computerisation (CNC) and the National Computer Board (NCB), established in 1980 and 1981' respectively, which provided institutional leadership to set up a software house and the Civil Service Computerisation Programme (CSCP). The *Malaysia ICT competitive positioning in the global economy* owes its initial momentum to the establishment of an Inter Agency Sub-Committee on Automatic Data Processing and Information Systems which was mandated with the co-ordination and review of computer acquisition by government agencies. Over the years this was superseded by the establishment of the national Committee on data Processing (NCDP), which became Malaysia's highest committee on computerisation, chaired by the Senior Deputy Secretary General in the Prime Minister's Department. To date Malaysia boasts of "Vision 2020", aiming to become an advanced industrialized nation by the year 2020. While the manufacturing industry was likely to be the main engine of growth for the foreseeable future, the services sector, driven by knowledge as a resource is seen to be the next replacement resource that is poised to transform Malaysia into the knowledge economy (k-Economy). South Africa's *ICT focus on pursuit of development goal and realisation of a knowledge society strategy* was characterised at the outset by a top level commitment of the Deputy President who committed in the early 1990s to transform South Africa into an information society. South Africa's top down commitment was supplemented by a bottom up intervention by the private sector taking a lead in local and regional initiatives, and international donors offering financial and human support, and NGOs becoming prominent in grassroots development-oriented activities. Information technology policy for Government was directed under the auspices of the Steering Committee on IT Policy which promoted the idea that the policy should have two major foci, namely, Government using IT as an enabler for its own internal administrative efficiency and effectiveness, and Government using IT as an enabler of external government service delivery to the people and for democratic governance. To make the transition to the ICT industry for development, a national project, known as The South African Information Technology Industry Strategy (SAITIS) was conceived with the goal to "*develop a strong South African information technology industry to contribute to sustainable economic growth, social upliftment and empowerment*". The South African model represents an explicit focus to use ICT in pursuit of development goals. For South Africa, this is probably a natural sequence, emerging as it does, from the era of the Apartheid regime. Significant in the model are the various initiatives aimed at meeting the needs of the citizenry, right from the affluent to the somewhat disadvantaged section of the national population. The model attempts to contribute to both broad-based economic growth by way of building up a viable ICT industry and specific development goals that are aimed towards the alleviation of poverty.

Future Research Directions

It is recommended that 'action research' be deployed to investigate the pragmatics of real-world critical success factors setting methodology in a diverse number *national* ICT initiatives worldwide. Adaptation to supra-national type projects or initiatives could pave the way for some generalisation with respect to applicability and scope of jurisdiction. It will also be useful to investigate on-going initiatives with the aim to articulating issues, parameters or factors that have been deployed as critical success factors and to what extent these success factors are able to validate the findings as presented in this chapter.

CONCLUSION

Critical success factors (CSFs) have been discussed in detail, based on a survey questionnaire approach. Our findings demonstrate that CSFs give a disciplined mode of approach with regard to initiatives aimed at building capacity for ICT in general and *e*Governance in particular. While the issues relevant or critical to building capacity for *e*Governance have variously been discussed elsewhere, the concept of bringing together these issues under the auspices of a single umbrella described by the CSFs is new and unique. The framework presented by the CSFs consideration is in no way to be construed as complete in itself. It is, however, presented as a practical checklist that countries on the roadmap to making some initial ingress into *e*Governance or *e*Strategy planning and implementation may find useful. Practitioners in developing countries, particularly late starters at *e*Governance initiatives are encouraged to use the CSF approach in their national initiatives..

REFERENCES

Blanche, M. T., & Durrheim, K. (Eds.). (1999). *Research in practice: Applied methods for the social sciences.* Cape Town: University of Cape Town Press.

BMi-TechKnowledge Group. (2002). *BMi-T Approach to eReadiness*, Johannesburg: BMi-T.

Chen, D. (2006a). *A Knowledge Economy Perspective on Malaysia: A comparative diagnostic.* Washington, DC: World Bank. Retrieved March 25, 2009 from http://info.worldbank.org/etools/docs/library/232719/D1_Malaysia_Benchmark.pdf

Ericon (2007). *Malaysia and the Knowledge Economy*. Retrieved March 25, 2009 from http://blog.beekens.info/index.php/2007/12/malaysia-and-the-knowledge-economy/

Goering, L. (2006). *Mauritius set to become first 'Cyber Island.*' Government Technology. Retrieved March 25, 2009 from http://www.govtech.com/gt/articles/98096

Information Technologies Group of Harvard University's Centre for International Development. (2002). *Readiness for the Networked World: A Guide for Developing Countries*. Cambridge, MA: Centre for International Development, Harvard University.

Malaysia, M. S. C. (2009). Retrieved March 25, 2009 from http://en.wikipedia.org/wiki/multimedia_super_corridor

Malaysia kEconomy, (2000). *Transcending The Divide*. Speech by the Prime Minister the Hon Dato Seri Dr Mahathir bin Mohamad at the Second World Knowledge Conference at the Ballroom, Mandarin Oriental, KLCC, Kuala Lumpur, Wednesday 8 March 2000, 5:00 PM.

Malaysia kEconomy, (2006b). *Knowledge-based masterplan (Malaysia)*. Retrieve March 25, 2009 from http://unpan1.un.org/intradoc/groups/public/documents/apcity/unpan013973.pdf

Malaysia kEconomy, (2007b). *Developing Malaysia into a knowledge-based economy.* Retrieved March 25, 2009 from http://www.epu.gov.my/new%20folder/development%20plan/opp3/cont_chap5.pdf

Malaysia kEconomy, (2007c). *Malaysia and the Knowledge Economy: Building a World-Class Higher Education System. Human Development Sector Reports*. Washington DC: World Bank. Retrieved March 25, 2009, from http://publications.worldbank.org/online

McConnel International (2002). *McConnel International Methodology, xxxx.*

National Computer Board. (1997). *Towards an Intelligent Island: NCB 10th Anniversary 1981-1991*. Singapore: NCB.

National Computer Board. (1997a). *IT2000–A Vision of an Intelligent Island.* Singapore: NCB.

National Computer Board. (1997b). *Transforming Singapore into an Intelligent Island.* Singapore: NCB.

Okot-Uma, R. W. O. (2000). *Electronic Governance – Reinventing Good Governance*. London: Commonwealth Secretariat. Retrieved March 25, 2009 from http://webworld.unesco.org/publications/it/EGov/wordbank%20**okot-uma**.pdf

Okot-Uma, R. W. O. (2003a). *Electronic governance masterplan for the government of the Republic of Mauritius*. London: Commonwealth Secretariat.

Okot-Uma, R. W. O. (2003b). *Capacity Building in ICT and Development, Background Paper commissioned for the Microsoft Government Leaders' Forum*, Johannesburg, 21-23 September 2003, Okot-Uma, R. W. O., (2005). *The Roadmap to eGovernance implementation: Selected Perspectives.* Retrieved March 25, 2009 from http://citeseerx.ist.psu/viewdoc/download?doi=10.1.1.84.1656&type=pdf; http://www.rileyis.com/publications/research_papers/rogers2.html

Taylor, R. (2008). *Mauritian cyber island dream.* Retrieved March 25, 2009 from http://www.news.bbc.co.uk/2/hi/programmes/click_online/7169467.stm

The International Records Management Trust. (2004). *Developing a module for Assesssing Electronic Records as a Component of E-Government Readiness Assessments*: London: Commonwealth Secretariat.

UK. *e*Strategy, (2000). *The UK vision as described in an April 2000 eGovernment Strategic Framework, was of modernised, efficient government, focused on better services for citizens and businesses and more effective use of information resources, source*. Retrieved May 1, 2003 from http://www.eEnvoy.gov.uk/ukonline/progress/estrategy/contents.htm

World Economic Forum. (2001/2002). *World Economic Forum Consultation on SADC eReadiness*. Geneva: World Economic Forum.

APPENDIX I

Building Capacity for Electronic Governance: Critical Success Factors

Table 4. Survey questionnaire

	Given Parameters of Critical Success Factors (Please read the Glossary of Terms, tabulated below, first)	**Indicate by a cross (X) your Choice of Likert 4-Point Scale (Please read the Guideline below first)**				**Indicate Your Ranking of the Factors of Utility (Please read the"Guideline below first)**
		1	**2**	**3**	**4**	**(1 to 17)**
1	Security					
2	Lifelong Learning					
3	Change Management					
4	Legal and Regulatory Framework					
5	Availability, Access (including open access, participatory access and universal access) and Affordability					
6	Marketplace					
7	Vision, Mission, Mission and Strategies					
8	Knowledge Management					
9	Human Resource					
10	Acculturation					
11	Telecommunications Infrastructure					
12	Internet Infrastructure					
13	Policy Scope					
14	Organisational Responsibility					
15	Top Level Commitment / (Prime) Ministerial Responsibility					
16	Policy Orientation					
17	Partnerships					

Table 5. Guideline for the assigning of the likert 4-point scale (1-4) and the ranking of the parameters of CSFs (1-17)

A: Assigning of the Likert 4-Point Scale (1-4) to the Parameters of CSFs	
Designated Level of Importance of Parameters	Equivalent Nominal Likert 4-Point Scale Assignment
NOT Important	1
WEAKLY Important	2
IMPORTANT	3
STRONGLY Important	4
B: Nominal Ranking of the Factors of Utility (1-17)	
Please use the last column of the Questionnaire Table above to rank the factors of utility in the order in which you perceive them as important to you as actor-participant (or as state-as-actor), from MOST important (rank-order 1) to LEAST important (rank-order 17).	

Table 6. An assessment questionnaire (a glossary of terms used in the questionnaire)

	Given Factors of Utility	Additional Information
1	Security	Building Capacity for eSecurity and Government Secure Intranets (GSI)
2	Lifelong Learning	Providing an IT-Enabled Environment for Life-Long Learning
3	Change Management	Perceiving the implementation of an eGovernance initiative as a Change Management project
4	Legal and Regulatory Framework	Developing a Legal and Regulatory Framework: Cyberlaw Capacity for IT/ICT
5	Availability, Access (including open access, participatory access and universal access) and Affordability	Becoming part of the Global Information Infrastructure thru Regulation, Competition, Private Investment and Access
6	Marketplace	Building a Marketplace for National ICT Capacity
7	Vision, Mission, Mission and Strategies	Enactment of a National Vision, Mission and Strategies for eGovernance
8	Knowledge Management	Building Knowledge Management for National ICT Capacity
9	Human resource	Developing a Competent IT Human Resource
10	Acculturation	Building an IT Culture and Promoting the Acculturation of IT/ICT
11	Telecommunications Infrastructure	Building the Telecommunications Infrastructure as an essential component of the National Information Infrastructure (NII)
12	Internet Infrastructure	Building the Internet Infrastructure as an essential component of the National Information Infrastructure (NII)
13	Policy Scope	Defining the core of the Information Technology Policy SCOPE for National ICT Capacity Building
14	Organizational Responsibility	The core of Organizational Responsibility (institutional entity based or committee-based) for driving a National ICT Initiative
15	Top Level Commitment / (Prime) Ministerial Responsibility	The core of Top Level Leadership for the Trigger and Drive for a National ICT Initiative '
16	Policy Orientation	Defining the core of the Information Technology Policy ORIENTATION for National ICT Capacity Building
17	Partnerships	Building Partnerships (PPP, SMART, etc) for National ICT Capacity

Chapter 12
The Use of Information and Communication Technologies for Health Service Delivery in Namibia:
Perceptions, Technology Choices, and Policy Implications for Sub-Saharan Africa

Meke I. Shivute
Polytechnic of Namibia, Namibia

Blessing Mukabeta Maumbe
Eastern Kentucky University, USA

ABSTRACT

Information and communication technologies (ICT) have transformed health service delivery (HSD) in developing countries although the benefits are not yet fully understood. This chapter examines the use of ICT for HSD in the Namibian context. To obtain insights into the extent and degree of the current ICT uses, the chapter begins by mapping a HSD landscape for Namibia. The reported ICT use patterns are based on a primary survey of 134 patients and key informant interviews held with 27 health service providers (HSPs) in Khomas and Oshana regions of Namibia. The results from the survey indicate that Namibian patients use diverse range of ICT to access health services including the traditional television and radio, and the more modern mobile phones and computers to a limited extent. HSPs reported the growing use of ICT in various functional areas such as admissions, clinical support, family planning, maternity, and emergency services. The chapter identifies key challenges and policy implications to enhance the uptake of ICT-based health services in Namibia. The relatively high penetration rates of traditional ICT such as televisions and radios coupled with a growing use of mobile phones presents new alternative opportunities for expanding HSD to Namibian patients in remote settings. The chapter will benefit HSP and patients as they decide on affordable technology choices; and policy makers as they design interventions to stimulate the use of ICT in HSD in Namibia. The results provide key insights for other Sub-Saharan African countries contemplating ICT integration in health services.

DOI: 10.4018/978-1-60566-820-8.ch012

INTRODUCTION

Information and communication technologies (ICT) have transformed the way health services are delivered in today's global society Sargeant (2005:305). For any society, however, it is important to understand how ICT are being deployed to support the delivery of health services to patients. In the case of Namibia, health service providers (HSP) in both the private and public health sectors must have the capability to use ICT as this will subsequently influence how they deliver services to their patients in the future. On the other hand, patients too, need to use relevant ICT to support and improve their access to health services. This chapter, therefore, examines the landscape for health service delivery (HSD) in Namibia. The Namibia Health Service Delivery Landscape (NHSDL) provides key insights and a better understanding of the current utilization and the future potential for ICT applications in health service delivery (HSD). The proposed NHSDL provides a comparative assessment of the emerging ICT use patterns in both rural and urban areas in Namibia. Patient's views and perceptions about ICT applications in HSD are described. Furthermore, the chapter highlights some considerations for improving access to ICT by both rural and urban-based patients in Namibia.

Namibia's health sector is distinctive and different from other sub-Saharan countries. What makes Namibia's HSD unique is the critical role that missionaries have consistently played in health care provision. Health services in Namibia are supplied by the government, missions and private providers (Namibia, Ministry of Health and Social Services, 2004b:5). The HSD system comprises two main components: a *public* and *private* health services sector (El Obeid Mendlsohn, Lajars, Forster and Brule, 2001:1). The government of Republic of Namibia (GRN) through the Ministry of Health and Social Services (MOHSS) supplies public health services, while the private health services are offered by private practitioners, hospitals and clinics. Mission health services are subsidized by the MOHSS and include hospitals, health centers and clinics. Mission health facilities are considered key part of the public health services.

In view of the fact that the healthcare sector is one of the fastest growing industries in the service sector, stakeholders in healthcare will be forced to make some adjustments to reposition themselves for the future. Those health service providers (HSP) that fail to introduce the necessary changes arising from increased use of ICT will either become obsolete or less competitive in the long-run. As a developing country, Namibia is confronted with the global competitive issues and this serves as a further motivation for this study. A competitive health service sector will play a major role in the GRN effort to meet the challenges of the Millennium Development Goals (MDGs).

Namibia is among the leading countries in Sub-Saharan Africa that have made significant strides in using ICT to transform government service delivery. Hesselmark & Miller (2002:40) indicates that Namibia's infrastructure has tremendously improved in the twelve years after independence, the number of telephone lines has doubled, the mobile network covers most of the population, the Internet can be accessed throughout the country, and 600 leased lines are in operation. Hesselmark & Miller (2002:40) further adds that ICT competence is widespread and that several large companies in the modern sector operate sophisticated enterprise software.

According to the MOHSS, Namibia has implemented a computerized health information system (HIS) that is currently being used in the public sector (Namibia, MOHSS, 2004a:26). Given new and emerging changes taking place in ICT provision, MOHSS is revising the current HIS with the view to modernize it and make it more efficient. The newly established Ministry of Information and Communication Technology has assumed the responsibility for (i) the overall national ICT policy development functions and (ii) establishment of a legal framework for the ICT industry (Namibia, 2008). The use of telephones, faxes, emails and Internet in the health sector is becoming widespread. However, there is need to assess the extent and degree to which the modern ICT are used to support HSD in Namibia's public and private health sector. Given the foregoing, the specific objectives of this chapter are to:

i. Describe Namibia's health service delivery landscape (NHSDL).
ii. Examine existing ICT uses in Namibia's health sector and identify major constraints in health service delivery (HSD).
iii. Discuss ICT policies and their implications for health service provision.
iv. Recommend measures to enhance the uptake of ICT in HSD in Namibia.

The rest of the chapter is organized as follows: the next section provides a review of the literature and that is followed by a brief description of the research methods in section 3. Section 4 presents the results of the patient survey and health service provider key informant interviews and that is followed in section 5 by a brief discussion of the policy implications and future directions. Section 6 concludes our chapter with suggestions for future studies.

LITERATURE REVIEW

The Use of ICT in Health Service Delivery (HSD) in Developing Countries

ICT has the potential to radically change the way health services are delivered to patients and improve the quality of these services by providing easy access to health care information in developing countries especially in Sub-Saharan Africa. The benefits of ICT as a technique to improve the quality of health services is widely acknowledged worldwide and health institutions are striving to identify ways to deliver their services efficiently and effectively (Hjelm, 2005; Chandrasekhar, 2001). The benefits of telemedicine (i.e. ICT applications in HSD) have been summarized as improved access to information, provision of new healthcare services, improved access to existing service, increase in care delivery, improved professional knowledge, better quality control of screening programs, and reduction in health care costs (Hjelm, 2005). The effective use of ICT can help overcome geographic isolation for the populations that live in remote rural areas, as it facilitates access, dissemination, utilization and exchange of information on combating debilitating diseases such as malaria, tuberculosis and HIV/AIDS (Yamuah, 2005:1). A study in a remote north-west district of Russia reported that a Mobile Telemedicine Unit with endoscopy, electrocardiography and digital photography capability was used for emergency and screening purposes (Uldal, et al, 2004). Mobile ICT is important in reaching patients in remote rural locations without access to primary care facilities. In Cameroon, mobile phones helped medical residents from University Cameroon deployed in remote villages reduce their isolation and maintain contact with mentors during their rural training (Scott.et. al., 2005).

Chandrasekhar (2001:851) explains that ICT has the potential to be used as a mechanism to increase the transparency and efficiency of governance, which in turn would improve the availability and delivery of health services. The author argues that the potential of ICT in the health sector can be that of acting as mediatory role between the HSP and beneficiary (i.e. patients). In the mediatory role, ICT is used as a medium to deliver health services to patients or can be used as tool to organize information more efficiently and effectively. Yamuah (2001:2) identifies similar benefits, which include the use of ICT to facilitate communication between different levels of delivery units (e.g. district hospitals, health centers, clinics and referral hospitals). At the referral level there is added value through effective resource management and planning, efficiency in processing transactions, and access to more reliable information. Health professionals would have the advantage of effectively and efficiently sharing the information with other health professionals (Yamuah, 2001:2). ICT such as email, telephone or mobile technologies facilitate communication between two or more health professionals by sharing information on various diseases that may afflict the patients.

In developing countries, innovative uses of ICT involve the dissemination by medical experts of disease surveillance, disease treatment, and disease control interventions which saves the patient both time and monetary costs in the form of transportation charges. Such advances in telemedicine allows for the exchange of health information and delivery of health care across distances by using telecommunications (Rao, 2001:223). With the growth of telemedicine, patients can have their regular consultations with medical specialists across a distance and they do not have to travel long distances for such health services. Zolfo & Lynene (2008), explain how telemedicine is being used in remote areas to assist physicians in clinical support and mentoring. A study conducted in Tanzania to investigate the use of ICT tools such PDA's to assist members of parliament in decision making confirmed that indeed MPs do need to use PDAs to support them in decision making processes, data management and enhance information accessibility (Kirondoma *et, a.,,* 2008*).* Such modern ICT provide opportunities to support health service provision or health researches especially in rural or remote areas without basic infrastructure. Telemedicine offers a wide range of benefits which include accessibility to health services, efficient and professional healthcare delivery, quality control of screening programs and reduced health-care costs (Hjelm et.al, 2005). Bynum, Cranford, Irwin and Banken (2006:39), observed that telemedicine can improve the quality of the diagnosis and management of patients in remote areas. Richards, King, Reid, Selvaraj, McNicol, Brebner and Godden (Richards's *et.al*, 2005), are of the view that the use of ICT in health has the potential to improve access to educational opportunities for professionals and access to care in remote areas.

The use of ICT may not deliver the expected health benefits automatically. Before ICT applications in health can become a real success numerous challenges that currently serve as obstacles to their effective utilization need to be resolved (Anderson, Vimarlund & Timpka, and 2002:159). These barriers that limit access to ICT have been defined as internal (first order) and external (second) barriers (Ertmer, Addison, Molly and Woods, 1999:54). Studies by Ertmer.*et al.* (1999:54) and Lim & Khine (2006:99) characterize *internal barriers* as those barriers that are extrinsic to the ICT user and this includes lack of access ICT, insufficient time to plan instruction, and inadequate technical and administrative support. All of the above can inhibit ICT access in health institutions. On the other hand, *external barriers* are those barriers that are intrinsic to the ICT user, and are more concerned with the unwillingness to change, or user's attitude and preferences about ICT use. These barriers are referred as "fear factors" (Yamuah, 2005:2; Ertmer *et al.*1999: 54).

Health stakeholders can be unwilling to use certain ICT due to certain beliefs or they can be resistant to change due to unfamiliar ICT (e.g. telephones, computers). Resistance to change can therefore hinder ICT use in health institutions for example if a HSP is used to record patient data in manual files, they might be resistant to using a computer for the first time if they have not been given proper training or administration support.

Hjelm (2005), identified four major constraints to ICT-based HSD and these include (i) breakdown in the relationship between health professional and patient, (ii) break-down in relationships between (and among) health professionals, (iii) potential threat of quality deterioration of health information and (iv) organizational and bureaucratic constraints. According to Yamuah (2005:2), the challenges that hinder ICT use in developing countries are (i) lack of electronic means to capture data, (ii) low literacy rates among health providers, (iii) infrastructure constraints such as low distribution of reliable telecommunications and grid power in community health units, (iv) lack of technical support and (v) high establishment costs for ICT projects given limited budget for health services. Sargean (2005:304), identified challenges for ICT use in health services as technological, educational, and social environment. These challenges hinder the widespread use of ICT by patients and staff members in health institutions. Furthermore, Huggins and Izushi (2002:114) added that the delivery of ICT learning in rural communities often has to start with quite a low level of ICT awareness and knowledge among target groups, the target groups being different stakeholders in the health sector. Overall, the barriers in using ICT to enhance HSD relate to the attitude and preference of users, access to certain ICT, and loss of social capital among some professional health workers. Hence, practical solutions to overcome these ICT challenges and enhance effective access to ICT and use are required.

Developing countries are lagging behind in new developments in ICT applications for health service delivery to patients. Despite the aforementioned challenges, telemedicine which is one of the fastest growing areas of ICT applications in the health sector offers numerous opportunities for expanding both urban and rural patient outreach (Baldwin *et.al,* 2002). Telemedicine which started back in the 1920s has been evolving ever since, and its use in developing countries is rising (Richards's *et.al*, 2005). Not only does the use of ICT in developing countries offer tremendous opportunities to enhance HSD, but it also offers new opportunities for poverty alleviation by improving life expectancy.

In Namibia, initiatives have been implemented in the education sector to encourage ICT use. The ICT policy for education is cited as one of the most successful endeavors by the GRN's in its continued effort to spread the use of ICT among its population. 21st century (Tech/na. 2007). The effectiveness, relevance, and potential to address national educational goals aligned to the MDGs goals could be duplicated by MOHSS in its vision to deploy ICT broadly in HSD in Namibia. In other countries such as Egypt, the Ministries of Health have established e-Health programs for purposes of rendering better health services to their society. The Ministry of Communication and Information Technology in Egypt has initiated the incorporation of ICT in health services. The services range from clinical consultation and administration to the provision of medical education to isolated areas. This initiative is made possible by different projects that are underway in Egypt. These include the telemedicine project, the health record system, the emergency medical service call center ambulance project, and the information system and national network for citizen health treatments by the Government (Egypt, 2005). Other African countries can follow the example of Egypt in introducing initiatives to improve ICT use for health services in the context of their countries.

METHODOLOGY

The chapter describes ICT use in the health service delivery in Namibia. The study describes how HSP and patients use ICT to deliver and access health services respectively. The chapter presents HSP and patient views about ICT uses in health service delivery (HSD) and discusses ICT-related health service delivery policy issues. The results are based on a primary survey of 21 health service providers (HSP) and 134 patients in the private and public health sector in Namibia conducted in 2005. The questionnaires covered three main sections; 1) existing ICT and how it is used to deliver health services to patients 2) the constraints in delivering health services, and 3) patient and HSP perceptions about ICT uses for health service delivery in Namibia, particularly in the Khomas and Oshana regions. The sample size was 144 patients and 21 health service providers. The data for the study was collected from the Khomas and Oshana region.

Khomas is representative of a typical urban area in Namibia while Oshana region is situated in the northern part of the country which is primarily a rural setting. The two regions were chosen in order to identify infrastructural and socio-economic deference's in ICT usage that could be attributed to geographic disparities. Khomas region is situated in the central part of Namibia and it is 37,007 square km^2, and this is where the capital city Windhoek, is situated. The region consists of various health facilities of which, one is a national referral hospital, one intermediate hospital, two health centers, seven clinics, three private hospitals, private clinics and thirty-three outreach points. The region has the highest population which stands at 264,616 compared to other regions in the country, and the population per hospital is 121,278 (Namibia, 2006). Access to ICT in the Khomas region might be higher than Oshana, as the community tends to have better access to resources especially infrastructure endowment. Oshana region is the smallest region (5,290 km^2) with a population of 167,797, hence it is densely populated (Namibia, 2004a:1). The Oshana region is the second largest population concentration in Namibia after Windhoek, but it lacks basic infrastructure and most of the services and facilities normally found in urban areas of this size (Anon, 2007a).

Secondary data sources that were utilized included publications such as journals, white papers and health policies some of which were still under construction. Government policies were examined to determine the enabling environment for ICT applications in HSD. In cases where the necessary documents were not easily accessible, GRN officials from the Policy and Planning Directorate were interviewed to provide additional insights on the potential of ICT deployment in Namibia's health sector.

The study is limited to the use of ICT in the health service delivery and it excludes use in other areas of the health sector such as pharmacies. The study explored ICT use by HSPs in the health facilities such as mission, private and public hospitals. The data collection was limited to the Oshana and Khomas regions, where a sample of stakeholders from each region was interviewed on ICT use in their health facilities and did not cover all the regions of Namibia.

The Namibian Health Services Delivery Landscape

The Oshana health directorate is divided into fifteen health facilities, one Intermediate Hospital (regional), four health centers and twelve clinics. The district also has twenty-eight outreach points. The intermediate hospital is the ultimate referral point in the Oshana region and it is responsible for providing essential back up services and support for clinics and health centers in the region. Health centers are smaller than intermediate hospitals and they are found in two types, namely, day care health centers and rural health centers (i.e., the latter is common to Oshana region). The population per hospital in the Khomas region is 169,147 (Namibia, 2006). The idea of a landscape development was first researched by Korpela et,al. 2004a. In their study, they explained that a health landscape is used to depict health service provision to patients in different health sectors and that the landscapes would differ from one country to the next. The 'generic' health service delivery landscape for the Namibia was the first step taken in developing a health landscape model for other regions and a 'generic' model for the whole country. A number of documents were specifically reviewed during the landscape development and these include national health accounts and other government publications. The Namibian generic landscape is depicted in Figure 1 and it illustrates the various stakeholders in the health sector and how they provide ICT related

Figure 1. The health service delivery landscape in Namibia, 2007. Source: Survey data, 2006

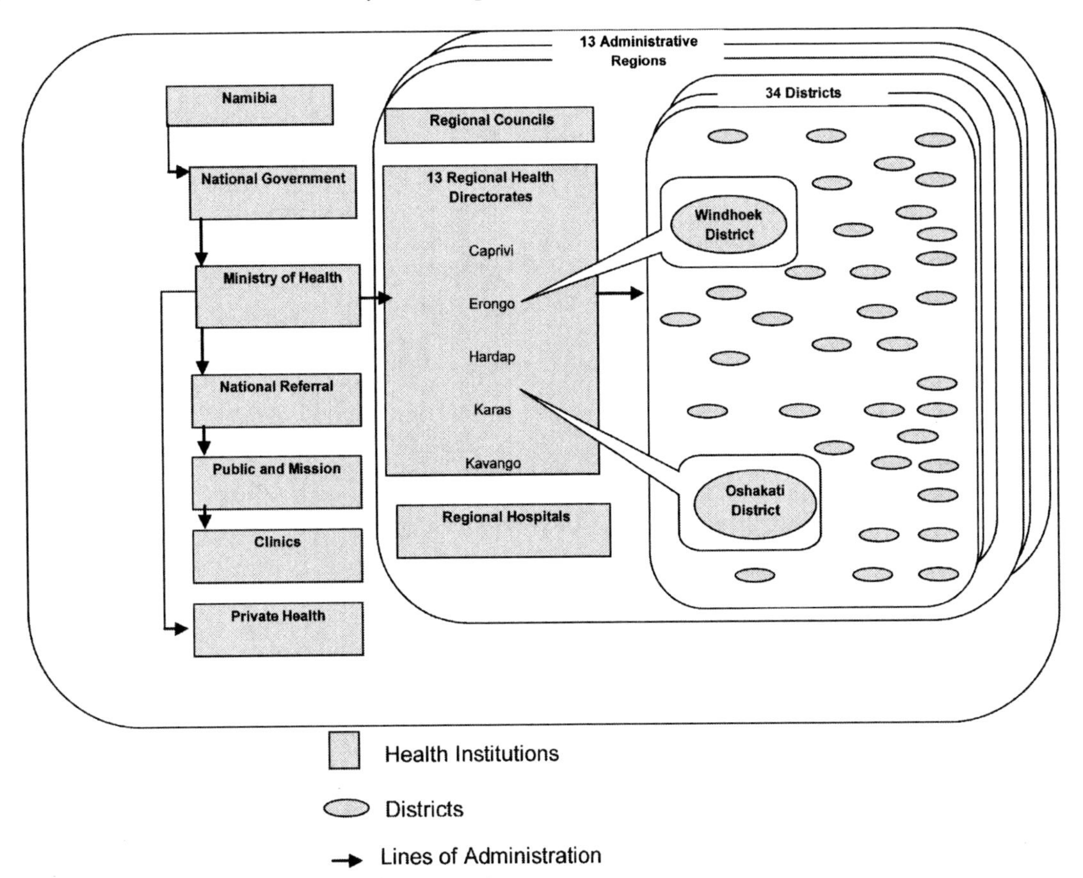

services to patients. The landscape further depicts inter-relations among key stakeholders.

The Ministry of Health and Social Services (MOHSS) is responsible for providing health and social services to the citizens of the country. There are 13 regional directorates, namely; Caprivi, Erongo, Hardap, Karas, Kavango, Khomas, Kunene, Ohangwena, Omaheke, Omusati, Oshikoto, and Otjozondjupa. These directorates are under the management of the MOHSS and within these directorates are 34 districts. The Khomas region has one directorate namely Windhoek. There is one national referral hospital country wide and it is located in the Windhoek district. The national referral hospital is under the management of the MOHSS which is at the national level. It is important to note that the description of NHSDL excludes other stakeholders in healthcare such as, pharmacies and medical aid companies because the study's purpose will primarily focus on use of ICT in the health service delivery for hospitals and health facilities (mission, public and private) in the Khomas and Oshana region. The entire NHSDL should encompass ICT utilization by both formal and informal health service institutions including traditional care providers.

RESULTS AND DISCUSSIONS

The ability to access affordable health services plays an important role in the effective delivery of health service especially to the poor communities. The use of ICT depends entirely on whether HSP or patients have access to affordable ICT. In addition to the cost factor, effective use of

Table 1. Summary of ICT use patterns and user perceptions in the Khomas region, 2007

	Personal computer	Email	Internet	Mobile phone	Radio	Television	Telephone	Fax machine
	%	%	%	%	%	%	%	%
ICT Access								
Current use	44	36	32	94	97	85	90	43
Previous use	72	47	47	97	99	99	97	67
Ownership	24	11	10	92	96	83	64	6
ICT Perceptions								
Expensive	6	n/a	7	0	0	0	0	4
Advanced/do not know how to use ICT.	18	n/a	32	1	0	0	0	14
Time consuming	4	n/a	3	0	0	0	0	0
Not user friendly	0	n/a	0	1	0	0	0	0
Unnecessary	4	n/a	4	0	0	0	0	8
Other	17	n/a	15	2	10	0	0	0
ICT Literacy								
YES	74	n/a	50	99	100	97	99	64
NO	25	n/a	49	1	0	1	1	32
ICT Importance								
Work	40	n/a	21	38	28	13	39	36
Leisure	43	n/a	33	93	94	92	83	38
Health	8	n/a	21	44	83	74	56	17
Education	47	n/a	32	53	76	69	51	19
Other	32	n/a	6	26	0	0	4	0

Source: Survey Data, 2006

ICT also depends on positive perceptions about deploying these modern technologies to deliver and access health services. The data on patients' ICT use patterns and perceptions is shown in tables 1 and 2 for the Khomas and the Oshana regions respectively.

ICT awareness in the Khomas region (Table 1) is relatively high as patients indicated they have access to most of the technologies. Most (99%) patients showed that they have used a radio and television before, even though for some patients they do not own these ICT at home. In case of television, some patients who did not own a television set at home indicated that they had access either in their neighborhood, or when they visit health facilities where they watch video cassettes on health education. Awareness of ICT also proves to be strong in mobile phone use, as 92 percent of patients that took part in the survey indicated that they own a mobile phone. Those that did not own them explained that they had at least one person in their household who owned a mobile phone. The high penetration of mobile phone use provides new avenues for the delivery of mobile health services.

There were various perceptions displayed by patients on ICT use. Some patients expressed the view that they cannot operate some of the ICT (e.g. computer, Internet). Most of the patients that expressed the view that they did not know

Table 2. Summary of ICT use patterns and perceptions in the Oshana region, 2007

	Personal computer	Email	Internet	Mobile/Cell phone	Radio	Television	Telephone	Fax machine
	%	%	%	%	%	%	%	
ICT access								
Current use	31	13	19	95	100	84	77	31
Previous use	40	15	21	98	100	100	97	63
Ownership	7	0	2	94	98	61	47	3
ICT perception								
Expensive	18	n/a	18	0	0	0	0	5
Advanced/do not know how to use	27	n/a	36	3	0	0	0	7
Time consuming	7	n/a	7	0	0	0	0	5
Not user friendly	2	n/a	2	0	0	0	0	0
Unnecessary	2	n/a	3	0	0	0	0	15
Other	32	n/a	44	3	0	0	0	32
ICT Literacy								
YES	39	n/a	21	97	100	97	36	22
NO	61	n/a	77	3	0	3	63	39
ICT Importance								
Work	23	n/a	10	39	24	18	42	31
Leisure	18	n/a	18	98	100	86	81	44
Health	7	n/a	13	39	92	79	45	7
Education	29	n/a	16	55	95	63	37	13
Other	11	n/a	2	12	0	0	0	3

Source: Survey Data, 2006

how to use a computer or Internet either felt that the technology was too advanced (i.e. functional illiteracy) or either they had no access to working computer at all. ICT literacy implies the ability for users to operate diverse ICT tools. The surveyed patients indicated a high ICT literacy rate in terms of their capability to operate the radio (100%), mobile phone, and a landline telephone (99%). The ability to use different ICT was also high for PC (74%) and average for Internet (50%).

Large proportions of patients indicated that they use mostly radio (83%) and television (74%) for health related services e.g. listening to health programs or watching health education programs on TV. Telephone (56%) trailed by mobile phone (44%) were the common ICT used for health-related purposes. The traditional ICT such as radio (76%) and TV (69%) are mainly used as mediums for educational purposes. Radio ranked high (i.e. 100%) as the ICT that all the patients in Khomas region could operate, and which could easily be adapted for health education purposes. This is consistent with the results of Kenny's study where he reported that radio was highly used (71%) by rural people in Nepal as their source of information (Kenny, 2002:150). The Internet is slowly being used for health-related services as only (21%) of patients use it for this purpose. The study did not ask respondents on whether or not they used email to obtain for health services.

On the other hand in the Oshana region (see Table 2), a lot of patients indicated relatively low access to ICT such as computers (31%) and the Internet (19%). The main reasons provided for not having access to these ICT were (i) lack of knowledge and (ii) absence of necessary infrastructure to enable access to specific ICT. Despite the low access to ICT, all the patients surveyed stated that they use radio (100%) for entertainment and for listening to health services and that (98%) owned radios in their household. Mobile phones (95%) are also highly used in the Oshana region. Patients in the Oshana region demonstrated a relatively high functional literacy, as most of them could operate radio (100%), telephone (99%) and mobile phone (97%). A summary of ICT use patterns and perceptions in Oshana are shown in Table 2. In the Oshana region, radio (92%) and television (79%), were the two leading ICT that are mostly used for health-related purposes while telephone (45%) and mobile phone (39%) came third and fourth respectively as being used for health-related services. ICT such as mobile phone (55%), radio (95%), television (63%), are currently being used for educational purposes in the study zone of Namibia. In the case of television, HSP mentioned that they provide health education to patients by playing videos with films that inform patients about dangers of spreading diseases such as HIV and AIDS etc and how to combat the disease outbreak.

During the survey, patients were questioned on their willingness to improve their ICT skills. Patients displayed a strong interest in learning how to improving their ICT skills which shows a positive perception towards ICT. On the other hand HSP had different opinions on how existing ICT in their institutions has helped them to enhance the services provided to patients. It is remarkable to note that there was complete consensus among HSP on the view that ICT helps them to access new health information. It was observed that use of computers helps to improve efficiency and effectiveness of health management information. This mostly applied to departments such as admissions, where patient data needed to be captured and stored accurately.

Ninety-one percent of HSP viewed ICT as helping them to interact with other HSP and other health institutions and higher levels of authority. Eighty-one percent of HSP strongly agreed that ICT does provide ways to improve health services even though there are costs involved in the introduction, maintenance and utilization of ICT. More training of staff on ICT use is required to achieve more effectiveness. Ninety-five percent of HSP strongly agreed that the GRN needs to play a more prominent role in the provision of ICT for HSD, as they felt that the public sector is lagging behind in ICT use as compared to the private sector. HSP felt that the public sector should learn from the private sector how ICT is used in the process of service provision to the patients. Results of specific uses of different ICT in HSD are presented in table 3 and discussed in the following section.

The results indicate that ICT are deployed across different functional areas in health service provision in Namibia. The functional areas where ICI is currently deployed are admissions, consultation and clinical support, maternity, family planning, and emergency purposes. Overall, telephone was the most commonly used ICT across all the various health departments. It is interesting to note that HSPs mentioned that TV was a common form of communication channel for health education as this corresponds with the patient's responses that they use TV as their main source of health information. Mobile phones usage was reported in admission and for emergency services especially when the cheaper landlines are not functioning. Pagers were commonly used in clinical support and emergency services. The personal computers and Internet were used in administrative support in admissions and clinical support respectively. The electronic patient record was used in clinical support to keep track of medications (ARVs) for HIV and AIDs patients.

Table 3. Major ICT applications in Namibia's health institutions, 2007

	Admission	Consulting/ Clinical Support	Maternity	Family planning	Emergency services	Purpose of use
Type of ICT used						
PC	√					Patient record's keeping Administrative duties
Internet		√				Search new health information Office duties
Email		√				Communication with higher levels (e.g. District)
Mobile phone	√					Used in cases where landline is not functional.
					√	-Contact emergency vehicles (ambulance) -Communication with other HSP (e.g. Polio campaigns)
Radio	√					Used for personal reasons
TV		√	√			Display videos on health education on diseases such HIV and AIDS, TB etc.
Telephone	√		√			Making appointments
		√				-Consultations -Ordering medication for pharmacies -Commutation with district office
				√		-To call ambulance -To call doctors in emergency cases -Communication with district office, in cases of urgent documents
					√	For referral cases to big hospitals
Fax Machine		√				For administrative duties e.g. sending fax to higher levels such as district
					√	For referral cases
Pagers		√				To page doctors on call
					√	To page doctors on call
Electronic patient record		√				Used to keep track of medication (ARVs) intake for HIV and AIDS patients.

Source: Survey data, 2006

Table 4. ICT use constraints in the Khomas and the Oshana regions, 2007

Constraints	Description
Budgetary constraints	Finances are not enough to buy all the necessary ICT to assist in the process of service provision to patients.
Lack of basic infrastructure to support health service delivery	Some health facilities especially those in the rural areas lacked basic infrastructure such as electricity and this was a constraint in using ICT such as personal computers.
Lack of basic technological skills	Some HSP providers do not have skills in operation ICT related tools such as Personal Computers

Source: Survey data, 2007

These results demonstrate the extent to which ICT have been deployed in different functional areas in Namibia. Despite making some tangible progress in ICT deployment for HSD, Namibia still confronted with some major challenges which are described in Table 4 below. These challenges were grouped into three main areas, (i) budgetary constraints, (ii) poor ICT infrastructure, and (iii) lack of basic ICT skills among health workers and patients. Budgetary constraints imply the shortage of sufficient public and private) funds required to acquire the necessary modern ICT in order to upgrade HSD systems in Namibia. Related to the problem of funding, is the weak back-borne infrastructure for ICT service delivery. The shortage of electricity hinders the effective utilization of ICT especially in remote rural areas. Without electricity, or alternative energy sources such as solar panels, the use of

computers and Internet will remain limited in geographically remote locations of the country. Even if the country were to allocate adequate funds for the purchase of needed ICT equipment and machinery, and rural electricity generation were to be expanded further, shortage of a critical mass of well trained ICT cadres will remain a major handicap for ICT use for HSD in Namibia. With new ICT constantly coming on stream annually, the development of human capital skills required to support and manage ICT-based health service delivery is central to its effective utilization and sustainability of such services. Although Namibia faces a number of constraints in delivering health services to patients, the foregoing results show a relatively high level of ICT use in both Khomas and Oshana region. This confirms that ICT use in Namibia has been growing and there is scope for further expansion and progress (Schware, 2003:3).

POLICY IMPLICATIONS AND FUTURE DIRECTIONS

In order to sustain the initial progress in ICT use in Namibia, the development of sound effective ICT policies is pivotal. The development of ICT policies plays a vital role in terms of providing timely interventions to stimulate ICT deployment and uptake. Namibia's health sector provides ample opportunities to create a huge market for ICT-based health services that will benefit both HSP and patients alike. According to Galloway and Mochrie (2005:41) policies interventions are needed to support rural ICT development and promote economic sustainability. Such policies should be developed with clear objectives and specific goals in mind, regarding expected achievements of the health institutions. Policies help determine the rate and direction of ICT uses for health purposes and benefits of policy development spills over artificial borders that exists between jurisdictions (Scott, 2004). In addition, key lessons can be drawn from policies, administrative and institutional arrangements, and ideas that work in a given setting. According to the

United Nations (2007) public policies are often best designed and implemented in close dialogue with various stakeholders. The UN report specifically mentioned examples of South Korea and India that close cooperation between public and private sectors was crucial to the success of their ICT Industry. Further, in those countries, private industry associations participated in ICT-related policy formulation. Sub-Saharan countries should draw some key lessons from the development of ICT related policies on HSD that have been successfully implemented elsewhere (UN, 2007).

In the East Africa, there is ICT policy coordination at the regional level. The approach aims to increase levels of regional cooperation and development. The ICT policy framework in East Africa is designed to foster sustainable development in various areas and this includes the health sector (UN, 2003:2). McFarlane, Murphy and Clerkin (2006:245) discussed a number of health information policies in Ireland as they relate to telemedicine services. They concluded that using ICT in telemedicine is a challenge that need to be tackled by taking cognizance of the complexities involved, and actively fostering levers to implementation and sustainability through policy, practice and research initiatives. With reference to Asia, Quibria, Ahmed,Tschang and Reyes-Macasaquit (2003:819) state that policies to promote ICT involve investment in education, infrastructure, creation of favorable institutions, fostering new institutional innovations and international cooperation.

Governments and health institutions in Sub-Saharan Africa should have clear policies and specific strategies on how certain ICT should be implemented (Schware, 2003:3). Maumbe and Owei (2007:1) argue that ICT policy in Africa is lagging behind other regions. In order to achieve long term success in e-government policy development and implementation, institutions should examine the socio-economic context, key constraints and formulate appropriate policies that address those challenges. However, governments cannot merely set up their own policies but they must adhere to the international standards. Namibia (2002:33) has adopted the position that government policies, standards, and procedures must be internationally benchmarked for best practices. Furthermore, policies and standards should also facilitate interoperability within an international, voluntary and consensus-based environment for standards setting. The international standards bodies such as International Telecommunication Union guide Namibian policies and standards formulations. According to (Maumbe and Owei, 2007:1) the area of ICT policy development has received less attention in Sub-Saharan Africa, and South Africa is not an exception despite making great strides in e-government development. A study conducted by Joseph and Andrew (2006), recommends that policies must be developed in such a way that they support and stimulate ICT use by rural people. Furthermore, policies should be crafted in such a way that it improves ICT access at low cost especially through public service. E-health policy response of countries has been classified as "none", "reactive" or "proactive" (Scott, 2004). While none is self-explanatory, reactive suggests policies that are crafted in response to some major internal shock or external pressure. Proactive, which what African countries should strive for, deals with policy development in advance of immediate need and guided by a context specific development strategy that is unique to given country and its socio-economic circumstances or resource endowments.

At the time of our study, Namibia's MOHSS did not have an ICT policy for HSD to patients as it was still in a development stage. But there is a national ICT policy that the MOHSS adheres to. According to the interview with the Directorate of Policy and Planning, the MOHSS adheres to policies established by the public service committees. With the recent establishment of a Ministry of Information and Communication Technology, Namibia is in a much better position to design a sustainable ICT policy for HSD. In addition, successful ICT policy initiatives in education and lessons drawn from other African countries could provide further insights into critical issues for consideration when crafting such a policy.

As already alluded, Namibia's Ministry of ICT is still in its infancy stage (Namibia, 2008). The new Ministry of ICT aims to establish future mechanisms for the collection of relevant information from operators and to conduct comprehensive research to identify relevant context specific issues, trends and other matters that may affect the sector. Further, the Ministry plans to establish a formal process for the periodic reviews of ICT policy documents with a comprehensive feedback mechanism from key stakeholders. Therefore, our chapter contribution to the book on e-government is expected to benefit the Namibian government and several other African governments involved in the design and implementation of sustainable ICT policies in the twenty first century.

CONCLUSION

This chapter examined the use of ICT in HSD in the Namibian context. The chapter described the NHSDL which shows the key health service facilities in the country and the existing and potential arteries for ICT utilizations. The results from the primary survey conducted with patients and also interviews held with HSP in Khomas and Oshana regions highlight the existing patterns of ICT use among patients and the health service departments or functional areas that have deployed ICT. The chapter outlines some of the major constraints in delivering ICT-based health services in Namibia namely, funding shortages, poor infrastructure and low ICT literacy. The high penetration rates for traditional ICT such as televisions and radios coupled with a growing use of mobile phones presents alternative relatively under-explored opportunities for HSD to patients. The increased use of mobile phones and mobile computer lat tops could help overcome geographic and infrastructural challenges associated with ICT use for HSD in Namibia.

The chapter highlights the need for the Government of the Republic of Namibia (GRN) to actively pursue the opportunities presented by ICT such as personal computers and mobile phones to complement traditional ICT-based HSD channels such as radios and televisions. The Namibia Ministry of Information and Communication Technology working in collaboration with the Ministry of Health and Social Services should develop "pro-active" ICT policies that will create an enabling environment for ICT use, and stimulate its uptake by both patients and HSPs. The results indicate a strong momentum for ICT use in HSD that can be further developed, strengthened and natured with the backing of a sound and effective policy development process. Without targeted policy development for ICT use in HSD, the current momentum will fizzle out, and one of the best opportunities to transform HSD in Sub-Saharan Africa (and thereby alleviate poverty) will be sadly missed.

The results in this chapter can be used to provide insights into similar efforts to incorporate ICT in HSD in other African countries. Future studies should examine ICT use in private or public health care focusing on specific patient groups (children, elderly, physically disabled members of society etc), functional areas (e.g. admissions, emergency services, maternity, clinical support, family planning etc), and the treatment of highly contagious disease outbreaks (e.g. HIV and AIDS, SARs, avian flu, swine-flu (HIN1), etc.). An alternative study could assess the willingness to use ICT such as mobile phones, personal digital assistants, and mobile lap-tops, as these are the likely future channels for HSD. Such a study should provide a comprehensive analysis of both the tangible benefits and the negative externalities associated with the use of such applications for HSD in both the short-term and the long-term.

REFERENCES

Anderson, A., Virmarlund, V., & Timpka, T. (2002). Management demands on information and communication technology in process-oriented healthcare organizations: The importance of understanding managers' expectations during early phases of systems design. *Journal of Management in Medicine*, *16*(2/3), 159–169. doi:10.1108/02689230210434907

Baldwin, L. P., Clarke, M., Eldabi, T., & Jones, R. W. (2002). Telemedicine and its role in improving communication in healthcare. *Logistics Information Management*, *15*(4), 309–319. doi:10.1108/09576050210436147

Bynum, A. B., Cranford, C. O., Irwin, C. A., & Banken, J. A. (2006). Effect of telemedicine on patient's diagnosis and treatment. *Journal of Telemedicine and Telecare*, *12*(1), 39. doi:10.1258/135763306775321407

Chandrasekhar, C. P., & Ghosh, J. (2001). Information and communication technologies and health in low income countries: the potential and the constraints. *Bulletin of the World Health Organization*, *79*(9), 850–855.

El Obeid, S., Mendlsohn, J., Lajars, M., Forster, N., & Brule, G. (2001). *Health in Namibia: Progress and Challenges*. Windhoek, Namibia: Raison Research and Information Services of Namibia.

Ertmer, P.A., & Addison, P, L., Molly, R, E. & Woods, D. (1999). Examining teacher's' beliefs about the role of technology in the elementary classroom. [Fall.]. *Journal of Research on Computing in Education*, *32*(1), 54–72.

Hesselmark, O., & Miller, J. (2002). A country ICT survey for Namibia. Retrieved from http://www.cyberzoo.co.za/download/milless/newdocs/Namrep%20v5.pdf

Hjelm, N. M. (2005). Benefits and drawbacks of telemedicine. *Journal of Telemedicine and Telecare*, *11*(2), 60–70. doi:10.1258/1357633053499886

Huggins, R., & Izushi, H. (2002). The Digital Divide and ICT Learning in Rural Communities: Examples of Good Practice Service Delivery. [May.]. *Local Economy*, *17*(2), 111–122. doi:10.1080/02690940210129870

Maumbe, B. M., & Owei, V. T. (2007). E-government Policy Development in South Africa: Current Status, Distributional Issues, and Future Prospects. In *Proceedings of the IST-Africa Conference*, 9th -11th May, Maputo, Mozambique.

Ministry of Information and Communication Technology. (2008). *Information technology Policy for the republic of Namibia 2008*. Windhoek, Namibia: Author.

Namibia. MOHSS (2004b). *The technical efficiency of District Hospitals in Namibia*. Windhoek: Directorate: Policy, Planning and Human Resources Development.

Rao, S. S. (2001). Integrated health care and telemedicine. *Work Study*, *50*(6), 222–229. doi:10.1108/EUM0000000006034

Sargeant, J. M. Medical education for rural areas: Opportunities and challenges for information and communications technologies. *Journal of Postgraduate Medicine*, *51*(4), 301-307. Retrieved June 20, 2007 from http://www.jpgmonline.com/text.asp?2005/51/4/301/19244

Scott, R. (2004). Investigating e-health policy-tools for the trade. *Journal of Telemedicine and Telecare*, (10): 246–248. doi:10.1258/1357633041424377

Scott, R. E., Ndumbe, P., & Wootton, R. (2005). An e-health needs assessment of medical residents in Cameroon. *Journal of Telemedicine and Telecare*, 11.

Uldal, S.B., & Amerkhanov, J., Bye, Manankova, S., Mokeev, A., & Norum, J. (2004). A mobile telemedicine unit for emergency and screening purposes: experience from north-west Russia. *Journal of Telemedicine and Telecare*, *10*(1), 11–15. doi:10.1258/135763304322764121

Yamuah, L. K. (2005). ICT in the African health sector; towards healthy nations with ICT wealth. *I4d magazine*. Retrieved June 20, 2006 from http//:www.i4donline.net/may05/africahealth.asp

Zolfo, M. & Lynen (2008). Telemedicine for HIV/AIDS care in Low Resource Settings: Proven Practices. In *Proceedings of the 2008 conference of the IST Africa, Windhoek,* 07-09 May 2008, Ireland, IIMC International Information Management Corporation, Ltd.

Section 4
Selected Readings

Chapter 13
ICTs as Tools for Poverty Reduction:
The Tanzanian Experience

Zaipuna O. Yonah
Tanzania Telecommunications Company Ltd., Tanzania

Baanda A. Salim
Sokoine University of Agriculture, Tanzania

ABSTRACT

This chapter attempts to enhance the understanding and knowledge of Information and Communication Technologies (ICTs) in relation to the Tanzania National ICT Policy as a case study. The authors extensively explore these pervading technologies as they impact on the education, commerce, social, cultural, and economic life of the poor Tanzanian people. The chapter looks at how Tanzania is coping with the issue of poverty eradication as one of the eight UN Millennium Development Goals (MDGs). It addresses the issue of digital divide and the role that ICTs can play in poverty reduction. Tanzania's efforts in embracing ICTs and the challenges facing the country in its efforts are also addressed. Overall, the chapter demonstrates that ICTs are a set of tools for knowledge sharing, which is a powerful means for poverty reduction. Furthermore, it is advisable to focus on information literacy rather than just focusing on computer literacy.

INTRODUCTION

In this chapter, an attempt has been made to enhance the understanding and knowledge of Information and Communication Technologies (ICTs) in relation to the Tanzania National ICT Policy as a case study. The authors extensively explore the evolving frontiers of these pervading technologies as they impact on education, commerce, social, cultural and economic life of the poor Tanzanian people. The chapter also attempts to show case examples of ICT national projects in Tanzania to demonstrate how the government of Tanzania, entrepreneurs, and some of the rural communities are appropriating ICTs to fit into their needs. The chapter reflects on perspectives, trends, and potential of using ICTs to develop innovative approaches and methods for poverty reduction in Tanzania.

The chapter is organized into six sections. The Background section looks at the general issues, putting Tanzania into perspective and how it is coping with the issue of poverty eradication as one of the eight U.N. Millennium Development Goals (MDGs). Targets for achieving poverty eradication are articulated in the National Poverty Eradication Strategy. The Development Vision 2025 aims at guiding Tanzania to achieve five goals by year 2025, namely, (1) high quality livelihood, (2) peace, stability, and unity, (3) good governance, (4) a well educated and learning society, and (5) a strong and competitive economy. This section further puts into perspective Tanzania's position on the role of ICTs in meeting the goals of the Vision 2025 as well as the MDGs. The framework for appropriating these ICTs for national development and poverty reduction is articulated in the National ICT Policy.

The second section attempts to answer the question: what is poverty and what causes it? It is noted that ICTs are now acknowledged to be a set of powerful tools for building the capacity for knowledge management and hence for building the capacity for poverty reduction. It also addresses the issue of digital divide. The third section addresses the role that ICTs can play in poverty reduction. It outlines the desirable characteristics and attributes of the modern ICTs useful for poverty reduction. This section further addresses issues of connectivity, affordability, and capability. The fourth section details efforts that Tanzania has made in embracing ICTs. It narrates on various projects being undertaken in Tanzania such as projects on increasing capacities and opportunities; projects on reducing vulnerabilities; projects on enhancing government capacity, efficiency, and accountability; and projects on participation, empowerment, and the strengthening of civil society.

The fifth section looks at the challenges facing the country in its efforts to embrace ICTs. It also looks at the prospects and try to project the trend of adoption of ICTs in the country in the next five or so years. In the last section of the chapter, some recommendations are drawn on "an entry point" into dissemination of ICTs to the rural areas of Tanzania. In this section, the authors try to lay out priority areas in harnessing the power of ICTs in bringing about development to the rural people of Tanzania.

BACKGROUND

Tanzania is one of the 191 United Nations member states that have pledged to meet the eight UN Millennium Development Goals (MDGs) by year 2015 (URT, n.d.). These goals are: (a) Eradicate extreme poverty and hunger, (b) achieve universal primary education, (c) promote gender equality and empower women, (d) reduce child mortality, (e) improve maternal health, (f) combat HIV/AIDs, malaria and other diseases, (g) ensure environmental sustainability and (h) develop a global partnership for development. In response to this pledge, Tanzania restated the UN-MDGs and came out with its own nine (9) development goals, herein to be referred to as TZ-MDGs, namely (URT, n.d.):

a. **Extreme poverty:** Halve the proportion of people living below the national poverty line by 2015.
b. **HIV/AIDS:** Halt and reverse the spread by 2015.
c. **Hunger:** Halve the proportion of underweight, under five year olds by 2015.
d. **Basic amenities:** Halve the proportion of people without access to safe drinking water by 2015.

e. **Primary education:** Achieve universal primary education by 2015.
f. **Gender equity:** Achieve equal access for boys and girls to primary and secondary schooling by 2015.
g. **Reproductive health:** Reduce maternal mortality ratio by three-quarters by 2015.
h. **Under-five mortality:** Reduce under five mortality by two-thirds by 2015.
i. **Environmental sustainability:** Reverse loss of environmental resources by 2015.

The focus of this chapter is on the first TZ-MDG, which is on eradication of extreme poverty. As is in many developing countries, poverty in Tanzania is characterized by low income and expenditure, high mortality and morbidity, poor nutritional status, low educational attainment, vulnerability to external shocks like natural disasters (e.g. drought, pests, diseases and floods), and exclusion from economic, social and political processes. Generally, poverty is particularly widespread in the rural areas, but is not insignificant in urban areas. There are also important regional and district differences in the levels and specific dimensions of poverty. Those most at risk of being trapped to live under poverty are young children and youths, the elderly, women, those in large households and those involved in subsistence agriculture, livestock production and small-scale fishing.

The primary targets for achieving the first TZ-MDG on poverty eradication are articulated in the National Poverty Eradication Strategy (NPES) (URT, 1997). The set timeline for this TZ-MDG, which is more ambitious than the international one (2015), is to halve extreme poverty (measured by income and expenditure) by 2010 and eradicate it by 2025, which corresponds to the articulation of the goals of the Tanzania Development Vision 2025 (URT, n.d.). The Development Vision 2025 aims at guiding Tanzania to achieve five goals by year 2025, namely: (1) high quality livelihood, (2) peace, stability and unity, (3) good governance, (4) a well-educated and learning society, and (5) a strong and competitive economy capable of producing sustainable growth and shared benefits. Overall, the Vision foresees that by the year 2025:

Tanzania should have created a strong, diversified, resilient and competitive economy, which can effectively cope with the challenges of development and, which can also easily and confidently adapt to the changing market and technological conditions in the regional and global economy.

The challenge is, therefore, to mobilize massive resources towards the realization of the Vision 2025 goals.

It is worth noting that the Government of Tanzania has already acknowledged that ICTs can be harnessed to meet the goals of the Vision 2025, as well as that of the TZ-MDG on poverty reduction. In broad terms, ICTs refer to any technique or knowledge used to create, store, manage, and disseminate information. They include simple information centres with notice boards, books, brochures, posters, and newspapers, simple content systems running on cheap (safely disposable) hardware, for example, audio and video cassettes, disconnected mailing systems, to locally browse-able content accessible through a range of electronic technologies such as telephone, fax, television, and radio. Modern ICTs include the Internet, e-mail, computers, mobile phones, digital cameras, online databases and portals.

In a sense, ICTs enable communication, a process that links individuals and communities, governments and citizens, in participation and shared decision making. This is done through use of a variety of ICTs to engage, motivate, and educate citizens of opportunities for development and poverty reduction, in this way promoting changes in peoples attitudes and behaviours and thus increasing their participation in the development or poverty reduction processes (Gillman, 2003).

The framework for appropriating these ICTs for national development and poverty reduction is articulated in the National ICT Policy (United Republic of Tanzania, 2003). In the policy framework, it is clearly shown that ICTs are crosscutting in nature, an attribute well shared with poverty. As such, therefore, ICTs can indeed be appropriated as tools for realizing effective communication processes to achieve, among others, distribution and sharing of knowledge and information for supporting poverty reduction initiatives.

In this chapter, an attempt is made to enhance the understanding and knowledge of ICTs and the evolving frontiers of these pervading technologies are explored as they impact on education, commerce, social, cultural and economic life of the poor Tanzanian people. It also attempts to show case examples of ICT national projects that demonstrate how the government of Tanzania, entrepreneurs, and some of the rural communities are appropriating ICTs as tools to fit into their needs and empower themselves to fight poverty. Further, an attempt is made to stimulate debate to reflect on perspectives, trends, and potential of using ICTs to develop innovative approaches and methods for poverty reduction.

WHAT IS POVERTY AND WHAT REALLY CAUSES IT?

Widespread and persistent poverty in Tanzania is the main development challenge since the 1990s. Currently, about 40% of the population lives below the basic needs national poverty line (World Bank, 2004), that is, a population surviving on less than US$2 per day. A debatable issue arises here: *what does poverty really entail?* Poverty is often viewed from an income-based perspective as lack of income and measured in terms of income and expenditure. In recent years, however, the concept has been enlarged. Poverty is now seen as a multidimensional concept that is the opposite of well-being, which includes more than income. Therefore, apart from the inability of poor people to meet basic needs (such as nutrition, clothing, and shelter), poverty also refers to disadvantages in access to land, credit, and services (e.g., health and education), vulnerability to violence, external economic shocks, natural disasters, powerlessness, and social exclusion (Shaffer, 2001; Williams, Sawyer & Hutchinson, 1999).

According to the Government Poverty Reduction Strategy Paper (PSRP) (URT, 2000) and the Poverty Monitoring Master Plan (PMMP) (URT, 2001), people living in poverty in our society are reidentified to include: illiterate people; unskilled labourers; self-employed micro-entrepreneurs; subsistence farmers; women; children; and population living in remote (rural) areas. Those in the group that are extremely vulnerable include children, persons with disabilities, youths, elderly, people living with HIV/AIDS, women, drug addicts, and alcoholics.

Poverty, the simplest indicator of alienation (Allen, 2002), causes a citizen to feel devalued, resentful, frustrated, and angry. Despite the vast advances that are being made in the spheres of science and technology, medicine, capital mobility, and so forth, income disparities are ever widening, both within countries and between nations. This imbalance in equity is characteristic of alienation—alienation between developed and developing nations; alienation between civilizations or cultures; alienations within our nation (regions, districts, etc.) and even alienation within families. Therefore, overcoming poverty, a consequence of alienation, will require the full participation of every member of the Tanzanian society. It is worthwhile to note that ICTs are now acknowledged to be a set of powerful tools for building our capacity to care, knowledge management, and hence for building the capacity for poverty reduction. In the literature, alienation based on modern ICTs is also popularly known as the digital divide (Weigel & Waldburger, 2004).

POVERTY REDUCTION

The poverty reduction challenge is about how to empower the poor with knowledge and skills, thus availing to them new opportunities to improve their livelihood (Weigel & Waldburger, 2004). In terms of income and expenditure this would mean an increase in income, and hence capacity to afford food, health services, and other basic needs.

The strategic approach is to use ICTs in a creative manner to level the playing field in economic, social, cultural, and political terms by reducing the rapidly growing gap caused by a very asymmetric architecture of opportunities between the rich and the poor. As a matter of emphasis, the role of ICTs in poverty reduction is not limited to reducing income poverty, but also includes non-economic dimensions, in particular, empowerment (Gerster & Zimmerman, 2003). This can be accomplished through a variety of strategies ranging from the *sustainable livelihoods approach*—by putting people first; *production-oriented growth strategy*—which focuses on pro-poor corrective measures; a *distribution-oriented strategy*—which emphasizes the redistribution of wealth within the nation; and a *rights and empowerment strategy*—which promotes knowledge about basic rights and empowerment (Gerster & Zimmerman, 2003;Weigel & Waldburger, 2004).

It was stated earlier that modern ICTs facilitate the creation, storage, management, and dissemination of information by electronic means much more easily, efficiently, and conveniently. Some of the desirable characteristics and attributes of these modern ICTs useful for poverty reduction include:

a. **Interactivity:** ICTs are effective two-way communication technologies, which have drastically changed the way individuals, organizations, and enterprises interact. This is made possible by the fact that modern ICTs embody broad communication and processes of economics, social, political activity, and organization to empower citizens. At the same time ICTs increases transparency within and among societies, fostering empowerment and accountability (Sharma, n.d.; Spence, 2003).

b. **Permanent availability:** The new ICTs are available 24 hours a day, seven days a week. For this reason they can be mainstreamed into our daily activities.

c. **Global reach:** When ICTs are embraced in any socioeconomic activity, geographic distances hardly matter anymore. This has changed the shape of the socioeconomic activities. One of the impacts is in conditioning and changing the structure of markets, hence pushing the world towards globalization. This is true for the case of the Internet, which enables individuals in any country to participate in markets or activities beyond the immigration constraints determined by geographic locations (Sharma, n.d.).

d. **Reduced per unit transaction costs:** Relative costs of communication have shrunk to a fraction of previous values and this effect has impacted on the cost of business transactions. Transaction costs have tended to rise with time, distance, and correlated variables, especially as the global economy has expanded and become more integrated. With modern ICTs, time and distance essentially do not matter. Therefore, the reduction in transaction costs can be larger in absolute and percentage terms than the rising trend. A good example is on long distance or international telephone calls vis voice over IP (VoIP) calls.

e. **Creates increased productivity and wealth or value:** ICTs are value adders and amplifiers in products and services. In addition to lowering transaction costs, modern ICTs can be very liberating by enabling wholly new possibilities of creating wealth/value. For example, by making services previously difficulty to trade or non-traded at all to become easily tradable—within countries and internationally. ICTs also are capable of untrapping the value of human resources or human capital (e.g., with local skills) by marketing such capital globally than just locally, for example, in the case of outsourced jobs and offshore opportunities (Proenza, n.d.; Yonah, 1999).

f. **Multiple sources of information and knowledge:** Modern ICTs embody a lot of knowledge, particularly if such ICTs include electronic networks. These embody and convey knowledge and in this sense provide important intermediate products and services and content in education, human, and social capital formation activities. In this context, it is a fact that the creation of knowledge and access to information is the basis of new prosperity. Access by the poor to high-tech opportunities empowers them with valuable knowledge and equips them with productivity skills, which is essential in any poverty reduction process. In this respect, poor people do not have a simpler set of living needs and aspirations, as may be assumed (Gates, 1999; Kao, 1996).

Despite these powerful attributes of ICTs, access to information using ICTs is determined by (a) *connectivity*—are the services available? (b) *affordability*—can the targeted poor users afford the access, and (c) *capability*—do the targeted users have skills required to support and utilise profitably the access? The user's skills relate to technical abilities, language, and literacy. Therefore, for ICTs for poverty reduction to succeed they need to be integrated and mainstreamed in the national development priorities and projects.

SHOWCASE EXPERIENCES

It is worthwhile to note that ICTs are impacting on all dimensions of life: education, health, quality of family, culture, leisure and arts, scientific and technological world. The way people do business globally is changing beyond imagination. ICTs are helping economies expand at an unprecedented rate and competitiveness has become the motto of the way. However, for ICTs, as tools of empowerment, to contribute to the poverty reduction initiatives, certain conditions have to be met. These relate to ownership, local content, language, culture and appropriate technology enabled by adequate functional basic amenities and connectedness, e.g. roads, electrical power, telephone, water, etc. Access becomes important only once these conditions have been met. Due to this requirement of connectedness most of the ICTs have an urban bias and discriminate against rural areas.

Since the 1990s, years of economic liberalization in Tanzania, there has been a large wave of investment in ICTs for development and some significant part of this has been aimed at poor people—both in terms of bringing ICT access to poor communities, and in using ICTs in many other ways which support poverty reduction. These investments have produced *many* documented successes, lessons learned, and experiences. Most of the documented materials on ICT projects are mostly sector and application specific covering multipurpose community access; access technologies amenable to poverty reduction; gender equality; education and human resource development; science, high-tech, and ICT-sector growth; business and livelihoods development and support; public sector, services, and poverty management; environmental and natural resource management; and transparency, accountability, and empowerment.

The many experiences that are discussed in the subsequent sections offer a lot of insight and value, though it is difficult to consistently maintain focus on poverty, and separate poverty reduction from broader economic development insights. However, it has to be noted that growth and development are necessary but not sufficient for poverty reduction; and also that pro-poor strategies and investments are as important for ICTs and knowledge-economy strategy as for other connected areas of social and economic development. This section uses some of the material from publicly available websites carrying content about Tanzania.

Projects on Increasing Capacities and Opportunities

One of the many deprivations that compound the misery of the poor and prevent them from rising out of poverty is their lack of access to adequate education, training, skills development, broader information, and knowledge resources that could help them improve their lives and livelihoods. One of the show case projects on information sharing is the *Sharing with Other People Network* (SWOPNet) managed by the Tanzania Commission for Science and Technology (COSTECH). It may be found at www.swopnet.or.tz. The SWOPNet portal electronically brings together all owners of ICTs projects under one roof. The site is essentially very useful for publishing outputs from different research and devolopment projects in the field of ICTs conducted within Tanzania.

It has been envisioned that education and knowledge helps the poor to improve their current livelihoods, address impediments and vulnerabilities that prevent them from seeking opportunities to improve their lives, and participate in new sectors of the economy that require advanced skills and, therefore, offer higher incomes. ICTs can help make educational institutions more efficient and responsive, both by improving communication flows within them and between them and their various constituencies; and by increasing their access to global knowledge and good practice in education. One of the most promising areas where ICTs can help improve education quality and outcomes is in teacher training. The Tanzania educational system is, in general, plagued by inadequate resources for teacher training and curriculum development leading to low quality of education. Creative combinations of the Internet (for content access and interactivity) and digitally stored training materials (including CD-ROMs) can dramatically increase both the *reach* and the *yield* per unit cost of teacher training efforts. The government of Tanzania is making efforts to build strategy to integrate ICTs in the Tanzanian secondary schools (Menda, 2005b). There is also a showcase ICT project owned by the Ministry of Education and Vocational Training (MEVT) geared at integrating ICTs in Secondary Education and Teachers Training colleges (TTCs). These are the preferred entry points for integrating ICTs into the secondary education in Tanzania (more at www.pambazuka.org/index.php?id=26807).

Projects on Reducing Vulnerabilities

Poverty and illness go hand in hand and feed upon each other. Poor people are particularly prone to disease and illness.

There are several ways that ICTs can help to address the health challenges facing developing countries. First, the capacity to monitor, respond to, and thus hopefully control disease outbreaks and address their causes, can be significantly enhanced by improving communication flows and the information-management capabilities of health care professionals at community, district, regional, and national levels. In many cases, time is of the essence in responding to disease outbreaks, and faster communication and information gathering can often make a dramatic difference in how well an outbreak is contained. More generally, the ability of health care providers to assemble and share timely information about health trends and needs enables a country's health care system to adapt more quickly and allocate resources more effectively.

The government of Tanzania is ready to adopt ICTs in the provision of health services (International Institute for Communication and Development, 2004; Menda, 2005a), a commitment made recently by the Permanent Secretary in the Ministry of Health and Social Welfare, Hilda Gondwe, when officiating at a national roundtable workshop on the application of ICTs in the country's health sector. The workshop was organized by Christian Social Services Commission (CSSC) and financed by the Catholic Organization for Relief and Development (CORDAID) and International Institute for Communication and Development (IICD) (The Guardian, 2006).

Several show case projects that show how ICTs are being appropriated to realise health information system, to collect health data at village levels, and on child vaccination programs are reported at http://www.swopnet.or.tz/ws_projs.html.

Projects on Enhancing Government Capacity, Efficiency and Accountability

Government officials, and the institutions in which they work, are hampered in many ways by poor information and knowledge flows. They often have weak access to even basic current data about the issues and trends in the country. Information flows poorly within most government departments because of a combination of weak communications infrastructure, hierarchical structures, and rigid bureaucratic cultures.

In addition, government officials have limited information on global good practice, and few opportunities for consultation and collaborative problem solving with colleagues elsewhere. At the same time, citizens, entrepreneurs, and particularly the poor, often have limited information about their rights and the services available to them, about the structure and functioning of government agencies, and about procedures for requesting services.

ICTs can help in a variety of ways to address these problems. ICTs can be used to reorganize and speed up administrative procedures, to increase the volume and speed of information both within government institutions and between them and the larger society, to train government officials in global best practices, and to allow greater collaboration and sharing of experience among government officials both within a country and across borders. The government of Tanzania is exemplarily active in appropriating ICTs to improve its e-readiness and the delivery of public services. The official online gateway of the United Republic of Tanzania is accessible at http://www.tanzania.go.tz. Information accessible from this site includes fully analysed population census data, and most of the government policies and presidential speeches are hosted at this site.

There are several other information gateway projects in the country, for example, the Tanzania Online Information Gateway at www.tzonline.org and the Tanzania Country Gateway project at www.tanzaniagateway.org. Collectively these Internet-based gateways give Tanzania a global presence. The gateways provide links to ideas and good practice, information about development activities and industry trends, funding and commercial opportunities. These portals provide and promote exchange and dissemination of information on development matters.

Another project is the Tanzania National Assembly (Parliament or "Bunge") portal at www.parliament.go.tz. Acts, bills, and useful Bunge documents are readily available from this site through its Parliamentary Online Information System (POLIS). This Internet-based global presence clearly demonstrates that the Tanzania Parliament is appropriating modern ICTs to transform the Bunge from an old-fashioned institution to a new, modern, paperless electronic parliament. Furthermore, the Tanzania Investment Center website at www.tic.co.tz acts as the official investor's guide to Tanzania.

ICTs can also play an important role in combating corruption and making government institutions more transparent, by reducing the opportunities and incentives for, and increasing the costs of, corruption. The most obvious role for ICTs is to "disintermediate" between the citizen and the services, procedures, and documents by automating and making widely accessible, many of the simpler procedures which have traditionally depended on the involvement of a local government officer. If a citizen can directly access a needed form, acquire required documents, permits and certifications, or register a new small business, using automated procedures, the opportunities for corruption are reduced. ICTs can also empower individual citizens and groups to hold government officials publicly accountable. These efforts are evident from the TIC Web site www.tic.co.tz especially links to the Business Registration and Licensing Agency (BRELA); www.necta.go.tz for the National Examination Council of Tanzania; and www.isd.co.tz for the insurance supervisory department.

Projects on Participation, Empowerment, and the Strengthening of Civil Society

ICTs can play an important role in informing and empowering citizens and strengthening the capacities of a wide range of civil society organizations and institutions. This is important not only in increasing the demand for good governance and strengthening the voice of citizens in government policy, but also for promoting both the stability and responsiveness of the political system and for the economy and society as a whole.

By facilitating new forms of many-to-many communication, collaboration, and information-sharing, both within a given country and among groups with similar interests and concerns across borders, ICTs can add to the vibrancy of civil society institutions and networks as a check on government. They can also act as a source of ideas and innovations, and an outlet for the interests, concerns, and desires for solidarity on the part of individuals and groups. This will reduce the alienation of the poor from feeling isolated, powerless, and neglected. A contrast can be found between a physical rally meeting to the famous eThinkTank user group at www.eThinkTank.org focused on issues concerning ICTs. Another portal is at www.hakikazi.org, a civil society meeting place for sharing ideas promoting the rights of all people to fully participate on social, technical, economic, environmental, and political (STEEP) issues. The portal carries a cartoon-based guide to popularize the PRSP (URT, 2000).

By definition, however, the poor have scarce resources, and the burdens of their daily lives often leave them little discretionary time to engage in activities designed to protect their interests and articulate their needs. Their limited education, and in many cases illiteracy, puts them at a disadvantage when faced with sophisticated ICTs that are not adapted to their most pressing needs, their modes of communication (including a frequent preference for oral communication), their cultural norms, and the social contexts in which they typically interact and pursue joint action. In such cases, community radio and video conferencing facilities (like that at the Tanzania Global Development Learning Center (TGDLC—www.tgdlc.go.tz) become very appropriate.

Projects/Challenges on Appropriating ICTs for Income Generation

Economic programs implemented by the Tanzania Government have been based on the philosophy that Tanzania is committed to a market economy whereby the private sector will take the lead in creating incomes, employment, and growth. On the other hand, the State will be a producer of public goods, play a regulatory role to level the playing field, and create conducive environment for the private sector to take the lead in driving economic growth. The private sector has started playing an ever increasing role in creating incomes and employment. Small and Medium Enterprises (SMEs) account for a large share of the enterprises active in Tanzania. In fact, SMEs are the emerging private sector and do form the base for private sector led growth (Hakikazi Catalyst, 2001; URT, 2002, 2003a).

According to Drucker, there are eight key areas that constitute any business (Drucker, 2001), namely: *marketing, innovation, human resources, financial resources, physical resources, productivity, social responsibility*, and *profit requirements*. Let us recall Gandhi's test for technology appropriateness: to "*find out how the last man would be affected by it*" (Gates, 1999). The authors, therefore, propose to apply Gandhi's test for the effectiveness of ICTs in each selected income generating activity by examining the impact of mainstreaming ICTs into the eight key business areas.

THE ULTIMATE CHALLENGE: TO FOCUS ON INFORMATION LITERACY AND NOT ONLY ON COMPUTER LITERACY?

The Tanzania Vision 2025 (URT, n.d.) would like to see Tanzania be a well educated and a learning society. What does this mean in relation to ICTs as tools for poverty reduction? It has been said earlier that ICTs are embedded in networks and services that affect the local and global accumulation and flows of public and private knowledge.

The authors hold the view that the government of Tanzania needs to focus more on information literacy rather than mere computer literacy. Information literacy is the ability to access, evaluate, and use information from multiple formats—books, newspapers, videos, CD-ROMs, or the Web. Information literacy is a set of competencies, skills that will grow with the people as the society evolves towards a knowledge society, even when current computer operating systems, search engines, or computing platforms and devices are obsolete.

To date, however, it has been observed that, in promoting the information society agenda, citizens and the government are still focused on technology. However, they must focus on increasing awareness of the potential applications; on improving the availability of and access to modern digital communications; and on encouraging people and organizations to use technology more efficiently and effectively. It also has to advocate for policies that encourage and allocate funding towards the development of skills to use technology and improve computer literacy. This focus has been, and is still, invaluable. To really move into the information age and get benefit from the potential of a knowledge based society, it needs a new focus, that is, on the content that flows through the ICTs, a focus on information and knowledge, and focus on how to create it, manage it, and use it. To do this a new focus on information literacy is needed. An information literate person is one who (American Library Association, 2000):

1. Recognizes the need for information and determines the extent of the information needed;
2. Identifies potential sources of information, accesses the needed information effectively and efficiently from sources of information including computer-based and other technologies;
3. Evaluates information and its sources critically and incorporates selected information into his or her knowledge base and value system;
4. Recognizes that accurate and complete information is the basis for intelligent decision making, develops successful search strategies, then organizes information for practical applications and uses information effectively in critical thinking to accomplish a specific purpose;
5. Identifies potential sources of information and understands the economic, legal, and social issues surrounding the use of information and accesses and uses information ethically and legally.

Therefore, there is a challenge: how to ensure that the Tanzanian society becomes information-literate? It is the ultimate challenge to the government, public sector, private sector, development partners, and the whole Tanzanian community though all of the society, that especially the poor people are flooded with information options on day-to-day basis.

CONCLUSION

Tanzania, like other developing countries, is confronted with the challenge of eradicating extreme poverty and hunger. It is acknowledged that the new opportunities which ICTs are opening up can be harnessed in Tanzania's efforts to eradicate poverty. The framework for appropriating ICTs for national development and poverty reduction is articulated in the National ICT Policy.

This chapter has attempted to enhance the understanding and knowledge of ICTs in relation to the National ICT Policy and has extensively explored the evolving frontiers of this pervading technology as it impacts on education, commerce, social, cultural, and economic life of the poor Tanzanian people. The chapter has also presented several show case examples of projects in Tanzania that demonstrate how the government of Tanzania, entrepreneurs, and some of the rural communities are appropriating ICT tools to fit into their needs. Some challenges have been identified and presented with the aim of stimulating debate to reflect on perspectives, trends, and potential of using ICTs to develop innovative approaches and methods for poverty reduction. It is emphasised that, since these ICTs are not solutions to social problems by themselves, they must be carefully chosen and implemented for each appropriate purpose. Overall, the chapter demonstrates that ICTs are a set of tools for knowledge sharing, which is a powerful means for poverty reduction. Further, it is shown that ICTs could only be tools of empowerment for those who have access to them. And that "I" (Information) and the "C" (Communication) are far more important than the various technologies which are just means to an end. Therefore, ICTs cannot turn bad development into good development; they can make good development better. It all has to do with focusing on information literacy as opposed to just focusing on computer literacy.

REFERENCES

Allen, I.G. (2002). *Can we eradicate poverty while tolerating alienation?* Retrieved June 29, 2006, from the Christian Mission for the United Nations Community, http://www.Christianmission-un.com

American Library Association. (2000). *American Library Association Presidential Committee on Information Literacy.* Retrieved June 29, 2006, from http://www.ala.org/acrl/nili/ilit1st.html

Drucker, P.F. (2001). *The essential Drucker.* Harvard Business School.

Gates, B. (1999). *Business @ the speed of thought: Using a digital nervous system.* Warner Books.

Gerster, R., & Zimmermann, S. (2003). *ICTs for poverty reduction: Lessons for donors.* Retrieved June 29, 2006, from http://www.commint.com/strategicthinking/st2003/thinking-187.html

Gillman, H. (2003). *Fighting rural poverty—The role of ICTs.* Paper presented at IFAD side event at the WSIS-Geneva. Retrieved June 29, 2006, from http://www.ifad.org/events/wsis/synthesis/index.htm

Gunawardene, N. (n.d.). *ICT for poverty reduction: Think big, act boldly.* Retrieved June 29, 2006, from http://www.teriin.org/terragreen/issue71/essay.html

Hakikazi Catalyst. (2001). *Tanzania without poverty—A plain language guide to Tanzania's poverty reduction strategy paper.* Retrieved June 29, 2006, from http://www.hakikazi.org/eng/

Hakikazi Catalyst. (2003). *Millennium development goals—No more broken promises*. Retrieved June 29, 2006, from http://www.srds.co.uk/mdg/nmbp-draft-04.pdf

IICD. (2004). *Telemedicine hampered by infrastructure and awareness—says Tanzanian health expert during seminar by SWOPNet*. Retrieved June 29, 2006, from International Institute for Communication and Development (IICD), http://www.iicd.org/articles/iicd-news.2004-09-02.7493425067

Kao, J. (1996). *Jamming: The art and discipline of corporate creativity*. Harper Business Publishers.

Mandela, N. (2005, February 3). *Make poverty history* (Speech during The Global Campaign for Action Against Poverty). London's Trafalgar Square.

Menda, A. (2005a). *ICT experts probe methods in Tanzania to train medics outside hospitals*. Retrieved June 29, 2006, from International Institute for Communication and Development (IICD), http://www.iicd.org/articles/iicd-news.2005-07-15.2614286290

Menda, A. (2005b). *Stakeholders build strategy to integrate ICT in the Tanzanian secondary school*. Retrieved June 29, 2006, from International Institute for Communication and Development (IICD), http://www.iicd.org/articles/iicdnews.2005-07-15.3965031729

Proenza, F.J. (n.d.). e-ForALL: *A poverty reduction strategy for the information age*. Retrieved June 29, 2006, from http://communication.utexas.edu/college/digital divide symposium/papers/index.html

Shaffer, P. (2001). *New thinking on poverty dynamics, implications for policy*. Retrieved June 29, 2006, from http://www.un.org/esa/socdev/poverty/paper_shaffer.pdf

Sharma, M. (n.d.) *Information technology for poverty reduction* (Proposal to Asian Development Bank, Manila, Philippines). Retrieved June 29, 2006, from http://topics.developmentgateway.org/ict

Spence, R. (2003). *ICTs, the Internet, development and poverty reduction—Background paper for discussion, research and collaboration*. Retrieved June 29, 2006, from http://www.mimap.org/

The Guarduan, (3rd February, 2006). *Govt ready to adopt ICTs in health services*. By Gardian Reporter. Retrieved from http://www.ipp.co.tz/ipp/guardian/2006/02/03/59218.html

United Republic of Tanzania (URT). (2000). *Poverty reduction strategy paper (PRSP)*. Dar es Salaam: Government Printer. Retrieved June 29, 2006, from http://www.tanzania.go.tz

United Republic of Tanzania (URT). (n.d.). *IDT/MDG progress—The United Nations and the International/Millenium Declaration development goals (MDG)—on United Republic of Tanzania*. Retrieved June 29, 2006, from http://www.undp.org/mdg/Tanzania.pdf

URT. (1997). *National poverty eradication strategy*. Dar es Salaam: Government Printer.

URT. (2001). *Poverty monitoring master plan*. Dar es Salaam: Government Printer.

URT. (2002). *Small and medium enterprise development policy*. Dar es Salaam: Government Printers.

URT. (2003). *National information communication technologies policy*. Dar es Salaam: Government Printer.

URT. (2003a). *The Cooperative Societies Act—No. 20 of 2003*. Dar es Salaam: Government Printers. Retrieved June 29, 2006, from http://www.parliament.go.tz

URT. (n.d.a). *The Tanzania development vision 2025*. Retrieved June 29, 2006, from http://www.tanzania.go.tz

Weigel, G., & Waldburger, D. (Eds.). (2004). *ICT4D—Connecting people for a better world—Lessons, innovations and perspectives of information and communication technologies in development*. Swiss Agency for Development and Corporation (SDC) and the Global Knowledge Partnership (GKP).

Williams, B.K., Sawyer, S.C., & Hutchinson, S.E. (1999). *Using information technology—A practical introduction to computers & communications* (3rd ed.). Irwin McGraw-Hill.

World Bank. (2004). *2004 world development indicators*. Retrieved June 29, 2006, from http://www.worldbank.org/data/wdi2004/pdfs/table2-5.pdf

Yonah, Z.O. (1999). Orienting engineers in exploiting applied engineering and information technology in Tanzania: Challenges, opportunities and practical solutions. In *Proceedings of the ERB Press Seminar on Engineers as a Resource for Sustainable National Development* (pp. 64-76). Arusha, Tanzania.

This work was previously published in Information and Communication Technologies for Economic and Regional Developments, edited by H. Rahman, pp. 305-319, copyright 2007 by IGI Publishing (an imprint of IGI Global).

Chapter 14
Impact of E-Government Implementation on Poverty Reduction in Rural India:
Selected Case Studies

Sandeep Kaur
PSG College of Arts and Science, India

N. Mathiyalagan
PSG College of Arts and Science, India

ABSTRACT

ICTs in general and e-governance in particular offer tremendous opportunities for improving demand-driven transparent and accountable service delivery targeting the underprivileged. The objective of this chapter is to examine the effects of E-government implementation in the context of widespread poverty in India through an extensive secondary data analysis on selected pro-poor initiatives in reducing poverty and improving rural livelihoods. Analysis also includes various contexts in which these ICT based interventions operate. Specific recommendations are made to involve the socially excluded groups in the design, implementation and access to e-government services. Governments to design appropriate public policies in implementing socially inclusive e-government strategies in the emerging information society draw the conclusion.

1. INTRODUCTION

The new information and communications technologies are among the driving forces of globalization. They are bringing people together, and bringing decision makers unprecedented new tools for development. At the same time, however, the gap between information 'haves' and 'have-nots' is widening, and there is a real danger that the world's poor will be excluded from the emerging knowledge-based global economy. (Kofi Annan, Secretary-General, 2002).

Mahatma Gandhi's vision of the true India in its villages has led several central and state governments to emphasize on grassroots governance for bridging a growing rural-urban digital divide. ICT impact on the poor is at an early stage, but the potential is being demonstrated at the micro, intermediate and macro levels (Hanna, 2003), thus providing enormous opportunities in remote areas in health care, education, other forms of public services.

India is home to 22% of the World's poor with 35% of its billion plus population living on less than US $1 per day; more than 900 million people surviving on incomes less than US $2 per day and poverty being more concentrated among SCs/STs (24 % of the total population of India - 252 million people). Thus, poverty reduction is considered fundamental for the achievement of international goals (Planning Commission, 10th Five Year Plan [2002-2007], Chapter 3.2, Page No 293). The Human Development Report (2005) ranked India 127th among 177 countries; 58th among 103 developing countries on the Human Poverty Index (HPI-1). This call for an extensive and continued efforts from government and international agencies in terms of sound macroeconomic policies; open trade relations; increases in human and physical capital; good governance; sound legal, incentives and regulatory frameworks; an adequately regulated and supervised financial sector; health, education and social services that reach the poor, women effectively; quality infrastructure and public services to promote rural development and livable cities; and policies to promote environmental and human sustainability, thereby, helping in the delivery of social and economic benefits across a broader base of the populace.

Indian ICT sector is witnessing a rapid expansion in telecom markets in the world with a target of 500 million telephone subscribers by 2010; PC penetration of 65 per 1,000 (from the existing 14 per 1,000) by 2008. **ICTs empower the people at the grassroots level to access information and service delivery effectively** is the underlying philosophy of the National e-Governance Action Plan of India (2003-2007) launched by the Central Government at a cost of Rs.12,400 crore ($1.3 billion) to connect 600,000 villages through 100,000 broadband-enabled multipurpose computer kiosks or CSCs by March 2008 through participation by states, government agencies and corporates. This will enable services like e-learning, e-teaching, ehealth, telemedicine, e-farming, e-tourism, e-entertainment and e-commerce in all of India's 600000 villages (Chandrasekhar, 2006).

Country has witnessed an explosive growth in E-governance projects to nearly $1,300 million in 2007 with increase in the government's IT expenditure to over Rs. 5,000 cr. in 2007 (Second Skoch E-governance report, 2005). UN E-governance Readiness Report (2008) ranked India 113th with an E-government Readiness Index Score of 0.3814 against a World Index of 0.4514. According to E-government Readiness Data (2005), India had a Web Index Measure (0.4783), Infrastructure Index (0.0435) and Human Capital Index (0.6195). India ranked 49 in terms of E-participation Index (0.2500). Indian ICT sector is marching towards achieving the MDGs in areas such as literacy, education, gender equity and employment to benefit larger sections of the population. But the basic service delivery challenges include limited access to social services; economic opportunities to rural poor; lack of efficient local service delivery; lack of budgets for services in rural and remote areas; lack of information about entitlements and availability of services; lack of effective communication channels; lack of accountability and transparency.

A country's overall progress in E-government closely correlates with its social, political or economic composition. Weak governance structure of a nation seriously obstructs poverty reduction towards achieving development goals (Country governance assessment for the Asian Development Bank, Asia Foundation, 2003-04). Countries such as India could benefit from E-government if literacy and basic infrastructure can be improved (UN Study, 2003). Only a few developing countries have implemented pro-poor E-governance strategies depending on the government's willingness to develop and design content relevant usable for the rural fabric.

Considering ICTs' potential to alleviate poverty, this chapter overviews the impact of ICT4D projects towards effective governance of rural areas in India by examining secondary data resources on selected pro-poor e-governance initiatives that have impacted the underprivileged and the marginalized in the milieu of extensive poverty in India. To map the effects among rural poor, this chapter studies six innovative E-governance projects that have contributed significantly to poverty reduction, irrespective of its success or failure, thereby resulting in income generation; increasing productivity; generating employment opportunities. Besides involving the poor in the design and execution of these initiatives, this chapter also emphasizes upon their social, technological, economic and political contexts of operations for enhancing rural livelihood strategies. Conclusion will help the government to free up public funding for use in the pro-policy implementation over a long-term sustenance.

2. LITERATURE REVIEW

The principal cause of poverty is the income and consumption in the globalized economy coupled with inefficient service delivery due to poor communication between the central government and decentralized units; and between urban and rural areas. Problems of accountability, large-scale corruption and abuse of public funds obstruct the transfer of allocated funds for education and health to the target beneficiaries. ICT through strengthening decentralized units and financial flows will make basic needs more accessible by the poor by reducing deprivation, inequality and regional bias in trade and development. A well planned ICT network will enhance participation and knowledge sharing among the poor, thereby empowering them by making commerce, knowledge and information accessible to all irrespective of their age, gender, religion or ethnicity.

New ICTs have tremendous potential to address the developmental challenges by reaching the un-reached and alleviating poverty in terms of increasing opportunity, enhancing empowerment, and improving security (The World Development Report 2000/01: Attacking Poverty). IT in India is still in its infancy in terms of socio-economic development due to lack of a long term vision and sustained efforts to support the government's plan and policy. Research reports reveal that the benefits of the globalization for the poor are particularly strong in the cases where inequality is stable or declining. Michael Bailey (2000) suggests long-term domestic strategies and policies that would enable poor farmers in developing countries to benefit from international trade, thereby contributing to rural poverty reduction and achievement of international development targets. International experience has demonstrated the direct relationship between agricultural growth and rural poverty reduction (Timmer, 1997; Hazell & Ramaswamy, 1991).

The Government's Working Group on Information Technology for Masses (2000) identified the following IT relevant applications that could have a profound effect on poverty reduction: e-governance; enabling literacy and education for masses; fulfilling local information needs of the people; and enabling a better economic condition of people. Rural penetration of ICTs enhances pro-poor access to markets, health, and education with a few success stories resulting in delivering desirable social, economic and government services to rural populations by increasing their efficiency and productivity; improving their livelihood and also helping them voice their concerns, demand their rights and participate in decision-making processes.

ICTs and E-government play a critical role in strengthening the linkages between policymakers and the poor and between the poor and the service providers as well as between the policymakers and the service providers (Bestle, 2004). Effective E-governance applications can minimize some of the vital causes of poverty (overpopulation, lack of education, uneven distribution of resources, deep-rooted corruption) through easy information access to the poor, thus resulting in transparency, accountability and responsiveness in service delivery. Such applications aid the poor to reach policymakers with feedback information about the progress contributing to the transparency and accountability in government services. Sustenance of such a system largely depends on a strong political commitment, administrative support and managerial cooperation to successfully implement E-governance for tackling poverty.

ICT has enabled governments to reach out to rural communities, meet citizens' expectations by providing E-government services, thereby improving their quality of life and empowering them through their participation in the political process, and also reaching out to them in order to deliver much-needed public goods and services (UNDP/APDIP, 2004). Effectual pro-poor e-governance delivery strategies necessitates uncomplicated accessibility and availability of user-driven services to proficiently elevate its quality; reduce socio-cultural, political, economic, technological and legal barriers; broaden the range of service providers; empower the users; decentralization and local governance. Such a stratagem always confine to the internal efficiency rather than service deliverance; is urban focused; pilots with established impact on poverty cutback being rarely simulated with minimal focus on the poor or MDG targets; lack understanding of the ICT demand in rural areas with varied access points for service, content, and the nature of needs of vulnerable groups.

Considered as "a powerful engine of rural development and a preferred instrument in the fight against poverty (World Bank, 2005 cited in IDRC 1999), telecenters are now emerging as vital catalysts for disseminating information and knowledge for developing socio-economic prospects for rural communities. ICTs are better served through telecenters to maximize time, effort and resources for bringing relevant information to rural villages, understanding their information needs and capabilities and to deploy future telecenter initiatives to benefit the poor (Duncombe, 2006; Chapman et al, 2004). These can, under certain conditions, help improve the living conditions of rural poor through better and more sustainable livelihood strategies (UN, 2004: 2-3). But accessibility of rural telecenter does not assure its use by the rural poor to elevate poverty. The incorporation and effectiveness of ICTs depend on the acceptance of individuals and organizations (Rivera, 2003). Though a few technology-mediated telecenter models have proven successful in empowering rural communities, the failure and massive under-use of some telecenters set up in developing countries have raised doubts over their relevance for poverty reduction and sustainability (Heeks, 2005). As stated by the Group of Eight Digital Task Force (G8 DOT Force) "efforts to increase access to ICTs should be rooted in a broader strategy to combat poverty" (Harris, 2004). The following conclusions from the current experiences with rural telecenters (Bhatnagar, 2004) are:

- Rural populations are willing to pay a fee for systems that have clear business potential and benefits for them.
- The uptake of electronic services depend on whether significant added value is delivered in comparison with existing ways of receiving information and services.
- Intermediaries are often needed to respond to the specific information needs of rural

citizens, and to interpret and disseminate knowledge from public documents.
- Poor telecom and power infrastructure in rural areas can affect the economic viability of rural kiosks.

Four pillars that need to be addressed for bridging the digital divide can be identified, as follows (Bhatnagar, 2004):

- Applications that draw a large clientele that pays for the service ensuring economic viability of the kiosks.
- Content that empowers rural citizens and enables the formation of communities.
- NGOs and grass-roots organizations that encourage and manage the community-building process.
- Technology that makes rural access inexpensive and robust.

The socio-economic impact of ICT on Poverty Alleviation brings forth enormous qualitative benefits such as e-education; e-marketing; better access to healthcare; e-governance; enhanced transparency; reduced corruption; direct and indirect job creation; increased income for village phone operators and telecommunications operators; marketing opportunities for individual producers and small businesses; access to information relevant to livelihood and production of work; increased access to education and training; empowerment of women through entrepreneurial activity; reduced costs of social and other communications; quality of life for individual users and encouraging the development of local poverty reduction indicators and targets.

Despite the developmental potential of E-governance, few developing countries have implemented pro-poor E-governance strategies depending largely on the Government's willingness to develop and design content relevant and usable for them. The 10th Five-Year Plan (2002-2007) in India had made scant reference to ICT intervention in poverty reduction due to paucity of government efforts at both central and state levels. A comprehensive UNDP-APDIP study by Roger Harris and Rajesh Rajora (2006) titled *"Empowering the Poor Information and Communications Technology for Governance and Poverty Reduction: A Study of Rural Development Projects in India"* examined the application of large-scale approaches to the use of ICTs for E-governance and poverty reduction. The study expressed concern over the factors that hold back the wider rollout of ICTs for poverty reduction in spite of all its technical and human resources at the state and national levels. It was found that such initiatives are not promisingly benefiting the rural populace. Successful projects were mostly expanded into more economically active areas with a stable technology and power infrastructure, leaving behind poorer sections of the population that are more in need. The study suggested a more generalized approach to information and service delivery system applicable for communities with highly varying socio-economic characteristics.

The United Nations Department of Economic and Social Affairs (UNDESA) estimates more than 60 percent of e-government projects in developing countries as failures due to poor understanding of the barriers in the design, implementation and development of E-government systems. These Barriers are the real or perceived characteristics of the social, technological, legal and institutional context that hinder E-government development either through hampering demand by the citizens and businesses for E-government services or through obstructing supply of E-government services by the public sector. E-government services have a variety of economic, social and cultural causes that hinder access to and the effective use of E-government services such as lack of confidence, illiteracy, training, trust, language and citizens' low expectations with respect to the government resulting in the limited use of online information and services (Helen Margetts & Patrick Dunleavy, 2002). E-government does not provide a timely opportunity in the decision-making process for people living in poverty. Being the most frequent users of government services, low-income groups require more information on social policies and programs, but experience greater difficulty in accessing the benefits of E-government. Thus, lack of considerable government efforts in making relevant information reach the poor deprives them from access to e-government services.

The digital divide is one of the crucial global barriers to E-government adoption. India records low Internet and personal computer (PC) penetration rates. World Information Society Report (2006, 2007) reports growth in Infrastructure in India from 0.04 in 2005 to 0.05 in 2006. Digital Opportunity Index (2006) ranks India 124 out of 180 countries surveyed in terms of eleven core ICT indicators. (http://www.itu.int/osg/ spu/statistics/DOI/doi-guide.pdfT.) Several developing countries have attempted to overcome the digital divide problems through programs promoting PC ownership and the provision of public Internet access points. Nonetheless, computer illiteracy, low Internet and PC penetration are still widespread. E-government in many developing nations inadequately represent or lack appropriate strategies and objectives to enhance digital inclusion. Enhancing the reach and effectiveness of services provided to socially excluded groups do reap considerable efficiency gains. Therefore, social inclusion must be a priority in the future development of E-government to reap full benefits of digital transformation.

3. RESEARCH PREROGATIVE

The potential of E-governance as a development tool hinges upon a minimum threshold level of technological infrastructure; human capital and e-connectivity for all. E-government strategies will be effective and "include all" people only if, at the very minimum, all have functional literacy and education, which includes knowledge of computer and internet use; all are connected to a computer and all have access to the Internet. The primary challenge of E-governance for development therefore, is how to accomplish this. Diverse Pro-poor E-governance initiatives are undertaken by Indian government. Six such projects are chosen for this chapter based on extensive secondary data resources fulfilling the following criteria:

a. Analysis of projects based on its duration, objectives, targets, implementers, stakeholders (supply-side in general and demand-side in particular) and current status
b. Contextual examination of poverty-driven e-government projects through STEP analysis (social, technological, economic and political)
c. Scaling up and sustenance of these projects articulating with the stakeholder needs and participation in government-to-citizen interfaces, thereby enhancing their livelihood

Reasons for selection of projects are shown in Table 1.

4. FINDINGS FROM STEP ANALYSIS OF SELECTED CASE STUDIES

Analysis of the selected projects demonstrates the value of projects targeting the poor, the outcome of which serves as models for replication into nationwide approaches and to attain the sustainability of rural telecentres. Development of socially inclusive E-government by identifying relevant context-specific, need-based applications is crucial to the success and sustainability of these initiatives. A suitable technology, the community's development capacity, national policy and infrastructure need to be addressed to make rural telecenters a viable option for poverty alleviation. The study found that e-governance is limited to e-government or e-services with little participation

Table 1.

Initiative	*Initiation*	*Objective*	*Agencies involved*	*Status*
AGMARKNET	-One of the farmer--centric Mission Mode Projects of Govt of India to computerize market-related information (market fees, market charges, better market access and price realization by the farming community) ensuring data reliability; operational and pricing efficiency in agricultural marketing system -Intended beneficiaries include farmers, traders, exporters, processors, planners / decision makers, researchers -Farmers benefit from new and improved agricultural practices, to have weather-forecast-based guidance for timely agricultural operations, to be alerted by satellite surveys of pests and diseases, and to access crop-output forecasting and marketing strategies for domestic and for export trade	-To enhance decision-making processes of the farmers in marketing their produce -To enable poor farmers in developing countries to benefit from international trade, thereby contributing to rural poverty reduction and achievement of international development targets	-Directorate of Marketing & Inspection (DMI); NIC;State Agri Marketing Boards / Directorates; and Mandis (APMCs)	-XI plan strategy predicts expansion of 3700 wholesale markets with an emphasis on ensuring regular reporting of prices and arrivals data by all market nodes -This network has Server Access nodes at 35 State Agricultural and Cooperative Departments, and 600 District Agricultural and Cooperative Offices
DISK	-IT-based initiative to streamline the milk production in the country and to organize milk producers at the grassroots level	-To automate milk collection procedures at the collection centres; reduce the time required in collecting milk; use electronic technology to weigh the milk and measure its fat content; automatically calculates the amount payable to each milk collector; maintain regular records at the Dairy Cooperative Societies (DCS) -To promote local entrepreneurship and support pro-poor market development by setting up information kiosks with strong collaboration at the local level	-GCMMF,localvillage cooperatives and active participation from farmers	-Nearly 600 such systems in operation in the Kheda district in Gujarat with over 700 locations in India -Out of 70,000 village societies in India, 3,500 milk collection centers have been computerized -More than 10.7 million farmers and laborers benefited

continued on following page

Table 1. continued

ITC e-Choupal	-Introduced by the International Business Division (IBD) of ITC Limited in June 2000 and covers over 35,000 villages in the states of Madhya Pradesh, Uttar Pradesh, Maharashtra, Karnataka, Andhra Pradesh, Rajasthan, Haryana and Uttaranchal providing millions of farmers with critical information on farming and to address the fallacies in procurement of agri products such as soybeans in a country where nearly two-thirds of the population's workforce is employed within the agricultural industry, thus bringing about enhanced knowledge sharing, convergence, transparency and trust -Intended beneficiaries are wheat, rice, and pulses farmers	-To demonstrate the role of computers and Internet access in rural farming villages in order to recognize markets and efficiently and effectively increase the agricultural system to benefit and empower the marginalized groups, such as illiterate farmers and women --To catalyze rural transformation to help alleviate rural isolation, create more transparency for farmers, and improve their productivity and incomes	-IBD Group is responsible for procuring, processing, and exporting of agricultural commodities; ITC Ltd finances the establishment of e-Choupals in villages to supplement the traditional mandi system in operation in rural areas	-6,500 eChoupals till date -ITC adding 7 new eChoupals a day and plans to scale up to 20,000 eChoupals by 2012 covering 100,000 villages in 15 states, servicing 15 million farmers - Transactions through these eChoupals may rise to about US $ 2.5 billion by 2010
IVRP	-Initiated in 1998 in Pondicherry, (a centrally administered territory in Southern India), by the MSSRF, a nonprofit organization for sustainable agriculture and rural development -Intended beneficiaries include rural communities for fostering development, social change, cultural values, solidarity, political awareness, community organization and participation	-To enable rural families to access modem ICTs through easy rural access to information related to health, credit, input price and availability, transportation, pest surveillance and agronomic practices -To create awareness in poor communities of the government programmes and entitlements that are available for them from a database of more than 100 such entitlements	-Funded, supported by the IDRC, Canada and the Canadian International Development Agency (CIDA); government partners the project and bears the initial cost of setting up information centers in villages -Implementers are MSSRF and local fishermen colony	-14 out of the 18 Village Knowledge Centres (VKCs) are functional now

continued on following page

Table 1. continued

TARAHaat	-Initiated in 2000 in the resource-deficient and commercially-poor city of Jhansi in the Bundelkhand region followed by kendras in the Bathinda district in the state of Punjab in 2001 to bring computer and Internet technology to rural regions and plans to use these technologies to create revenue streams leading to financial viability for itself and its franchisees -Social objectives include creation of sustainable rural livelihoods; enrichment of the rural Indian economy through improved information flows; education; direct job creation; address issues pertaining to women's health, education, governance, and resource conservation, primarily by providing relevant and easily accessible information on its Web site -Intended beneficiaries are rural folks and students	-Empowerment through access to information and improved access to government services to rural consumers via the Internet to uplift the social rural fabric in India	-Partnership between Development Alternatives (DA), an NGO focused on promoting sustainable development in India, and its rural marketing arm, Technology and Action for Rural Advancement (TARA) as a franchise-based business enterprise	-Has 22 centres presently
Warana Wired Project	-"Bring the world's knowledge at the doorstep of villagers through the Internet" is the stated goal of the Warana "Wired Village" project, being launched in 1998 by the Government of India, the Government of Maharashtra and Kolhapur Sugar Cooperative on a cost-sharing basis demonstrating not only an increase in the efficiency and productivity of the sugar cane cooperative, but also to provide a wide range of information and services in accelerating socioeconomic development of 70 villages around Warana Nagar in the Kolhapur and Sangli districts of the state of Maharashtra -Intended beneficiaries include villagers to access sources of both local and global knowledge and help provide transparency in administration	- to utilize IT to increase the efficiency and productivity of the existing cooperative enterprise; provide agricultural, medical, educational, communication facilities to villagers at facilitation booths in their villages; provide distance education to both primary and higher educational institutes; and establish a Geographic Information System (GIS) of the surrounding 70 villages leading to greater transparency in administration especially in land related matters	-Jointly funded and supported by government of India (50%), government of Maharashtra (40%), Warana Cooperative (10%)	-The sugarcane cooperative is serving about 70,000 farmers across 75 villages

by the beneficiaries in decision-making processes. Few have empowered socially and economically underprivileged groups. Benefits to the women from such initiatives are very few and little effort is made to encourage its use by women as most kiosks are operated by men. Content and services are more geared towards needs of men rather than women. Most initiatives suffer from problems such as power cuts or lack of adequate power, and low-quality connectivity.

Economic liberalization and globalization has resulted in extensive agricultural marketing in India with70% of its population live in villages and depend on agriculture as their main occupation. **AGMARKNET** has succeeded in filling a gap by providing access to information at reasonable cost. This case demonstrates the need to sensitize farmers to respond to new challenges in agricultural marketing by using ICT as a vehicle of extension. Some of the constraints are connectivity in rural areas; training the stakeholders; ensuring data updation in real time frame coupled with erratic power supply; frequent transfers of market personnel trained for reporting data; nontransparent auctions in markets. Regional language content needs to be generated for the benefit of farmers by regularly analyzing data for needed market intelligence.

In order to streamline the production of milk in the country by organizing milk producers at the grassroots level, **computerized milk collection centers** demonstrate that useful content and services from the government and other institutions can be delivered effectively to poor rural populations through information kiosks. It has increased the efficiency and effectiveness of the services delivered to rural farmers. Rural farmers are now willing to invest in technology as useful content and services from the government and other institutions can be delivered to poor rural populations through information kiosks. The effort required for motivating rural communities to use IT, and the logistical problems of maintaining communication links and hardware and software can be mitigated through extensive knowledge creation of local dialects and conditions. The role of small private sector and the committed volunteer is also extremely important. In the dairy sector the district unions are willing to spend because they stand to gain as the system described above increases the efficiency and effectiveness of the services delivered by them to rural farmers. Even though the existing application was very successful, careful analysis provided several opportunities to enhance the value of the system by extending the data base and the capacity to analyze the data. A larger part of the effort went into understanding the needs of the farmers and collection center staff, and in building the content on the portal. The effort required for motivating rural communities to use IT, and the logistical problems of maintaining communication links and hardware and software generally keep the large companies away from such rural ventures. Successful venture of this initiative is attributed to the use of appropriate technology. Building useful content in local languages is absolutely necessary. To enhance the efficiency and effectiveness of the services delivered to rural farmers, it is necessary to build partnerships among development organizations, telecom companies, small IT service companies, and government agencies. The role of the committed intermediaries is also extremely important as most of the current success of rural kiosks has been built around the enthusiasm of the same. Intensive training in operating this system and maintenance of the IT systems need to be enhanced. Although AMCS is priced competitively, the cost and poor financial position is the biggest stumbling block in fast replication. Knowledge of local dialects and conditions can help to mitigate these problems. It is recommended to maintain a database of milch cattle, decision support systems, information on veterinary related issues and innovations to benefit the poor.

E-Choupal is a catalyst in rural transformation by providing easy access to agro information; developing local leadership; creating a profitable

Table 2. STEP analysis of e-governance initiatives for poverty reduction in India

Initiative	Social	Technological	Economic	Political
AGMARKNET	-Location-specific e-Government model for the Poor to foster agricultural growth; poverty reduction -Computerization of market related information such as market fees, market charges, etc., ensure data reliability and efficiency in agro markets, better market access and price realization by the farming community -Help farmers to improve their labor productivity, increase their yields, realize a better price for their produce, cut transaction cost -Farmers benefit from new improved agro practices, weather-forecast-based guidance for timely agro operations, alert by satellite surveys of pests and diseases, crop-output forecasting	-Server Access nodes at 35 State Agricultural and Cooperative Departments, 600 District Agricultural and Cooperative Offices to make rural infrastructure reach upto 6.5 lakhs villages - NIC and IT through DISNIC-AGRIS Project, and AGRISNET seeks to reach all agricultural districts and blocks through its massive Gateway Networks	-Existence and dissemination of complete and accurate marketing information is the key to achieve both operational and pricing efficiency in the marketing system.	-Need for long-term domestic strategies; policies to would enable poor farmers in developing countries benefit from international trade, thereby contributing to rural poverty reduction and achievement of international development targets -Improves the present agro marketing by linking vital agricultural produce markets in the country, the State Agricultural Marketing Boards & Directorates and DMI for effective information exchange. The dissemination of market information is a common function of Agricultural Produce Market Committee (APMC) that maintains the statistics of arrival, sales, prices etc. Prices prevailing in the market are displayed on the notice boards through various media and supplied to State & Central Government -Highly scalable, planned through bottom-up process, and implemented through active involvement and collaboration of agricultural market committee

continued on following page

Table 2. continued

Dairy Information & Services Kiosk	-Over 700 locations in India -Increased transparency; faster processing; immediate payments to farmers -Remarkable business success for private entrepreneurs -Farmers benefited in earnings due to accurate measurement of fat content and weight -Cooperative societies benefited with a fewer employees to manage daily accounts; automatic daily printing of payment slips helps easier communicate with the farmers -Active grassroots participation circulates thousands of liters of milk nationwide everyday; Number of farmers selling to their local cooperative milk collection centre varies from 100 to 1,000; daily milk collection varies from 1,000 to 10,000 litres; collects 700,000 - 800,000 litres of milk every day from milk societies in Anand and Kheda districts -Farmers contribute to the dairy development across the country, subsequently to the national economy -Milk producers no longer exploited by middlemen -Illiterate villagers access information on animal husbandry practices, health and hygiene of cattle, veterinary care, cattle insurance; download government forms, receive documents, order supplies and agricultural inputs from manufacturers -Women's Dairy Cooperative Leadership Program (WDCLP) extended to 50 district unions that include 2,062 cooperative societies; 90,000 women participants. Women's Dairy Cooperative Societies (WDCSs) and Women's Thrift Groups (WTGs) enhanced women's participation --Accurate and up-to-date records reduced fraud or corrupt practices -Reduction in under-payment to farmers, shorter queues at the milk collection centres save considerable time -Increased productivity through extensive education program to reach millions of farmers and dairy workers	-Maintenance of daily computerized accounts reduced chances of errors; fraud -Poor ICT infrastructure, frequent power breakdowns major impediments in its functioning -Building useful content in local languages is absolutely necessary -Necessary to build effective partnerships among development organizations, telecom companies, small IT service companies, and government agencies -Role of intermediary between the computer screen and the rural farmer is extremely important -Larger part of its success is attributed to the use of appropriate technology -Connectivity to a Dairy Portal serve information-transactional needs -DISK and dairy portal have improved delivery of artificial insemination, veterinary services, functional education about the dairy sector	-Cost a major impediment as milk vending system costs around $2,000 per centre, adapted milk-testing machine costs $500 as compared to $4,000 from the original manufacturer -Socio-economic empowerment through a steady, secure flow of income to more than 10.7 million landless laborers and marginal or small farmers -Promoted local entrepreneurship; supported pro-poor market development -Poor financial position of most of DCSs is the biggest stumbling block -Computerized profit and loss accounts and balance sheets also enforce accountability in the system.	-Increased accountability evident from training sessions, workshops, computerized centers. -Cooperative organizational structure enabled the daily milk collection from millions of illiterate, poor, vulnerable farmers across the country -Local organizational capacity of WTGs mobilize the savings of the women members to facilitate loans for animal maintenance or for the purchase of new livestock

continued on following page

Table 2. continued

ITC e-Choupal	-Established a network of 5,100 local computer kiosks in villages to provide access to real-time information to 6,50,000 farmers; plans to cover 15 states, 1,00,000 villages, 10 million farmers in 20,000 kiosks in coming years -Enhances rural social fabric by improving farmers' productivity; crop quality; livelihoods; e-commerce -Solved the problem of farmers' low income; difficulty in accessing credit limits on agro opportunities by partnering with financial institutions -Provides non-cash loans for farm inputs; loans to sanchalak; direct loans to farmers based on sanchalak recommendation; insurance & risk management services etc -Farmers involvement in local content creation helps customize information, develops local leadership quality -Delivers real-time information; customized knowledge to farmer's decision making ability in securing better quality & price, higher incomes -Farmers reimbursed for transport to the procurement hub of e-choupal -Direct marketing channel to eliminate intermediation, multiple handling, reducing transaction cost, making logistics efficient	-Links agricultural research centers; universities; other agencies in developing technologies to overcome the rural technological gap -Technology helps resolve distortion of information; distance to markets	-Impacts the agro chain from seed to money in the bank; incomes from farming risen by about 10% in 2004 alone; support services risen by over 38% since 2000 -Economic empowerment of farmers by connecting to markets and information on agro inputs; products; services to enhance farm productivity; scientific farming practices; market prices of crops (home-abroad); weather broadcast in local (Hindi) language	-Inefficiencies in serving the farmers; agriculture trading companies; the Mandi system under the Agricultural Produce Marketing Act (APMA) -Faces regulatory barriers from prohibited procurements from outside the mandis; waiving the mandi tax on the produce procured through the eChoupal -Government operates these kiosks as per the nature and requirement of the Act to benefit both the market and farmers -ITC continues paying the tax rather than risking the relationships with the government and the mandis
IVRP	-Provides e-knowledge delivery to the poor' -Awareness in poor communities of the govt programmes, entitlements from a database of more than 100 such entitilements -Promotes development, social change, cultural values, solidarity, political awareness, community organization, participation -Information on crop prices and the prices of inputs help farmers to improve their negotiating power against price-fixing middlemen -Creation and updation of relevant local content on agro practices help farmers market their produce more profitably -Enhances Entrepreneurial activity of women and *Dalit* landless labourers -Procurement of paddy seeds and timely paddy price information result in saving time and money to farmers -Eliminated exploitation of agricultural labourers -Gender sensitivity in the context of participation, content creation, and assessment of information needs facilitates the participation of women (about 50%) in operating the centers and disseminating information -Women gain information on issues relating to reproductive health, income-generating activities	-Proactive intermediaries in the village KCs have contributed to its success. -ICTs intervention in a rural setting has empowered the marginalized groups, such as women and the underprivileged. -Building content is a resource-intensive activity and has implications for the sustainability of such projects	-MSSRF helped in mobilizing financial resources for the project, coordinating the villagers to identify their own information requirements and building an information system responding to their information needs	-Citizens' charter data base provides information from government departments, such as current activities and procedures to obtain various entitlements resulting in greater transparency in governance, increasing accountability of officials, minimizing bureaucracy and corruption

continued on following page

Table 2. continued

TARAHaat	-Leading supplier of online information; services; market opportunities; computer-enabled education to the rural Indians -Created sustainable rural livelihoods -TARAhaat.com primarily transmits information; interactive online services such as health, government schemes, livelihoods, agriculture, education, e-governance, entertainment, matrimony, astrology, land records, grievances redressal, complaints to local government officials, local jobs, daily local commodity prices -Improved employment prospects for students; generated confidence among rural children about their future prospects -Women empowerment as both users and operators -Farmers benefited through weather forecasting, procurement services, sales negotiations; younger generation benefit through career counseling, entertainment, educational, career opportunities	-Connectivity and bandwidth are continuing challenge to the venture -Diesel-powered generator as part of the franchise infrastructure adds considerably to costs and maintenance needs	-Need for the franchise model to harness the entrepreneurial drive; local knowledge through a stronger, more resourceful, more locally-responsive network of kendras	-Local governance reforms towards greater democratic participation and clear regulations on Citizen's Charters are an important prelude to ICT applications
Warana Wired Project	-Computerization in the dairy and sugar industries has brought transparency; improved service delivery; permitted efficiency gains to accelerate socioeconomic development of 70 villages around Warana Nagar in the Kolhapur and Sangli districts of the state of Maharashtra -Provision of agro-marketing information, health, educational, communication, government procedure, automated assistance in completing applications for government certificates (ration cards, land records, birth and death certificates),crop information, bus and railway services, grievance redressal benefit the villages -Generates employment opportunities for local population as computer operators -Increased the efficiency and productivity of the existing sugar cane cooperative enterprise, directly impact livelihood -Provides distance education to both primary and higher educational institutes -Computerization result in time and money savings by the farmers on administrative transactions	-Level of human development is relatively high in the sugar cane belt of Maharashtra -Good infrastructure and local organizational capacity is praise worthy	-Socio-economic challenges are still predominant faced by the WGC in terms of low sugar prices and lack of coordination, efficiency and transparency among the cooperatives -Fertilizer stock inventories are now better managed with savings of about US$750,000 to the cooperative -Empowered community adopting ICT to bring efficiency and transparency gains to its main economic activity	-Establishment of a GIS of the surrounding 70 villages lead to greater transparency in administration especially in land related matters -Lack of local staff participation in software development, implementation process -Irregular updation of information on sugar cane growing and agricultural prices since 1998 by local staff -Provide transparency in administration -Access to government services through the Internet benefit people with currently grievances about government services solved through the local village committees (panchayats). Thus, its been perceived useful in terms of sending grievances, downloading government forms, and accessing land records

distribution; alleviating rural isolation; improving productivity and income; creating transparency for farmers. E-Choupal case have shown that provision of ICT services in rural areas enhances the quality of life of the rural Indians and make profit for the service providers, both these companies and their local partners. Local content with appropriate software development, regular updates of prices, local ownership and participation ensure long term sustenance. This system has tackled the challenges posed by the unique features of Indian agriculture, characterized by fragmented farms, weak infrastructure and the involvement of numerous intermediaries. ICT interventions enabled ITC to directly negotiate product sale by making farmers obtain information on mandi prices, good farming practices and place orders for agricultural inputs like seeds and fertilizers. There is a recorded rise in their income levels, improvement in quality of output and a fall in transaction costs. Customized knowledge is disseminated to the farmers free of cost. This case has successfully eliminated non-value adding activities by differentiating product through identity preservation; value added products traceable to farm practices;e-market place and support services to future exchange. Critical factors in its apparent success are managerial competence to execute a complex project and manage cost and ITC's extensive knowledge of agriculture in order to retain existing production system, including maintenance of local partners, the company's commitment to transparency, and the respect and fairness with which both farmers and local partners are treated. Infrastructure creation and maintenance to eliminate market inefficiencies by creating valued services delivered to hitherto underserviced markets will drive its long term sustenance. Some of the constraints are poor telecom infrastructure in villages, ICT illiteracy, low trust on electronic system, selection of an appropriate sanchalak, improper knowledge about rural market, vicious circle of intermediaries, improper and complex user interface on e-choupal, lack of rules and regulation related to electronic choupal, mistrust about inspection, testing and weighing of produce on centres.

IVRP demonstrates the perceived role of ICTs in sustainable agriculture and rural development through e-knowledge delivery to the poor. Replication of such projects require the backing of facilitating conditions as the rural population lack necessary expertise to demand and design a knowledge center on its own. Analysis reveals that villages are not interested in the technology, but in accurate, dynamic and customized information. Building content is a resource-intensive activity and has implications for the sustainability of such projects.

TARAhaat has succeeded in empowering the rural poor; inspiring confidence and higher aspirations among rural children; enabling farmers to gain market information and substantially higher prices for their crops besides harnessing local entrepreneurial energies. Human capacity constraints, pose a challenge. Illiteracy; limited technology experience and low disposable income among its intended users pose continuing challenge to its business strategy. TARAhaat's operational challenges broadly relate to infrastructure development at the franchise level, product development at the enterprise level, and management skills required at both levels. The franchise model should harness the entrepreneurial drive and local knowledge of franchisees through a stronger, more resourceful, and more locally-responsive network of kendras. Understanding and delivering products as per the needs of rural customer base is vital for its long term sustenance. TARAhaat's unique combination of commercial and NGO characteristics is mandatory and depends on societal and environmental factors for its success and replicability. Effective training and management support to its network of franchised TARAkendras will help in providing standardized services. Lack of government support; lack of motivation of farmers; miniscule returns from the rural kiosks compared to the investments made by private entrepreneurs; lack of appropriate content and services; heavy dependence on e-commerce activities pose serious challenges to its organizational, social and financial sustainability. Poor back-up arrangements coupled with lack of backward integration has made this financially viable rural network redundant. This case strongly recommends that the financial viability for the project partners and private entrepreneurs is crucial for kiosk expansion. Functional relationships need to be built with government departments through which numerous e-government services can be delivered to the masses. e-Education and e-health services needs to be outsourced. It is recommended that more community practitioners than technologist be employed in the project.

Warana kiosk case demonstrates the need for community information need as a crucial factor while conceptualizing e-government applications to directly impact the rural livelihood. Effective coordination with the local self government agencies has impacted G2C service delivery through these kiosks. Some of the constraints are lack of women's involvement wherein women generally visit information kiosks to obtain sugar factory services; but only men using the Internet. Warana information kiosks are mostly accessed by farmers who own their land. Non-relevant content for a majority leads to further marginalization wherein the poorest, landless labourers are excluded in their access to e-government as they were unconnected with sugar cane growing and harvesting. Warana has improved the standard of living of the community, especially of rural youth by providing employment opportunities. Providing necessary incentives can enable grassroots operators to become ICT champions in their villages, easing access to information for farmers, providing training to children, and creating new economic opportunities through software development. This case calls for motivated and skilled grassroots intermediaries in bringing e-government to rural communities.

5. TOWARDS SOCIALLY INCLUSIVE/ PRO-POOR E-GOVERNANCE STRATEGY IN INDIA

Implementing pro-poor e-governance is a complex phenomenon. Most e-government for development projects fail, particularly large scale e-government projects with high costs, due to enormous gaps between project design and on-the-ground reality, known as "design-reality gaps" (Heeks, 2003), besides the lack of a supporting organizational and institutional reform as well as poor strategic planning and implementation. Pro-poor strategy in India enhances opportunities for local service delivery to the poor as part of the national poverty reduction strategy and a way towards achieving the MDGs. Successful implementation of E-government initiatives for the poor depend on building trust, a sense of ownership, ensuring transparency, accountability and commitment and building networks. Strategic collaborative planning focuses on how different stakeholders can interact more effectively to add value, ensure that there are no gaps in service delivery, minimize duplication and enhance synergy (ESCAP 2004).

Capacity building, participatory planning for civil servants, local government and communities is vital in designing, implementing and utilizing pro-poor E-governance services. The intended impact on the poor and the vulnerable need to be examined by regular monitoring and evaluation of the existing projects and programmes as well as planning future interventions followed by a feedback mechanism for stakeholders and beneficiaries of the projects. This helps to create and sustain the accountability, transparency and effectiveness of service delivery. Visions and goals should be clearly articulated in terms of socio-economic development and quality of life enhancement for all members of society, including the marginalized poor. Stakeholders are a valuable resource for a sustained E-government function.

6. POLICY RECOMMENDATIONS FOR SOCIALLY INCLUSIVE E-GOVERNMENT

Technology is most effectively deployed in pursuit of some predefined developmental outcome, rather than as an outcome itself. A well-conceived project will be more likely to deliver useful benefits to its target audience in a sustainable manner. Confidential information provided to the study team from a recent project for the Organisation for Economic Co-operation and Development (OECD) found that E-government initiatives targeted on, or particularly useful to, socially excluded groups achieved the highest returns on investment amongst E-government initiatives for citizens. Considering lack of E-government strategies for socially excluded, following recommendations are drawn:

- For socially excluded groups to benefit from digital transformation, local government needs to seize the opportunities that ICT provides by:
  - Making better use of information and data to understand the distribution and magnitude of social exclusion problems and to use this knowledge to underpin social inclusion strategies
  - Greater legal clarity and promotion of the benefits of information sharing between public organizations
  - Making better use of information, ensuring that key staff who work with these individuals/groups are informed and trained so they understand and therefore use their knowledge more effectively to target initiatives and to provide personalized services
- Central government objectives and goals need to emphasize the role of technology and E-government in meeting the needs of socially excluded groups. Action is required to ensure:
  - Social inclusion is a priority in the future development of the E-government agenda
  - Central government develops clearer objectives to support and guide the use of technology to enhance social inclusion
  - Goals for E-government emphasize the provision of better quality services for users as well as increased efficiency for service providers

ICT driven poverty reduction and E-governance projects need to remain financially sustainable. The closer that it comes to self-financing sustainability, the better incentives need to be generated to stimulate people living in poverty to use and benefit from E-government.

7. CONCLUDING REMARKS

This chapter examined the effects of E-government implementation in the context of widespread poverty in India through an analysis of six ICT initiatives in rural areas. Findings call for E-government strategies to target poverty for efficiently serving those who lack access to relevant information and are less likely to participate in government. Lack of integrated E-government strategies will provide less opportunity for these communities to indirectly or directly participate in government, consequently widening the digital divide affecting the development of democracy and its institutions. Equitable access and use of E-government services requires a broader perspective by strengthening the strategies to include socially excluded. Government agencies should creatively extend the use of their information networks and replace current perspectives with a more decentralized focus that translates into the broader reach of their services. Government organizations should implement innovations to explore how citizens use and perceive Internet.

High rate of access to technological devices (e.g., cellular phones) in rural areas must be considered by government to integrate strategies that incorporate the use of other devises to encourage increased use of e-government services to make E-governance a reality in India.

REFERENCES

2007 Asia - Telecoms, Mobile and Broadband in India: http://www.researchandmarkets.com/reportinfo.asp? report_id = 481013

ADB (2003). *Toward E-Development in Asia and the Pacific: A Strategic Approach to Information and Communication Technology*, www.adb.org, accessed 27 May 2007.

Akula, V.B. (2000). *Putting Technology to Work for Poverty Alleviation: A Draft Proposal for $151,030 to Develop Smart Cards for Microfinance,* Swayam Krishi Sangam, Hyderabad.

Asia-Pacific Journal on Information, Communication and Space Technology: Reviews and Updates. First Issue – 2006, Economic and Social Commission for Asia and the Pacific

Baramati Initiatives (2001). *SKS-Smart Cards: Case Study*. www.baramatiinitiatives. com/cases/case10.htm.

Best, M., & Maclay, C.M. (2002). *Community Internet Access in Rural Areas: Solving the economic sustainability puzzle*. The Global Information Technology Report 2001-2002: Readiness for the networked world, ed., Geoffrey Kirkman (Oxford University Press, 2002). Available online at http://www.cid.harvard.edu /cr/gitrr_030202 .html (May 20, 2007)

Bhatnagar, S. (2000). Empowering Dairy Farmers through a Dairy Information & Services Kiosk. Washington, DC: World Bank.

Bhatnagar, S. (2004). *E-government: from vision to implementation – a practical guide with case studies*. New Delhi: Sage Publications (pp. 17, 53, 54, 73).

Bhatnagar, S., & Schware, R. (Eds.) (2000). *Information and Communication Technology in Rural Development. Case Studies from India*. Washington, DC: World Bank.

Bhatnagar, S., & Vyas, N. (2001). Gyandoot: Community-Owned Rural Internet Kiosks. Washington, DC: World Bank.

Cecchini, S. (2001). *Back to Office Report: Information and Communications Technology for Poverty Reduction in Rural India, Mimeo*. Washington, DC: World Bank.

Cecchini, S. (2002). *Back to Office Report: Evaluation of Gyandoot and Bhoomi and International Conferences on ICT for Development, Mimeo*. Washington, DC: World Bank.

Cecchini, S., & Raina, M. (2002, April). Warana: *The Case of an Indian Rural Community Adopting Information and Communications Technology in Information Technology in Developing Countries, 12*(1). www.iimahd.ernet.in/egov/ifip/apr2002/apr2002.htm.

Cecchini, S., & Shah, T. (2002). Information and Communications Technology as a Tool for Empowerment in Empowerment and Poverty Reduction: A Sourcebook. Washington, DC: World Bank. www.worldbank.org/poverty/empowerment/

Centre of E-governance. (2003, June). IIM Ahmedabad, Evaluation Studies by the Centre for E-Governance. Indian Institute of Management, Ahmedabad (CEG-IIMA). *Information Technology in Developing Countries, 13*(1). Available online at http://www.iimahd.ernet.in/egov/ifip/jun2003/ article6.htm

CRISP Group (2003, July). National Informatics Centre, Department of IT, Ministry of Communications & IT, Government of India, Information

Needs Assessment for Rural Communities- An Indian Case Study. Available online at http://ruralinformatics.nic.in/files/4_12_0_229.pdf

Critical Success Factors for Rural ICT Projects in India: A study of n-Logue kiosk projects at Pabal and Baramati. Submitted in partial fulfillment of the requirements for the degree of Master of Management By Vivek Dhawan, IIT Bombay.

Edwin, L. (2003, November). Project Leader, OECD E-Government Project. Challenges for E-Government Development. *5th Global Forum on Reinventing Government*, Mexico City.

Gartner estimate rates India as the fourth largest IT spender. The Economic Times, 13 July 2003. http://www. infochangeindia .org/archivesl.jsp?secno=9&monthname=September&year=2003&detail=T

Handbook on Implementing E-Government with basic introduction to E-Commerce: Developed by D.Net - Development through Access to Network Resources.

Harris, R.W. (2004). Information and Communication Technology for Poverty Alleviation. *E-primers for the Information, Economy, Society and Polity*, Kuala Lumpur (UNDP-APDIP), (pp. 21, 35).

Heeks, R. (1999). *Information and Communication Technologies, Poverty and Development, Development Informatics Working Paper Series, Paper No. 5.* www.idpm.man.ac.uk/idpm/di wp5.htm

Heeks, R. (2003). *Most e-Government-for-Development Projects Fail: How Can Risks be Reduced*? University of Manchester, United Kingdom: IDPM. Accessed from http://www.egov4dev.org.

http://dinkaunc.blogspot.com/2004/09/impact-of-ict-on-poverty-reduction.html

International Telecommunication Union, Global Survey on Rural Communications. (2004, June). *ITU News, No. 5.* www.itu.int/itunews, Accessed 28 May 2007.

James, J. Pro-Poor. (2000). *Modes of Technical Integration into the Global Economy in Development and Change, 31*, 765–783.

Kaur, S. & Mathiyalagan, N. (2007, December 28-30). Adoption of e-governance: Social dimensions of e-government: Poverty perspective. In *Proceedings of the 5th International Conference on E-Governance,* Hyderabad, India.

Keniston, K. (2000). *The Four Digital Divides.* Available online at http://web.mit.edu/~kken/Public/ PDF/Intro_Sage_1_.pdf (May 11, 2007)

Keniston, K. (2002). *Grassroots ICT Projects in India: Some Preliminary Hypotheses ASCI Journal of Management, 31*(1&2). Available online at http://web.mit.edu/~kken/Public/PDF/ASCI _Journal_Intro __ASCI_ version_. pdf (May 11, 2007)

Keniston, K. (2007). *IT for Masses: Hope or Hype*? Available online at http://web.mit.edu /~kken/Public/PDF/EPW_paper.pdf (May 11, 2007)

Kenny, C., Navas-Sabater, J., & Quiang, C. (2001). *Information and Communication Technologies and Poverty in World Bank Poverty Reduction Strategies Sourcebook.* www.worldbank.org/poverty/ strategies/ict/ict.htm

Margetts, H., & Dunleavy, P. (2002). Cultural Barriers to EGovernment, published by the National Audit Office, April 4, 2002 (HC 704-III) in conjunction with the Value for Money report "Better Public Services Through EGovernment."

McNamara, K.S. (2003). *Information and Communication Technologies, Poverty and Development; Learning from Experience.* A background paper for the InfoDev Annual Symposium. The World Bank.

Milestones in India's Internet Journey (2007). http://www.amitranjan.com/category/5/

Mission 2007 India http://www.mission2007.org/

OECD (2005). *The contribution of ICTs to Achieving the Millennium Development Goals (MDGs).* Forthcoming. Good Practice Paper on ICTs for Economic Growth and Poverty Reduction. http://www.oecd.org/dataoecd/15/38/34662239.pdf, accessed 23 May 2007.

Recommendation (2004*). Record of the Committee of Ministers to member states on electronic governance.* Adopted by the Committee of Ministers on December 15, 2004 at the 909th meeting of the Ministers' Deputies.

Singh, S. H.(2000, September 26). Ways and Means of Bridging the Gap between Developed and Developing Countries. High-Level Panel on Information Technology and Public Administration, United Nations, New York.

Sociocultural Barriers to E-Government: An Analysis from a Poverty Perspective: María Inés Salamancal, FLACSO Chile, DIRSI (Regional Dialogue on the Information Society), funded by the IDRC.

Sood, A.D. (2003, May-June). The Kiosk Networks: Information nodes in the rural landscape, I(1). Available online at http://www.i4donline.net/issue/may03/aditya_full.htm

Spence, R. (2003). *Information and Communications Technologies (ICTs) for Poverty Reduction: When, Where and How*? OECD International Development Research Center (IDRC). http://network.idrc.ca/uploads/user-S/ 10618469203RS_ICT-Pov_18_July.pdf

The E-Government Divide: Mutual Effects of Digital Divide and E-Government. (2003, October 12-19). *World Congress on Engineering and Digital Divide*, Tunis.

UNDP (2006). *Human Development Report.* India.

UNDP-APDIP (2004). *ICT for Development: A Sourcebook for Parliamentarians* (p. 39), New Delhi: Elsevier.

World Bank (2003). *ICTs and MDGs: A World Bank Perspective*. www.worldbank.org, accessed 26 May 2007.

World Bank Group (2004). *World Development Report 2004: Making Services Work for Poor People* (p. 7). Washington, D.C., World Bank.

World Bank. (2001). *World Development Report 2000/01: Attacking Poverty.* Oxford University Press. www.worldbank.org/poverty/wdrpoverty/.

WEB SITES

www.ejisdc.org/ojs2/index.php/ejisdc/article/view/462/230

http://www.eapf.net/ casestudies/in/agmarket.asp

www.digitalopportunity.org/article/view/113259/

http://www.unapcict.org/ecohub/case-studies/anand-milk-collection-centres-anand-gujarat

http://web.worldbank.org/WBSITE/EXTERNAL/TOPICS/EXTINFORMATIONANDCOMMUNICATIONANDTECHNOLOGIES/EXTEGOVERNMENT/0,,contentMDK:20486020~menuPK:1767268~pagePK:210058~piPK:210062~theSitePK:702586,00.html

http://poverty2.forumone.com/library/view/14655/

www.apdip.net/resources/case/in16/view

http://www.digitaldividend.org/case/case_echoupal.htm

http://www.expresscomputeronline.com/20070702/management02.shtml

http://www.thehindu.com/2007/11/28/stories/2007112862051900.htm

unpan1.un.org/intradoc/groups/public/documents/UN/UNPAN028607.pdf

www.igidr.ac.in/~susant/PDFDOCS/Thomas2003_commoditiesmkts_wb.pdf

download.microsoft.com/download/0/5/d/05dd56a0-1fec-4490-84da-ff2c30974152/Agmarknet.doc

www.patelcenter.usf.edu/assets/doc/IPSA-Singh.doc

India targets PC penetration of 6.5% by 2008: http://mungee.org/archives/2005/05/19/india-targets-pc-penetration-of-65-by-2008/.

India: e-governance market to have an explosive growth by 2007: http://www.digitalopportunity.org/article /view/125575/1/

This work was previously published in E-Government Development and Diffusion: Inhibitors and Facilitators of Digital Democracy, edited by G. Sahu, Y. Dwivedi, and V. Weerakkody, pp. 80-100, copyright 2009 by Information Science Reference (an imprint of IGI Global).

Chapter 15
The Development of an Organic Agriculture Model:
A System Dynamics Approach to Support Decision-Making Processes

Črtomir Rozman
University of Maribor, Slovenia

Andrej Škraba
University of Maribor, Slovenia

Miroljub Kljajić
University of Maribor, Slovenia

Karmen Pažek
University of Maribor, Slovenia

Martina Bavec
University of Maribor, Slovenia

Franci Bavec
University of Maribor, Slovenia

ABSTRACT

This article describes the problem state of organic farming development and procedures for modeling by the means of system dynamics, with emphasis on the organic products market. The modeling principles are described in the following steps: problem state formulation, development of causal loop diagrams, model development, scenario analysis and formulation of acceptable strategies. Basic structures developed by the system dynamics principle are presented. The concept of archetypes in the field of organic agriculture modeling is described. The simulation scenarios are formulated as a case study for the Slovenian organic agriculture.

INTRODUCTION

Organic farming is the most environmentally valuable agricultural system, and has strategic importance at the national level that goes beyond the interests of the agricultural sector. The development of ecological agriculture countrywide can be regarded as a long-term process of economic activities, land use, population growth, material energy-information flows, and interactions between humans and nature that satisfy regional sustainable development demands. It emphasizes the sustainable use of internal resources (in the ecological, economic, and social senses), rather than external (or extra-regional) flows, to support long-term agricultural development. In particular, it emphasizes intensive use of renewable resources, which allows the recycling of matter, which is a basic requirement of ecological farming. This alternative agricultural paradigm may provide the link between the objectives of sustainable resource use and sustainable regional development.

The consequences of policies are long-term and irreversible. In this light, the system approach for evaluation of development policies for organic farming must be developed. The decision support and system analysis in organic farming has been described by Pažek et al. (2006), Rozman et al. (2006), and Pažek et al. (2006). System dynamics methodology (Forrester, 1961; Sterman, 2000) can be used as an alternative to econometric and mathematical programming approaches when modeling agricultural systems for policy evaluation (Bockerman et al., 2005; Elshorbagy et al., 2005; Saysel et al., 2002). For our research, the most important work are presented by Shi and Gill (2005) and Kljajić et al. (2002, 2003). Shi and Gill are developing a simulation model based on system dynamics (SD) for the ecological agricultural development of Jinshan County, China. Kljajić et al. produced an integrated SD model for the development of the Canary Islands, where the main interactions between agriculture, population, industry and ecology were taken into consideration. The preliminary research in the area of system dynamics SD simulations of organic farming development was conducted by Rozman et al. (2006), Rozman et al. (2007), and Rozman et al. (2008). Yet this research did not incorporate the aspect of the food market fully.

The aim of this study was to develop an SD-based model for the development of organic agriculture in the Republic of Slovenia, with emphasis on the aspect of food markets. The present article will describe the problem state of organic farming development and the procedures for modeling it by the means of SD, with an emphasis on organic food market development. The modeling principles are described by the following steps: problem state formulation, model development, scenario analysis and formulation of acceptable strategies. Basic structures important for organic agriculture developed by the SD principles are presented. The simulation scenarios are presented as a case study for the Slovenian public's assessment and decision.

THE SYSTEM STRUCTURE AND MODEL DEVELOPMENT

There are approximately 80,000 farms in Slovenia, both conventional and organic. In the year 2006, only 1,728 farms were in the organic farm control system. Even though a subsidy has been offered to the farmers, the proportion of the organic farms is still low, not higher than 5% of all farms. The short term strategic goal is to reach 10% or 15% by the year 2015. This was determined by the state action plan, ANEK (Majcen & Jurcan, 2006). The long-term plan considered the feasibility of a complete conversion from conventional farming to organic farming. Our previous research (Rozman et al., 2007) demonstrated that conversion to organic farming relied on subsidies to provide the main means of conversion from conventional farming to organic farming, and that the negative feedback loops in the systems were dominant. The system was thus kept in the unwanted equilibrium state that prevented the achievement of the strategic goal (15% of farms were to be organic). However, subsidies are not the only driving force in the system; even more important are other activities that promote organic farming. The SD model used for the simulation of organic farming development considered key variables such as the numbers of conventional and organic farms, conversion, subsidies, the promotion of organic farming, marketing, market development, education, the organization of a general organic farming support environment, system self-awareness, and delay constants of process changes.

The conversion to organic farming is, however, also influenced by organic food market development (demand for organic food on the market) and the productivity of organic farm production. Initially the demand for organic products is higher than the supply, resulting in high prices of organic food and initiating higher conversion to organic farming. However, the supply increase would cause the expected prices to fall, and that would negatively influence the conversion to organic farming. This loop keeps the system in an equilibrium state preventing achieving higher shares of organic farms. Therefore, we upgraded the model with additional variables in order to fully incorporate the mechanism of the organic food market.

Figure 1 shows the Causal Loop Diagram (CLD) of the organic farming marketing system. Here, the emphasis is put on the demand-supply mechanism. Loop Ⓐ represents the negative feedback loop in which an increase in organic food capacity leads to a higher production ratio, which represents the ratio between desired production and actual production. A higher production ratio yields a lower market price, which decreases organic farms production, since the price is not as attractive for producers as previously. As the organic farms production decreases, food supply also decreases, leading to a stabilizing effect on the organic food capacity. This is the control mechanism of production and prices. As the prices get lower due to higher production, the intensity of production moderates. Loop Ⓑ represents the negative feedback loop in which an increase in organic food capacity leads to increased customer demand. This leads to increased sales, which again diminishes the organic food capacity, in the sense of stock depletion. Loop Ⓒ represents a negative feedback loop. As the production ratio increases, this leads to a decrease in market price, which is a characteristic manifestation of negative feedback loops.

Figure 1. Causal loop diagram of organic farming marketing system

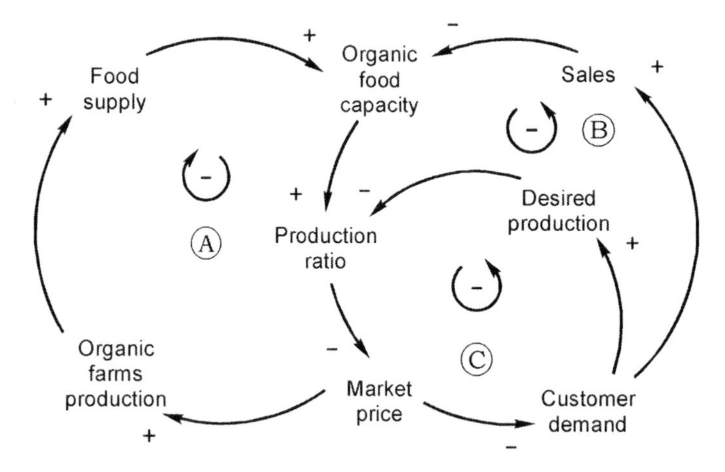

The structure of the SD model is shown in Figure 2. There are six levels of elements applied in the model. The variable "conventional farms" represents the number of conventional farms; the variable "organic farms" represents the number of organic farms. These are the farms that are in the control system of one of the control organizations. The other two important levels in the model are the number of customers who buy organic food and the number of customers who buy conventional food. Those two variables represent market shares of organic and conventional food. The "market price" is also modeled as a level element in order to demonstrate the inherent time delays of long-term price changes. There are two flows in the SD model: conversion from conventional to organic farms and conversion of consumers of conventional food to organic food. Figure 2 shows the model diagram and key connections between major variables in the system. The model consists of 28 variables and 39 links (see Figure 3).

SCENARIOS AND SIMULATION RESULTS

The model can be used for strategic planning at the national level. Table 1 shows the parameter values for the seven scenarios performed on the developed model. The initial scenario is set with the productivity of organic farms equaling 12. This productivity represents the number of organic customers that can be supplied by one farm. Scenarios 2 and 3 assume that there is an

Figure 2. Model diagram

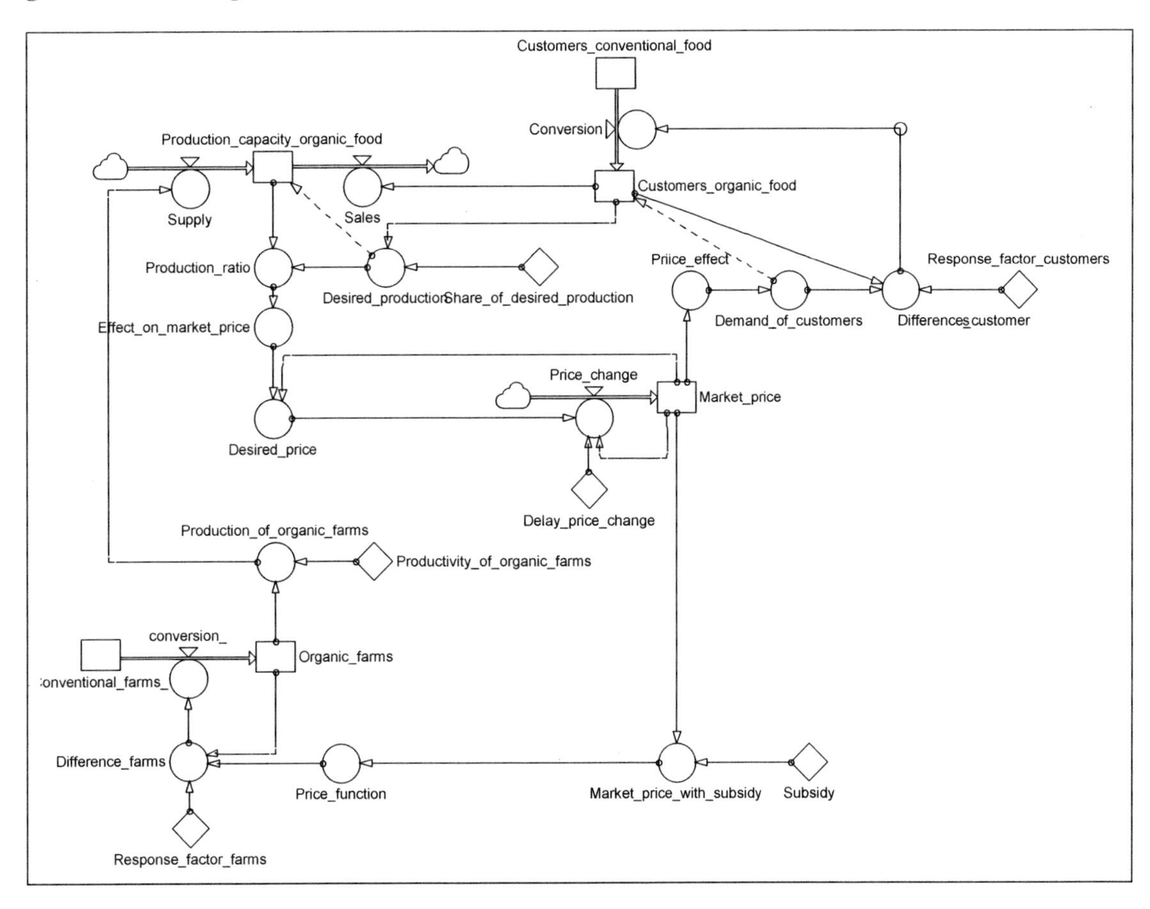

increase in subsidies. Scenarios 4 and 5 simulate the effect of an increase (decrease) in the changing of price delays. Scenario 6 assumes an increase in organic farm productivity at an initial subsidy level, while scenario 7 assumes that there are no subsidies for increased organic farm productivity. The initial value of the levels elements are set as follows:

Figure 3. Model equations

```
init C      onventional_farms_ = 78000
flow        Conventional_farms_ = -dt*Conversion_farms
init        Customers_conventional_food = 1976000
flow        Customers_conventional_food = -dt*Conversion
init        Customers_organic_food = Demand_of_customers
flow        Customers_organic_food = +dt*Conversion
init M      arket_price = 50
flow M      arket_price = +dt*Price_change
init O      rganic_farms = 2000
flow        Organic_farms = +dt*Conversion_farms
init P      roduction_capacity_organic_food = Desired_production
flow        Production_capacity_organic_food = -dt*Sales
            +dt*Supply
aux         Conversion = Difference_customers
aux         Conversion_farms = Difference_farms
aux P       rice_change = ((Desired_price)-Market_price)/Delay_price_change
aux         Sales = Customers_organic_food
aux         Supply = Production_of_organic_farms
aux D       emand_of_customers = Priice_effect+1
aux D       esired_price = Effect_on_market_price*Market_price
aux D       esired_production = Customers_organic_food*Share_of_desired_production
aux         Difference_customers = Response_factor_customers*(Demand_of_customers-Customers_organic_food)
aux         Difference_farms = Response_factor_farms*(Price_function-Organic_farms)
aux E       ffect_on_market_price = GRAPH(Production_ratio,0.5,0.1,[2.07,1.77,1.52,1.32,1.14,1,0.88,0.78,0.7,0.63,0.57"Min:0;Max:2;Zoom"])
aux         Market_price_with_subsidy = Market_price+Subsidy
aux         Price_function                                                                                      =
GRAPHCURVE(Market_price_with_subsidy,0,10,[0,86.2,221.3,554.3,1307.5,2725.6,4661.3,6402.2,7470.5,7980.4,8194.6"Min:0;Max:10000"])
aux         Priice_effect                                                                                       =
GRAPHCURVE(Market_price,5,10,[997746,518040,232332,95918,37357,13908,4991,1736,588,194"Min:0;Max:1000000;Zoom"])
aux P       roduction_of_organic_farms = Productivity_of_organic_farms*Organic_farms
aux P       roduction_ratio = Production_capacity_organic_food/Desired_production
const       Delay_price_change = 7
const       Productivity_of_organic_farms = 18
const       Response_factor_customers = 0.8
const       Response_factor_farms = 0.5
const       Share_of_desired_production = 5
const       Subsidy = 0
```

Table 1. Scenario parameters

Parameter Scenario	Subsidy	Delay price change	Productivity of organic farms
SC1	3	7	12
SC2	5	7	12
SC3	7	7	12
SC4	3	10	12
SC5	3	3	12
SC6	3	7	18
SC7	0	7	18

number of organic farms = 2,000
number of conventional farms = 78,000
number of organic customers = 24,000

Figure 4 shows the results of the seven scenarios with respect to the parameter "number of organic farms". The highest values are achieved at scenarios 3 and 2 with increased subsidies, followed by scenarios 1, 4, and 5. The increase of

Figure 4. Number of organic farms

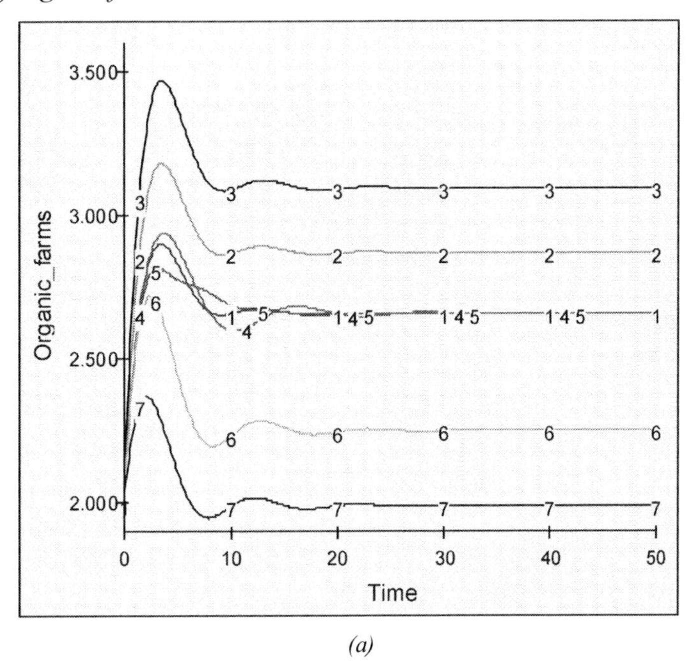

(a)

Number of "organic customers"

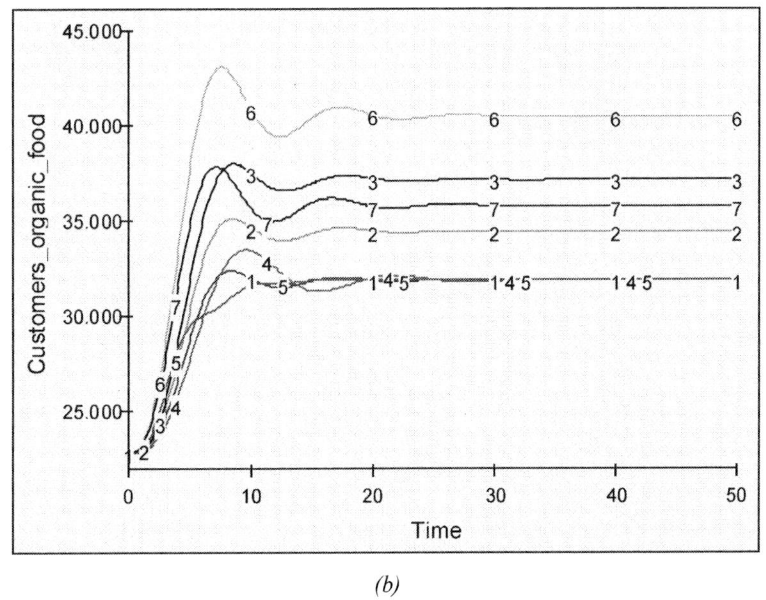

(b)

organic farm productivity results in lower values (scenarios 6 and 7). The small decline after the initial fast growth can also be observed caused by the negative feedback loop in the system: an increased supply of organic food on the market negatively results in lower prices (confirmed also in Figure 5) and influences conversion of farms negatively.

A slightly different response for the seven simulated scenarios can be observed for the parameter "number of organic customers". The highest values are achieved in scenario 6 (an increase of organic farm productivity with an unchanged level of subsidies).

The market price of organic food is declining in the long term for all the scenarios. The highest decline is at an increased level of farm productivity (scenario 6) and at the highest subsidy levels (scenario 3), followed by scenarios 7 and 2. The increase in productivity of organic farms by 50% (from 12 to 18) increases the supply, resulting in lower market prices. Similar results are achieved at higher subsidy levels.

Figure 6 shows the conversion dynamics for each scenario. Negative numbers mean that some organic farms convert back to conventional farming.

The oscillations in the initial phase of the response are caused by the immanent delay structure of the considered system. In order to reach the desired production, the production system has to adapt. This represents the delay in the system response. The second delay is present in the price adaptation mechanism. The prices change according to the market perception, which is represented by the parameter "delay price change" (initially set at a value of 7). Further simulations reveal that lower values in "delay price change" result in less intensive oscillations (Figure 7). Figure 7 demonstrates the response of the "number of organic farms" to the "delay price change" parameter set at 1 (line 1) and at 15 (line 2).

The results clearly show that the subsidies from this perspective are the major driving force stimulating the conversion to organic farming. How could that be changed? According to the results of the preliminary model (Rozman et al., 2007), subsidies are not the only driving force in the system; even more important are other activities that promote organic farming. Thus, future

Figure 5. Market price

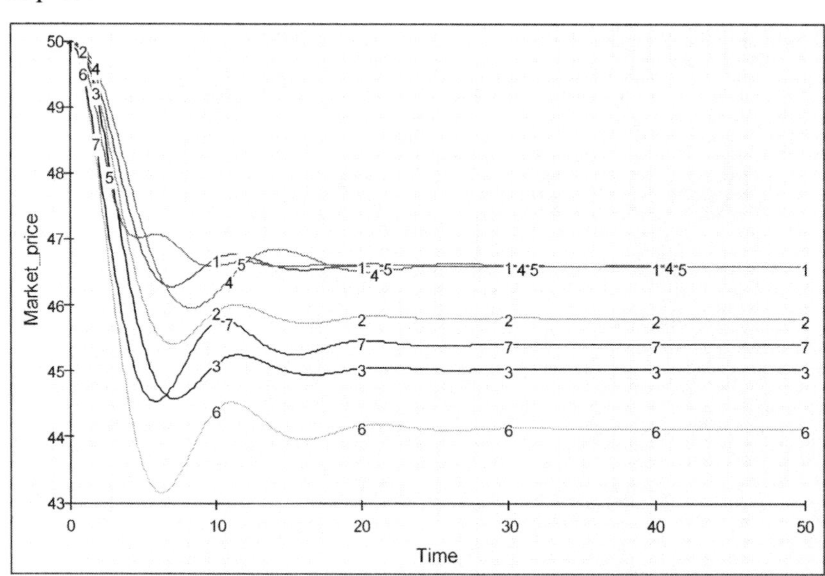

Figure 6. Conversion dynamics

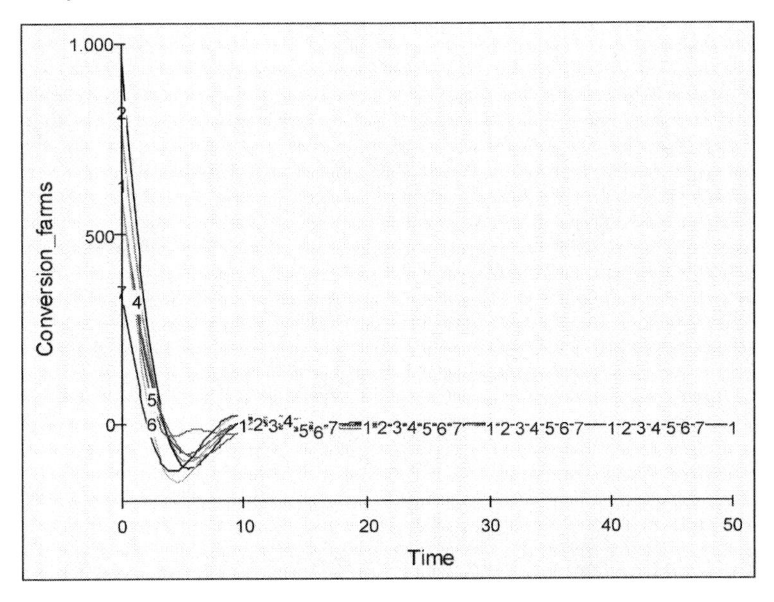

Figure 7. Initial oscillation when changing parameter "delay price change"

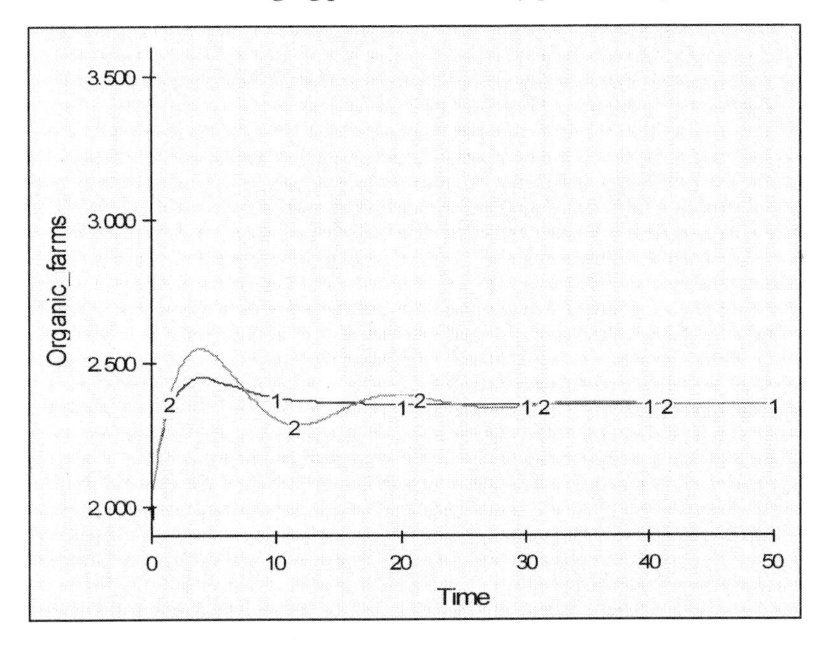

research is needed in the field of the incorporation of the model presented here with our preliminary model (Rozman et al., 2007).

Besides that, in its further development, the system dynamics model will be incorporated into a simulator which will enable group decision-making (Škraba et al., 2003) regarding the key control strategies for organic farming defined by the scenario. An important factor in the process is learning about the system structure and key system interdependencies. The of simulator application will enable the estimation of major weaknesses in the organic farming system's structure and actions needed in order to provide an effective system response; i.e. the proper ratio of the conversion. The fully developed system will enable the cooperation of different experts in making rational decisions based on a larger knowledge database compiled by expert teams. An important contribution in this respect is application of developed simulators by decision teams. The coordination of team views on the problem state is of the highest importance since only coordinated and rational strategies can enhance the organic farming conversion process.

CONCLUSION

After performing several simulation scenarios, the following findings were obtained:

- Conversion to organic farming relies on subsidies, which provide the main means of conversion from conventional farming to organic farming.
- The current output parameter; i.e., the number of organic farms; is caught in an unwanted equilibrium value due to balancing feedback loops in the system.

These results confirm our previous results (Rozman et al., 2007), which suggested that subsidies are the main driving force in the development of organic farming. Further strategic actions should consider the dynamic response of the system and the feasibility of the stated system's target values. Consideration of the interaction of the main feedback loops indicated in the system that determines the system performance provided the means for proper definition of the strategy.

The presented combined methodological SD framework for the analysis of the development of organic farming could provide additional informational support to agricultural policy makers and bring additional clarity to their decisions. It could therefore play an important role in the further development of organic farming, in particular in assistance and advisory capacities in policy planning. The developed model represents a novel application to a new product's spread, and in its dynamic hypotheses it represents novelty at a global scale. By taking into consideration the importance of the proper control of food production, the model represents an important contribution to current global cybernetics issues. The problem addressed in this article relates to such important topics as world food production, which is subject to stochastic demand and supply rules. It must also be mentioned that organic food production also affects the production capacity of national agricultural production systems, with important impacts on a global scale. Further research is needed in the field of SD modeling in order to properly evaluate the applicability of the proposed model.

REFERENCES

Bockermann, A., Meyer, B., Omannc, I., & Spangenberg, J. H. (2005). Modeling sustainability: Comparing an econometric (PANTA RHEI) and a systems dynamics model (SuE). *Journal of Policy Modeling, 27*(2), 189-210.

Elshorbagy, A., Jutla, A., & Barbour, L. (2005). System dynamics approach to assess the sustainability of reclamation of disturbed watersheds. *Canadian Journal of Civil Engineering, 3*(1), 144-158.

Forrester, J.W. (Ed.). (1961). *Industrial dynamics.* Cambridge, MA: MIT Press.

Kljajić, M., Legna, C., & Škraba, A. (2002). System dynamics model development of the Canary Islands for supporting strategic public decisions. In *Proc. of the 20th International Conference of the System Dynamics Society* (p. 16). Palermo, Italy: The System Dynamics Society.

Kljajić, M, Legna, C., Škraba, A., & Peternel, J. (2003). Simulation model of the Canary Islands for public decision support: Preliminary results. In *Proc. of the 20th International Conference of the System Dynamics Society*. Albany, NY: The System Dynamics Society.

Majcen, M.H., & Jurcan, S. (Eds.). (2006). *Action organic farming development plan in Slovenia to year 2015 (ANEK)*. Ljubljana: Government of the Republic of Slovenia. ISBN 961-6299-73-5.

Pažek, K., Rozman, Č., Borec, A., Turk, J., Majkovič, D., Bavec, M., & Bavec, F. (2006). The Use of multi criteria models for decision support on organic farms. *Biological agriculture & horticulture, 24*(1), 73-89.

Pažek, K., Rozman, Č., Turk, J., Bavec, M., & Pavlovič, M. (2005). Ein Simulationsmodell für Investitionsanalyse der Nahrungsmittelverarbeitung auf ökologischen Betrieben in Slowenien. *Bodenkultur 56*(2), 121-131.

Rozman, Č., Pažek, K., Bavec, M., Bavec, F., Turk, J., & Majkovič, D. (2006). The Multi-criteria analysis of spelt food processing alternatives on small organic farms. *Journal of sustainable agriculture, 2*, 159-179.

Rozman, Č., Škraba, A., Kljajić, M., Pažek, K., Bavec, M., & Bavec, F. (in press). The system dynamics model for development of organic agriculture. In *D. M. Dubois* (Ed.), *Eighth international conference on computing anticipatory systems*, HEC-ULg. Liege, Belgium: CASYS'07.

Rozman, Č., Škraba, A., Kljajić, M., Pažek, K., Bavec, M., & Bavec, F. (2008). The Development of an organic agriculture model: A system dynamics approach to support decision-making processes. In *Symposium on Engineering and Management of IT-based organizational systems* (pp. 39-44). Baden-Baden, Germany: The International Institute for Advanced studies in System Research and Cybernetics.

Saysel, A.K., Barlas, Y., & Yenigum, O. (2002). Environmental sustainability in an agricultural development project: a system dynamics approach. *Journal of Environmental Management, 64*(3), 247-260.

Shi, T., & Gill, R. (2005). Developing effective policies for the sustainable development of ecological agriculture in China: The case study of Jinshan County with a systems dynamics model. *Ecological Economics, 53*(2), 230-246.

Sterman, J. (2000). Business dynamics: Systems thinking and modeling for a complex world. Boston, MA: McGraw-Hill.

Škraba, A., Kljajić, M., & Kljajić-Borštnar, M. (2007). The role of information feedback in the management group decision-making process applying system dynamics models. *Group decision and negotiation, 16*(1), 77-95.

This work was previously published in the International Journal of Decision Support System Technology, Vol. 1, Issue 3, edited by G. A. Forgionne, pp. 46-57, copyright 2009 by IGI Publishing (an imprint of IGI Global).

Chapter 16
Role of Telecommunications in Precision Agriculture

James M. McKinion
USDA-ARS, USA

ABSTRACT

Precision agriculture has been made possible by the confluence of several technologies: geographic positioning systems, geographic information systems, image analysis software, low-cost microcomputer-based variable rate controller/recorders, and precision tractor guidance systems. While these technologies have made precision agriculture possible, there are still major obstacles which must be overcome to make this new technology accepted and usable. Most growers will not do image processing and development of prescription maps themselves but will rely upon commercial sources. There still remains the challenge of storage and retrieval of multi-megabytes of data files for each field, and this problem will only continue to grow year by year. This chapter will discuss the various wireless technologies which are currently being used on three proof-of-concept farms or areas in Mississippi, the various data/information intensive precision agriculture applications which use wireless local area networking and Internet access, and the next generation technologies which can immensely propel precision agriculture to widespread use in all of agriculture.

INTRODUCTION

Multispectral image-based precision agriculture technology is beginning to have widespread use in row crop production agriculture in the United States, particularly in the cotton belt. Companies such as InTime, Inc.1 (InTime, 2007), located in Cleveland, Mississippi, USA, are providing image-based products from which clients have access to scout maps derived from multi-spectral images. InTime uses Geospatial System's Inc. multispectral image cameras mounted on fixed-winged aircraft to obtain their image information (GSI, 2007). Specific scout maps target different plant or soil characteristics such as overall plant biomass, differences in soil type, and differences in fertilizer nitrogen uptake. The scout maps are used to determine rates of plant growth regulators, insecticides, defoliants, herbicides, or fertilizer to apply to the plants. Utilizing InTime's Web-based Crop-Site, growers and consultants can easily transform their scout maps into vehicle/controller prescription files that allow for chemical rates to be varied automatically with minimal operator inputs.

The information products generated by such activity, as described previously, can easily be expressed as multi-megabyte sized data files, especially when geographic information system (GIS) technology is used. In almost all situations, multispectral-image based maps are geo-referenced with pixel sizes in the 0.5 meter to 1 meter range generating large data files for the applications derived from these maps. The application maps are also generally geo-referenced for use by the geographic positioning systems (GPS)-based controllers on the farm equipment. For a 500 ha field, application maps can easily be generated which are from 1 to 5 megabytes in size or larger.

Many controller manufacturers today use PC cards or similar technology which has to be hand-carried from the farm base of operations to the equipment in the field and inserted into the controller (Raven Industries, 2008). Conversely, after the application has been made by the equipment (planting, fertilizer application, pesticide application, etc), the PC cards have to be manually collected and taken to the operations base to have the as-applied map uploaded into the farm computer. While many medium and small farms are not affected by this information movement process since the farm manager/owner is also the equipment driver, larger farms which have many pieces of equipment are often scattered over 50 km from one side of the farm land to the other, or even greater distance. The distances involved from the farm base station to the equipment in the field presents an operations problem for growers since when they are involved with precision applications not only must they travel to and from the fields where the equipment is located they must also find the equipment in the field which can often be two or more kilometers across to deliver and pick up the PC cards containing information. This type of operation can easily take a person three or more hours per day just to deliver and/or pick up the data cards.

When time critical operations are involved in delivery of pesticide application maps to the spray equipment controller, this problem becomes even more exacerbated. Our research in early-season plant bug control has shown that from the time the multispectral image is taken by the airplane to the time the spatially variable insecticide is delivered to the spray equipment no more than 48 hours should elapse (Willers et al., 1999). The optimal time is no more than 24 hours. Obviously, there is a better way to solve this time constraint/labor problem than using people for hand carrying PC cards.

Wireless local area network technology is a practical solution to movement of information to and from farm machinery which use GPS-guided precision application controllers. Low-cost wireless network solutions are available and are beginning to see widespread use in the United States. Commercial cell telephone networks are beginning to offer medium–speed Internet access via their cellular telephone towers.

Another technology which is becoming very widely used in the United States is the use of precision guidance on farm application machinery. This technology has proved it worth in labor savings alone by removing the tedium from equipment operators so that they can monitor the application operation to ensure seeds or chemicals are being applied as the equipment moves across the field, allowing the operator to stay longer in the field without the exhaustion which accompanied pre-precision guidance operations. Another tangible benefit from precision guidance cultivation is the recovery of lost row acreage which has been estimated as high as 10% of large field area. This recovery for cotton production on larger plantations has been estimated to pay for the precision guidance system in as little as one year. Many precision guidance systems use the 900 MHz public spectrum to send the signals from the GPS base station to the farm equipment to allow up to centimeter accuracy in the placement of farm application machinery in operation.

This chapter will discuss the various wireless technologies which are currently being used on three proof-of-concept farms or areas in Mississippi, the various data/information intensive precision agriculture applications which use wireless local area networking and Internet access, and the next generation technologies which can immensely propel precision agriculture to wide spread use in all of agriculture.

CURRENT STATE OF AFFAIRS

All farms engaged in row crop production agriculture have extensive information available on their farm operations, crop inputs in terms of seed used and fertilizers applied, soil types, soil fertility, pesticides applied and weather information. Some farms are more organized than others and have all this information developed in computerized databases. Modern production agriculture requires that the aforementioned information and much more be available for not only preseason and in-season decision making, but also for many regulatory purposes required by state and federal government agencies and by seed companies which provide genetically modified varieties used in current practices, especially in cotton production.

The National Cooperative Soil Survey began more than a century ago. Soil databases began to be accumulated by the US Department of Agriculture for the entire US in earnest in the 1940's, 50's and 60's in terms of soil maps whereby soil types were identified and classified by agents of the Soil Conservation Service. These agents would walk the fields, take soil cores manually, and identify broad areas on soil maps by soil type. Today, one can obtain detailed soil maps for each county in a state and generally rely on the placement of these soil types within crop fields. Since August 2005, soil maps and associated data and information have been available through a geographic information system called the Web Soil Survey (WSS). The newest version of the WSS is at http://Websoilsurvey.nrcs.usda.gov.

In the past, soil fertility has been obtained by taking representative soil cores from each major soil type within a field, where soil type was obtained from the maps mentioned previously, and sending the soil samples to a state or private laboratory for analysis to determine N, P and K content.

Yield histories on a field or worst case farm basis have been obtained for management purposes. The advent of private and commercial access to the Global Positioning System (GPS) has changed not only the type of record keeping but has made the level of record keeping much more detailed. Most of the data records were paper records of field histories, planting data, fertilizers applications, chemical applications, fuel usage, equipment usage, and yield data which were kept on a field or farm basis. Only in the last 15 years have these data been migrated from the paper environment to the computer database environment.

Early use of telecommunications on farms was primarily limited to farm headquarters communicating to equipment operators and farm personnel using citizens-band radios, licensed two way radios, and paging equipment.

GPS AND PRECISION STEERING

In the 1970s and 80's the US Military began launching a series of 24 satellites into low orbit above the earth with a pattern of orbits such that most of the time at least 4 to 6 satellites are in line-of-sight to an observer on the earth's surface

(Daly, 1993). These satellites emit radio signals with extreme precision time information so that with appropriate receiving equipment, one can determine one's location on the surface of the earth. This system is known as the Global Positioning System or GPS. Today using what is known as differential signal correction one can determine one's position within plus or minus one meter of the true location with hand-held devices. Going even further, when a secondary radio system is used with GPS signals and the secondary system's location is precisely known, position accuracies within plus or minus one centimeter can be determined. This later system is known as a survey-grade GPS system and is the basis for auto-steering technology which is used in precision guidance of farm machinery.

In the 1970's and 80', software began to be developed to take advantage of satellite imagery so that precision maps could be obtained and manipulated. As GPS became available, this feature was integrated into the software systems so that ground features collected manually using GPS sensors could be added to the maps. This allowed the soil maps which were collected by the Soil Conservation Service and the US Geological Survey to be digitized and converted into spatially sensitive computer maps. Several break thoughs happened during this time which affected the use of the data contained in the old soil maps. Land-leveling was found to be advantageous especially concerning the application of irrigation technology (Walker, 1989). Land-leveling required the use of laser technology (one-way communication) to automatically raise or lower machinery used in the leveling process. This process also changed the morphology of the soil types in the field so that field had to be re-sampled to determine their new soil type. GPS technology was used to map the boundaries of the soil types in the fields to much higher level of accuracies. Millions of acres of land were land-leveled. Even farms that were not land-leveled began to have their field re-sampled using GPS to determine more accurate soil-type boundaries within the fields.

Within the last 10 years, precision guidance or auto-steering systems have become widespread in use (Lessiter, 2006). These systems make use of GPS and an on-site beacon system typically using the 900 MHz unlicensed spectrum to broadcast location and timing signals to field machinery used in precision operations. A typical auto-steer system will have two to three satellite GPS receivers mounted on the field vehicle and a receiver antenna for the geo-referenced beacon system, which can be as far as 50 km from the machinery. These systems typically use frequency hopping radios and a fairly low transmission rate such as 125 to 250 kbps. According to the physics of signal propagation, the lower the transmission rate the further the signal propagates. Again this type of system is a one-way transmission from the satellites and the beacon down to the receivers on the field equipment. Signals from the auto-steer system allow tractors, sprayers, combines, and harvest equipment to navigate a predetermined path with centimeter accuracy. Auto-steer systems allow extended hours of operation by operators without the fatigue associated with non-auto-steer vehicles. Operators can pay more attention to the operation of the equipment to assure the operation is being carried out properly, i.e., seed planting equipment or chemical applicators are working properly and not clogged. Another huge benefit of using auto-steer systems is the recovery of lost acreage. Conventional row crop operations lose area because the operator can not precisely steer the equipment 100 % of the time so rows are made with less precision, typically loosing 10% of the field. This 10% can now be recovered and made productive. Large cotton plantations typically can pay for the cost of auto-steer systems within the first year of operations just from the recovery of lost acreage and the increase in productivity from that area.

REMOTE SENSING VIA AIRCRAFT

Another facet of the precision agriculture technology package is the availability of timely images of row crops within the growing season. Because of the timeliness factor, satellite imagery is of very limited to no use at all, because from the time the image is acquired to the time it is made available for decision making months have elapsed. Imagery acquired via fixed-winged aircraft has become commercially available widely in the US over the last five years. Companies such as InTime, Inc can provide 24 to 48 hour response time to image requests making this technology available for rapid response pest management as well as longer term crop management decisions. Images acquired for use in crop and pest management typically involve four layer images using red, blue, green, and infrared filtered images which are spatially and geo-rectified. From these four layers, image analysis software such as Imagine and ARCView can be used to construct prescription maps which can be used to make decisions about fertilization, irrigation, pesticide applications, application of plant growth regulators, and harvest aid chemical applications. Since most farms are by definition remotely located and do not have access to the Internet, these images and application maps currently have to hand-carried to the farm headquarters for distribution to farm managers and crop and pest consultants. The lack of high speed digital data communication is a major bottleneck in the promotion and acceptance of image-driven precision agriculture technology. This scenario, however, is beginning to change.

PRESCRIPTION FARMING

Even without the help of remote sensing and imaging technologies, precision agriculture has been in use for over 15 years and is becoming wide spread in use. The advent of the low cost micro-controllers based on the microcomputer has made this possible. With computer technology becoming available on the farm beginning in the middle 1980's, farm record keeping was a natural advance from paper records to digital database technology. As computers became more and more powerful and hard discs became ever larger in size, the computer became more than just a good way to keep records to satisfy income tax requirements and EPA regulations on chemical and pesticide usage. Software became available in the early 1990's which allowed farm managers the capability to specify planting density, chemical application rates, and fertilizer application rates which corresponded to their historical records of soil type, soil fertility, cation exchange capacity, field yield data, and weather conditions. The microcontrollers on the field machinery coupled with GPS sensors could then make the seed, chemical, or fertilizer applications required by the specific site requirement of the area within each field (DuPont et al., 1999). Harvest equipment using sometimes the same microcontrollers and GPS sensors could record the yield of the field as the harvesters moved across the field generating a spatially-referenced yield map, which then could be used to help plan the next season's prescriptions. The only missing link at this time is the aerial imagery to gauge how the crop is progressing within the season so that mid-course corrections could be applied to maintain or enhance yields.

The primary method for conveying prescriptions from either the farm managers computer where prescriptions were generated or from a third party commercial prescription generator's computer was then and still is to a large extent today the use of PC Cards. The PC Cards would have to be hand-carried to each specific controller located of the farm machinery and inserted into the appropriate slot whereby the application program would then be uploaded and the machinery would carry out the prescription operation.

Today, growers have the controllers and equipment which can carry multiple varieties of seeds and plant these varieties according to soil conditions by soil fertility by soil types or by specific area (AgLeader Technologies, 2008). The GPS signal coupled with the controller and application software determine which variety to plant and how many seeds per row foot to plant. Chemicals applied at planting can be applied at different rates. During the growing season spray equipment can apply as many as six chemicals concurrently by different rates assuming there is no incompatibility between chemicals.

Aerial chemical applications are also beginning to use variable rate technology with GPS sensor technology and high speed controllers changing the chemical application rate as the aircraft speeds across the field.

For many growers the use of PC cards does not cause any problems. However for large farms which have thousands of acres and often are noncontiguous, the PC card limitation is a problem. The farm manager or other skilled personnel have to hand-deliver PC cards to each piece of equipment when precision operations are called for. Many times the farm equipment is in operation and has to be located. All of this takes time away from skilled personnel who could be more productive performing other tasks. If the wrong card is delivered to the wrong piece of equipment, that equipment will have to wait until it receives the right card; again lost productivity or worse the wrong chemical is applied. Major controller vendors have yet to move away from proprietary closed systems to open systems which have telecommunication capability.

Most growers today do not have the training or capability to perform image analysis and application map generation and do not wish to invest in equipment, software or training to be able to do so. They instead will rely on third party prescription generators to perform these operations. Companies exist which can perform all of the image acquisition, image analysis, and application map generation in-house as a turnkey operation. Other companies exist which can acquire the aerial imagery and hand this imagery off to other companies which perform the image analysis and application map generation. There are also some large farm operations which can perform their own image analysis and application map generation.

Another important aspect of prescription farming is ground truthing. The first component of ground truthing is to record actually what was applied or done where. Because of soil conditions, equipment failure, weather, etc, some field operations may not be carried out as required by the application map. In this case, the controller records what is actually done at each time frame and location and this is stored in the controller as an as-applied map. It is vitally important to retrieve this information from the controller and archive the data at the farm headquarters computer because when it comes time to analyze what was done and what was the outcome this information may explain any anomalies.

The second part of ground truthing occurs during the growing season. When imagery is used to generate application maps, the crop and pest management consultants and their helpers have to stay intimately involved in the decision making loop. While imagery has been shown to reduce by as much as 40% of the observations required by consultants to confirm, extend, or compact zones of treatment predicted by the imagery, these still need to be confirmed by boots on the ground. The imagery maps, often in the form of Normalized Difference Vegetative Index (NDVI) maps (Schowengerdt, 1997), can be used to preselect optimum observation or scouting points within fields. The observer navigates to these points using hand-held GPS sensors and confirms the observation or reports discrepancies. These data are then relayed back to image analysis expert who combines the observations and generates the final application map. This operation is another strong requirement for high speed bidirectional data transmission to the farm and within the farm.

In today's farm world most of the events where we are discussing the movement of information or data, it is movement by hand-carrying the data or information to a central site where the data can be loaded in to a base operations computer or downloaded into a microcontroller from a PC Card. The entire US is covered by satellite Internet access which is unsymmetrical in nature. There are fairly high speed downlinks which can reach as high as 1 mbps but the uplink is much slower and peak out at 150 kbps. All of these speeds are relative to the amount of traffic on the shared link and can be much less than the top speeds. As mentioned before, by definition, the great majority of farms are remote and do not have access to ground-based high speed Internet and will not for the immediate and medium term future. The cost of running fiber optic cable is too high and copper will not meet the needs. Telephone modems are not reliable enough to transmit hundreds to thousands of megabyte-sized image files.

CURRENT AND FUTURE TELECOMMUNICATION CAPABILITIES

Current Services

Satellite capabilities have the possibilities of being significantly upgraded, but have the main limitation of being point-to-point and not covering mobile operations without expensive servo equipment to maintain antenna pointing direction. While providing the link to the Internet for the farm operations centers, satellite is not the answer for communicating with farm machinery and pest and crop consultants in the field.

Unlicensed broadband applications have been in use for over 22 years in the US. Most of the applications have become know as WIFI and use the unlicensed 900 MHz and 2.4 GHz bands with some application in the 5 GHz bands, though these are mainly used as links to the 2.4 GHz and 900 MHz sites. Unlicensed WIFI has two major problems for providing the Internet access to farm operations center; these are range and line-of-sight. Rules for use of WIFI were set by the FCC (FCC, 1985). These rules specified the frequencies to be used and the power of the signal to be broadcast. The main problem for 2.4 GHz broadband is the power of the signal is limited to 1 watt so as not to cause problems with adjacent licensed bands above and below 2.400 to 2.483 GHz. The physics of the problem is simple. When you are trying to send high speed data signals, the more data you send at the same power density the shorter the range will be. Further, at 2.4 GHz, the signal propagation is blocked completely by buildings, trees and foliage so line-of-sight is required for relatively long distances. The 5.0 to 5.8 GHz spectrum behaves comparatively the same as the 2.4 GHz spectrum. The 900 MHz spectrum behaves much more favorably than either 2.4 GHz or 5.0 GHz spectrum, but there is less of it. The 900 MHz spectrum ranges from 900 to 928 MHz. Because the frequency is lower, the signal propagates much better at the same 1 watt power density. Non-line-of-sight signals can be reliably transmitted up to a 10 km radius using horizontally polarized antennae. After that, line-of-sight is required. Using line of sight conditions, a central antenna can broadcast reliable signals to a transceiver with a yagi antenna up to 65 km away. Because of the spectrum limitation and FCC rules, the maximum user data rate is about 2 mbps and can be set symmetrical. For more information on WIFI systems and 801.11 standards see Reid (2001), Geir (2002), and Ohrtman et al. (2003).

Because 2.4 GHz WIFI is widely available and costs are very low, a WIFI system can be used to set up high speed data communications link between the farm base of operations and equipment in the field using multiple overlapping WIFI radios broadcasting from the farm's edges inward, much like cellular telephone towers overlap and hand off signals as one travels from cell to cell. Such a system for demonstration purposes was set up at the Good Farm in Noxubee County, Mississippi, USA (McKinion et al., 2004). The base station radio used was an Alvarion BreezeAccess access point (AP) radio with an omni antenna located on the highest structure at the headquarters site. The AP radio provided the link to BreezeAccess subscriber units located at three corners of the contiguous 800 hectare farm as shown in Figure 1. A typical subscriber unit (SU) repeater station is shown in Figure 2. At the repeater station, the BreezeAccess was directly connected to BreezeNet AP radio which then broadcast its signal into the farm property using a sectorial antenna covering a range of almost 4 km. On each farm tractor, combine or picking machine, a BreezeNet subscriber radio combined with an omni antenna mounted on the top of the equipment completed the link from the farm headquarters to the equipment in the field. A user data rate of 2 mbps bidirectional was accomplished.

For farms with noncontiguous fields with treelines in between, a different solution was needed. On a large cotton plantation in the Mississippi Delta in Bolivar County, MS, we tested another demonstration system on Perthshire Farms. Because line-of-sight was an issue with various treelines bordering numerous fields of the 5,000+ hectare farm, a Waverider Model 3001 900 MHz AP radio with an onmi antenna mounted on a 60 meter tall guyed tower was used to communicate to eight repeater stations. A Waverider EUM Model 3000 900 MHz subscriber unit completed the link to the farm headquarters with the remainder of the system being the same as the Good Farm. The 900 MHz radio allowed non-line-of-sight

Figure 1. Map of Good Farm located in Noxubee County, Mississippi, USA showing location of wireless local area network base station (light circle) and three repeater stations (dark outline circles) using 2.4 GHz frequency hopping spread spectrum (FHSS) radios to provide seamless coverage of farm cotton acreage. Vehicle and ground personnel use WIFI radios to complete link for scouting applications with uplink being a FHSS radio

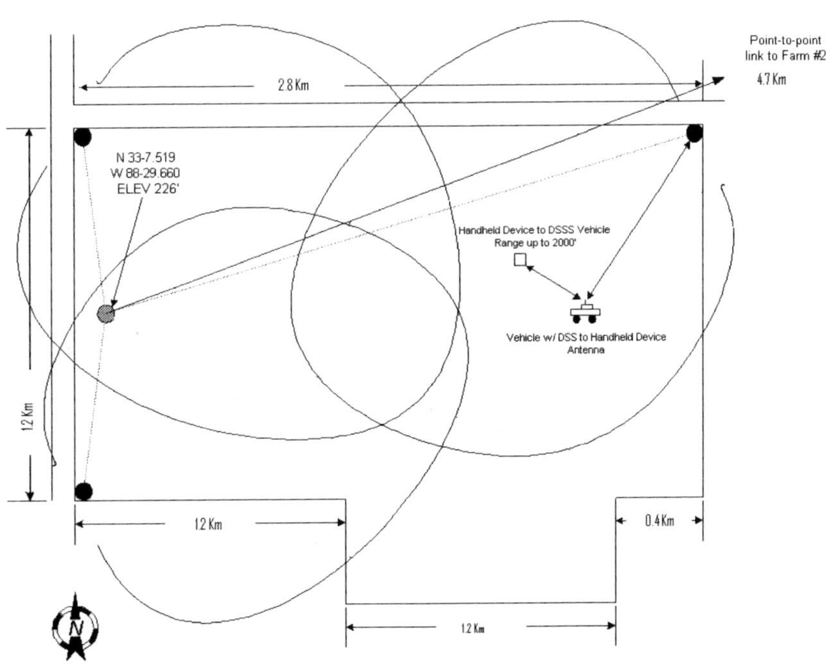

operation up to 15 km. In addition to providing high speed bidirectional communication between headquarters and farm machinery, the 900 MHz digital radio system was also used to connect computer systems located at two cotton gins, one close to the headquarters and another 7 km away, allowing rapid communication and system backup from the gin computers to the headquarters computers. Figure 3 shows the geographical layout of Perthshire Farms. The user data rate on this system was 2 mbps up and down.

Both of these farm demonstration systems were connected to the Internet using satellite links while very workable but did not allow fast uplinks to transmit large datasets (from hundreds of megabytes to several gigabytes in size) back to image analysts. To address this problem and to explore wide-area networking, a third radio system was established to demonstrate the effectiveness of wide-area telecommunications (McKinion et al., 2007). In Noxubee County, Mississippi on Prairie Point Road approximately 15 km east of Macon, a 100 meter tall microwave tower owned by Teletec Communications, LLC of Columbus, Mississippi was used to place three 120° sectorial antenna configured to broadcast horizontally polarized signals in the 900 MHz band using Motorola Canopy 900 radios. A diagram depicting the sectorial layout of the area of coverage from the 100 m tall tower is shown in Figure 4. Each antenna broadcast at a different frequency for each sector to prevent interference. Non-line-of-sight signal propagation was achieved up to 14 km in radius from the tower. Line-of-sight connections were made up to 60 km from the tower essentially providing coverage to stationary antennae located in Noxubee County, southern Lowndes County, and northern Kemper County in Mississippi, and western Pickens County in Alabama. Internet access to the system was provided by Teletec using high gain point to point dish antenna with the microwave tower being the receive location and Teletec's 125 meter tower being the originating site located in Columbus , Mississippi, 45 km away. Internet access was provided as a full duplex 18 mbps service to the three sector antenna array at Prairie Point. A user data rate of 2 mbps up and down was achieved. A typical user radio link is shown in Figure 5.

Cellular telephone providers recently have extended Internet services called EV-DO and WCDMA which provide medium speed Internet

Figure 2. Picture of repeater station with sectorial antenna at top, square panel antenna communicating to base station in the middle of the picture and solar panels for power at the bottom of the tower

Figure 3. Map of Perthshire Farms, Bolivar County, Mississippi, USA showing size of area covered

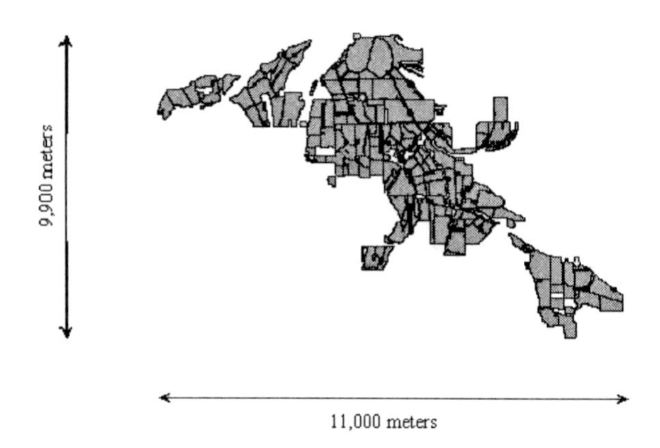

Figure 4. Coverage map indicating the non-line-of sight coverage area advertised by Motorola. 5 km is typical non-line-of sight coverage while we actually were getting 9+ km under non-line-of-sight conditions

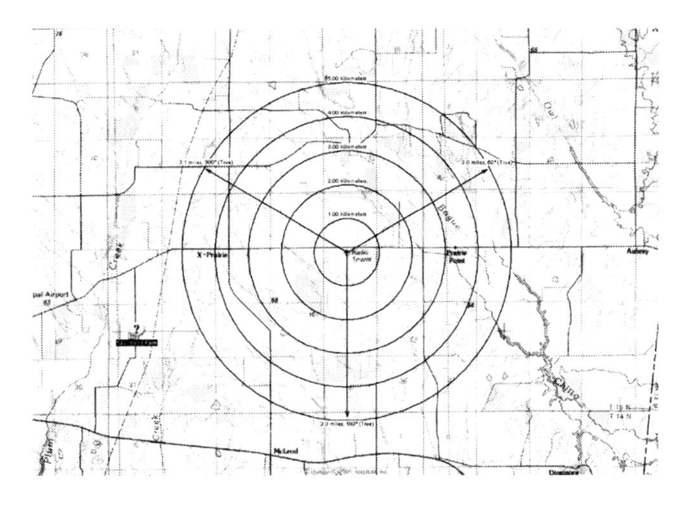

Figure 5. Outdoor antenna mounted at end user site 9 km east of the base station. Antenna and radio are shown on the left with the antenna on top with the radio mounted below. The NEMA 4 box contains the surge protected power strip mounted on the left side interior of the box along with the power converter which supplies 24 VDC power over Ethernet (POE) to the radio, and antenna lightning protector shown mounted on the lower right

and data communication capabilities which serve mobile applications. Cellular services such as CDMA 1xRTT and GPRS, transmit data at speeds less than 200 kbps in one direction and provide customers with access to mobile data applications such as text messaging, e-mail, and ring tone downloads. Wireless *broadband* networks, on the other hand – such as CDMA 1x EV-DO (EV-DO), Wideband CDMA (WCDMA) with High Speed Downlink Packet Access (HSDPA), and Wi-Fi – transmit data at speeds greater than 200 kbps in at least one direction and provide access to the applications available on the slower networks as well as services that require greater bandwidth, such as video programming, music downloads, and high-resolution. Commercial high speed data communications for farm operations are becoming more available, but current data rates are still fairly low with 800 kbps being the standard. The end user also has to face equipment charges and usage charges. Cellular telephone companies are beginning to extend their data communications systems (EV-DO and WCDMA) to rural areas where they already have cellular telephone service.

One very useful service is the coupling of digital paging systems to critical automated systems in farm applications which send a paging signal and brief message telling the recipient that trouble has occurred and what system has been affected. This system could be greatly helped by being included in a digital communication system so that the recipient could not only recognize the nature of the trouble but take steps remotely to fix the problem or stabilize things until someone could come on site to fix the problem.

Licensed broadband Internet access is a strong possibility which promises high speed Internet access and wide area of coverage. This service is known as WIMAX and has been specifically designed for wireless high speed Internet use. Many vendors now have equipment certified by an independent laboratory in Spain as conforming to the WIMAX standard and interoperable with any other vendors certified equipment. While WIMAX may bring high speed Internet to the rural US, most holders of licensed bands are focusing on populated areas to get the highest return on their investment, and the likelihood of rural applications is still several years away.

WIMAX systems promise much higher data rates and greater areas of coverage than WIFI, EV-DO and WCDMA. Because of these two properties WIMAX should be a less expensive commercial service WIMAX was designed from the ground up to be a wide area data communications system and takes advantage of numerous technical advances such as steerable beam forming to achieve range and non-line-of-sight capability as well as much higher user data rates. WIMAX is beginning to be rolled out but because of economics will be available only in the populated areas first.

Future Services

Higher broadcast power satellites have been proposed using highly directional antenna to service smaller areas of the earth with much higher data rates than currently available (Mir, 2008). For most no-time-critical applications, this would be suitable. But where real-time response is needed, satellites present too great a time delay because of distance involved for the space the signal travels.

The ultimate solution is for every home and business in the US to have access to fiber optic cable, but this service will be a long time coming, if ever, because of the cost involved.

A recent Public Notice published by the Federal Communications Commission (http://wireless.fcc.gov/spectrum/index.htm?job=proceedings_details&proid=369) could have a very large impact on rural America's access to high speed digital communications and the Internet. The FCC has proposed that unused bandwidth in the commercial television channels from channel 2 through channel 51 be allowed to be used as unlicensed spectrum with certain limitations as early as January, 2009. For rural areas in particular and for high population density areas of the country, TV whitespace, unused TV channels, promise very high speed data and Internet access over large distances with signal penetration capability properties which all other broadband licensed and unlicensed spectrum do not possess. Where WIFI signals are blocked by buildings and trees, this is TV spectrum, and we all know that TV signals penetrate buildings and propagate through dense foliage and trees. This spectrum will be available for free usage. Each TV channel uses 6 MHz of spectrum. Assuming a data transmission efficiency of 5 bits/hertz, a single channel could be used to transmit 30 mbps of data. For a rural area like Mississippi or Alabama, the most channels broadcasting in either public or commercial channels in an area is about 6 channels plus channel 39 which is reserved for space usage. This means there would be 43 channels available for high speed digital wireless communication, or in terms of capacity, over 1290 mbps with only using omni antennae and not making use of current spectrum reuse technology at all! Limitations on the use of TV whitespace may preclude the use of channels next to broadcasting stations in the lower TV bands, channels 2 thru 13. Data radios will have to have the capability of listening to a channel to ascertain no one is broadcasting on it before the radio attempts to use that channel, otherwise the radio will have to change channels and negotiate the listening process again. All digital radios using TV whitespace will have to be registered with the FCC identifying spectrum to be used and physical location. All radio manufacturers will have to certify and prove to the FCC that their radio equipment will not interfere with public and commercial broadcast stations. These radios will use low power as ordered by the FCC. System builders have said that they can meet all of these requirements. A summary of current wireless technologies is presented in Table 1.

Table 1. Comparison of telephone and wireless local area network protocols; GPRS and EV-DO are used alongside cellular telephone networks while the remainder were designed as wireless data network protocols

Protocol	Upload (mbps)	Download (mbps)	Mode
GPRS	0.02	0.08	Full Duplex
EVDO, Rev a	1.80	3.10	Full Duplex
FHSS, 802.11	2.00	2.00	Half Duplex
WI-FI, 802.11b	10.00	10.00	Half Duplex
802.11a,g	54.00	54.00	Half Duplex
WIMAX, 802.16	70.00	70.00	Half Duplex
802.11n	280.00	280.00	Half Duplex

CONCLUSION

Things have advanced greatly from the early use of telecommunications when the only thing being used was the telephone. Next came unlicensed and licensed radio telephone service. Low powered walkie-talkies were useful for small distances but CB radios and licensed two-way radios quickly became the standard for on-farm communication. Some one way communication was used in land leveling operations which used laser technology so that centimeter accuracy could be obtained in elevation establishment in fields for drainage and irrigation flow. When GPS became available for commercial use, another one way communication technology was rapidly adopted. Hand held GPS units were used to mark field boundaries on computer maps which were spatially registered. This confluence of technology, GPS system and digital mapping, allowed precision farming to begin. GPS also allowed the advent of precision guidance of farm machinery, also called auto-steer. GPS was crucial for the development of microcontrollers placed on farm equipment so that as the field equipment proceeded across the field, the GPS sensor told the equipment where it was so that the microcontroller could vary the rate of the application in response to its control program. These last precision agriculture milestones have all used one-way communication. The time has come to close the loop. As commercial and private high speed networks become available to the farm community, all farm data and information traffic can and should become two-way communication. The potential is there for significant savings in manpower and amplification of effort to improve farm productivity. No longer would data cards have to be delivered to controllers and installed by hand. This could all be handled from the farm operations center by trained personnel making best use of their time and effort. Application maps delivered to the wrong piece of equipment could be totally prevented. All equipment could be tracked in real time at the farm operations center. This means that as supplies are being used and applied in the field, operations can track and anticipate delivery of additional supplies to optimize use of farm machinery and personnel.

The availability of systems for use in TV whitespace spectrum could and should have a major impact in rural America. The potential is enormous with bandwidth capacities exceeding that of current state-of-the-art 750 MHz digital cable systems and being equal to or exceeded by only fiber-optic-to-the-curb systems. If this system is put into place in rural America, the digital divide between rural America and urban America will cease to exist.

REFERENCES

AgLeader Technologies (2008). http://www.agleader.com/products.php?Product=pfadvantage. (Verified January 23, 2008)

Daly, P. (1993). Navstar GPS and GLONASS - global satellite navigation systems. *Electronics & Communication Engineering Journal, 5*(6), 349-357.

Dupont, J. K., Willers, J. L., Seal, M. R., & Hood, K. B. (1999). Precision pesticide applications using remote sensing. *Proc. 17-th Biennial Workshop on Color Photography and Videography in Resource Assessment, Reno, NV.*

FCC. (1985). http://www.access.gpo.gov/nara/cfr/waisidx_01/47cfr15_01.html (Verified January 23, 2008)

Geir, J. (2002). *Wireless Lans, Implementing High Performance IEEE802.11 Networks.* Indianoplis, IN: SAMS Publishing.

Geospatial Systems, Inc. (2008). http://www.geospatialsystems.com/imaging-products/ (Verified February, 6, 2008)

Intime, Inc. (2007). http://www.gointime.com/ (Verified January 23, 2008)

Lessiter, F. (2006). Auto-Steer's New Frontier – Precise implement steering. *Farm Eguipment.* http://www.gpsfarm.com/_resources/uploads/files/Farm%20

Equipment%20Sept%202006%20AFTracker.pdf. (Verified January 23, 2008)

McKinion, J. M., Turner, S. B., Willers, J. L., Read, J. J., Jenkins, J. N., & McDade, J. (2004). Wireless technology and satellite internet access for high-speed whole farm connectivity in precision agriculture. *Agricultural Systems, 81*(2004), 201-212.

McKinion, J. M., Willers, J. L., & Jenkins, J. N. (2007). Wide area wireless network (WAWN) for supporting Precision Agriculture. *Proceedings of the 2007 Beltwide Cotton Production Research Conferences, New Orleans, LA.* Unpaginated CDROM.

Mir, R. M. (2008). http://www.cs.wustl.edu/~jain/cis788-97/ftp/satellite_data.pdf. (Verified January 23, 2008)

Raven Industries. (2008). http://www.ravenprecision.com/Manuals/pdf/016-0171/147A.pdf. (Verified January 23, 2008)

Ohrtman, F., & Roeder, K. (2003). *Wi-Fi Handbook: Bulding 802.11b Wireless Networks.* New York, NY: McGrawHill.

Reid, N. P. (2001). *Broadband Fixed Wireless Networks.* New York, NY: McGrawHill/Osborne.

Schowengerdt, R. A. (1997). *Remote Sensing: Models and Methods for Image Processing.* New York, NY: Academic Press.

Walker, W. R. (1989). *Guidelines for Designing and Evaluating Surface Irrigation Systems.* FAO, 1989: ISBN 92-5-102879-6, http://www.fao.org/docrep/T0231E/t0231e01.htm#preface (Verified January 23, 2008)

Willers, J. L., Seal, M. R., & Luttrell, R. G. (1999). Remote Sensing, Line-intercept Sampling for

Tarnished Plant Bugs (Heteroptera: Miridae) in Mid-south Cotton. *The Journal of Cotton Science, 3*, 160-170.

KEY TERMS

CDMA: Code division multiple access, a coding methodology for digital radio transmission which improves the amount of information carried by each hertz of frequency. WCDMA is *broadband* CDMA, a further refinement.

DSSS: Direct sequence spread spectrum digital radio transmission methodology which spreads the radio signal over a range of contiguous frequencies to achieve noise immunity.

EV-DO: Evolution Data-Optimized = high-speed mobile data standard used by CDMA=based networks

GIS: Computer based geographic information system technology which records and/or manipulates map data using a map coordinate system to reference map values by pixels.

GPRS: general packet radio service is a standard for wireless communications that allows packets of data, such as e-mail and Web content, to travel across a wireless telephone network and to the Internet.

GPS: The collection of 24 satellites in low earth orbit which transmit precise time information so that receivers can precisely located themselves on the surface of the earth.

Geo-Referenced Map: A computerized map in which each pixel has a value and a geographic information system location reference.

FHSS: Frequency hopping spread spectrum digital radio transmission methodology in which the radio signal is spread over a narrow range of spectrum and the signal hops in a pseudo-random order to other narrow ranges all within a defined overall range to achieve better noise immunity than DSSS, greater range than DSSS and inherent signal security.

Omni and Sectorial Antennae: An omni antenna broadcasts/receives in a 360° pattern completely covering an area while a sectorial antenna broadcasts/receives in a more narrowly defined direction such as 30°, 60°, 90° , or 180° patterns

Precision Agriculture: The practice of planting, applying chemicals, and recording harvest yields based on GIS and GPS information directed by a prescription map using variable rate controllers/recorders on farm machinery.

Prescription Map: A geo-referenced map which contains rate information so that variable rate controllers can apply the appropriate application to the appropriate location using real time GPS sensor information.

TV Whitespace Spectrum: Each television channels occupies 6 MHz of spectrum and in rural Americal there are typically only 6 to 8 channels in use from the set of channel 2 thru channel 51; the unused chanels in this spectrum are called whitespace.

WI-FI: A DSSS radio system based on industry standard IEEE 802.11b which uses the 2.4 GHz signal spectrum, also called wireless fidelity has a range of 10 km line-of-sight and a bandwidth of 11 Mbps.

WIMAX: Wireless metropolitan area network standard IEEE 801.16 which is the new wireless broadband with a range of up to 80km with a bandwith of up to 75Mbps and is the successor to WI-FI.

ENDNOTE

[1] Mention of trade names is for information purposes only and does not imply a recommendation or endorsement by the USDA-ARS.

This work was previously published in the Handbook of Research on Telecommunications Planning and Management for Business, edited by I. Lee, pp. 836-850, copyright 2009 by Information Science Reference (an imprint of IGI Global).

Chapter 17
World Wide Weber:
Formalise, Normalise, Rationalise: E-Government for Welfare State – Perspectives from South Africa

Nicolas Pejout
Centre d'Etude d'Afrique Noire (CEAN), France

ABSTRACT

Many of African States are focusing on ICTs and developing e-government infrastructures in order to fasten and improve their "formalisation strategy". This philosophy drives the South African State in its impressive efforts to deploy an efficient and pervasive e-government architecture for its citizens to enjoy accurate public services and for this young democracy to be "useful" to them. By focusing on the South African case, people will be able to understand the role of ICTs as tools to register, formalise and normalise, supporting the final objective of Weberian rationalisation. The author will consider the historical process of this strategy, across different political regimes (from Apartheid to democracy). He will see how it is deployed within a young democracy, aiming at producing a balance between two poles: a formal existence of citizens for them to enjoy a "delivery democracy" in which they are to be transparent; an informal existence of citizens for them to live freely in their private and intimate sphere. In this tension, South Africa, given its history, is paradigmatic and can shed light on many other countries, beyond Africa.

Numerous governments, particularly those of developing countries, have to deal with challenging economic, socio-economic and political *realities*. More challenging is to deal with *unrealities*, i.e. realities that do exist but that governments can't manage because they don't know about them. These realities are real but informal: one typical example is moonlight work. They all do exist but have no official, formal, legal-administrative and statistical existence. They are "parallel" to the official-formal world of public action and stay "underground", in the shadow of public policies.

This problem of "informality" is particularly experienced by governments in developing countries. They face tremendous difficulties in terms of public action upon realities that they don't and can't know of. That is due to a lack of measuring resources and public management capacities. Various examples are: the absence of a satisfactory statistical machinery, the ineffectiveness of a formal civil status (for instance, the registry of birth), the inefficiency of tax rolls...

For governments to act upon realities, they need to know them and therefore to reveal and measure them. In other words, they need to formalise them so as to be able to control them. Governments have to normalise human activities, i.e. to put them into norms, into measurable and controllable frameworks. This explains, for instance, the importance of statistical machineries into the construction of nation-states (Desrosières, 1993).

This quest of formalisation has been growing with the strengthening of nation-states throughout time. It has been using various tools, the last generation of which being Information and Communication Technologies (ICTs). Governments consider these technologies as powerful instruments to formalize and normalize realities. The use of ICTs to rationalize reality and therefore public action is set along the deployment of electronic government (e-government). The ultimate objective is to make a society (individuals as well as groups) highly visible—some might say transparent—to the power in place. By formatting knowledge for the State, e-government is supporting a move towards genuine rationalisation: ICTs enable an extreme degree of accuracy and sophistication (data mining) so that everything and everyone can be labelled, measured, compartmentalised.

Obviously, such power of knowledge, based on the knowledge of power (Foucault, 1997), can threaten democracy: full transparency of individuals to the State is impossible, due to the absolute necessity of protecting the private sphere. Nevertheless, the development of the welfare State requires the administration to know most of personal data, so as to provide relevant services, for instance well-measured pensions or health care (Gilliom, 2001). This is all the more true when the welfare State is getting ICT-intensive, making the most of e-government to provide e-services. For such provision with efficiency and cost-recovery, the State needs to be scientific, somehow omniscient. That is why transparency of the society to the State is necessary (Lyon, 2003), but to a certain extent beyond which democracy is at risk.

Most of governments in African countries are confronted with informal realities, particularly in hard socio-economic contexts. They don't have enough resources—financial, human, ...—to know of realities that they nevertheless need to tackle with. That is why some African States are focusing on ICTs and developing e-government infrastructures in order to fasten and improve their "formalisation strategy": by getting to know their society better, they can act upon it better (Cheneau-Loquay, 2005). South Africa is one of these and certainly the most advanced on the continent in that regard. The South African State is indeed deploying a remarkable e-government architecture for its citizens to enjoy accurate public services (Péjout, 2004; Péjout, 2007). The challenge is not just that of administrative efficiency but is also highly political: this young democracy needs to be "useful" to its citizens.

This chapter will highlight the role of ICTs as tools to register, formalise and normalize realities. We will first show how formality is key to the e-welfare State. We will see how "the World Wide Web meets Weber": ICTs are ideal the ideal tools to implement the Weberian administrative organization (Weber, 1978). This ICT-based administration must serve an e-welfare State that has to guarantee the effectiveness of the social contract. We will then consider the historical process of the South African e-strategy, from Apartheid to democracy. We will see how it is deployed within a young democracy, aiming at producing a balance between two poles: on the one hand, a formal existence of citizens—even a transparency to the State—so that they can enjoy the deliveries of democracy; on the other, an informal existence of citizens so that they can live freely in their private and intimate sphere. This tension is magnified by the intrusion of e-government. South Africa, given its history, is paradigmatic and can shed light on many other countries, beyond Africa.

FORMALITY IS KEY TO E-WELFARE STATE

ICT are powerful tools for e-government to be deployed in the most comprehensive way. The condition for this scenario to happen and for e-government to mean something for everyone in everyday life is: what can e-government deal with? Indeed, e-government functionalities can only be put into place if they get the clearest picture of the reality to be managed or to be transformed.

According to Lautier (2004), the word "informal" was first proposed in 1971 by Hart (1973) to describe the complementary revenue that is necessary to face wage stagnation, inflation, insufficient kin solidarity and limited access to credit opportunities in developing countries. The term was then popularized by the International Labour Office (ILO) in 1972 in its report "Incomes and Equality – A Strategy for Increasing Productive Employment in Kenya". Whereas Hart was taking "informality" for a set of practices, the ILO is using it to depict a situation, the "informal sector", the "informal economy". As Lautier (2004) shows it, searching for a definite list of criteria to define informality is a vain activity. He rather focuses his analysis on the economic informality, that of economic activities that are characterised by the following: law infringement, relatively small-scale, under-employment, poverty, survival strategies…

Because we agree with Lautier (2004) on the impossibility of a comprehensive definition of informality, we shall push for a narrower understanding of this notion by focusing on the *administrative informality*. It gathers all human activities and products that do not exist in the eye of the State, that are not part of any statistical apparatus and thus can not be acted upon because they simply don't exist for the administrative machinery.

The construction of e-government requires the constitution of vast and complete databases. An "ignorant" e-government, that does not satisfy its appetite for information and data, can not run efficiently. Furthermore, these data must be usable by the technical architecture in place and thus must be standardised, formatted into the frameworks of the administrative system that collects, manages, produces and diffuses these information. In this regard, any e-government strategy must develop "policies of formalisation" (Lautier, 2004). Using the notion of formalisation refers to four objectives: making some realities official in the eye of the administration; formatting realities into specific statistical and administrative frameworks; informing authorities to produce knowledge; controlling these now well-known realities.

Aiming at satisfying its statistical appetite in order to ensure the exhaustivity of its knowledge and of its control, the South African State is implementing several ICT-based initiatives. Citizens, their identities and their activities must be visible to the State. Some might say they must be "transparent" to the Leviathan. Let's note that whereas the total population of South Africa was about 45,4 million in 2002 (StatSA, 2002), only 28 million South Africans had an identity card in May 2003 (Buthelezi, 2003). Three million unidentified children were still out of the social security system in 2004 (Mapisa-Nqakula, 2004).

The existence of people in the eye of the administration means that they can enjoy their "second-generation" socio-economic rights and access the facilities of the Welfare State. For instance, the *Vital Registration Programme* is using an online data collection and compilation system in order to register births with no delay in hospitals and clinics. The *Child Support Grant* is thus correctly paid to eligible households. The ID card is truly an enabling documentation. This formalisation of people's identity is also a condition of their citizenship : thus, according to the Home Affairs Minister, Mapisa-Kqakula, the attribution of ID documents to San communities in 2004 enables them to live "their full participation as citizens" (Van Der Berg, 2004). Geographical data are also crucial: the South African Post Office (SAPO) and the firms Spatial Technologies and MapInfo Corporation are involved in the ICT-based cleaning up of all street addresses of the country (Minnaar, 2004b).

Another example is the possibility for South African women to check their marital status online by typing in their ID number in a specific area of the Web site of the Department of Home Affairs. In July-August 2004, about 200 women discovered they were registered as married but… to men they didn't know. The "*Check Your Marital Status Campaign*" enabled 5 000 women to check their status (Burrows, 2004a). This campaign has enabled the prosecution of five Home Affairs officials who facilitated 1 500 fraudulent marriages between South Africans and illegal immigrants and 200 faked registrations of birth and unlawfully nullified marriages (Engelbrecht, 2007a).

This policy of administrative formalisation is essential to the tight administration of the South African population. In this regard, formalisation is feeding "normalisation" as understood by Michel Foucault (1997) when he looks at the "society of normalisation" and the generalisation of disciplinary projects mixing knowledge and power. Here comes in the notion of "normalisation" as proposed by Guillaume (1978, p.8): normalisation is a compromise between order and disorder that can "lock up individuals, assign them to specific places and dictate their representation in the past, the present and the future". For instance, the officers of the *Johannesburg Metropolitan Police Department* can now get the personal data of any driver by entering into their cell phone the car registration number, its serial number, the engine number or even a spare part's number; the information is sent by SMS to the main server (CPSI, 2003, p.58).

The ICT-based formalisation applies not only to identities but also to activities, with the perspective of (economic) transactions between the State and citizens. The tax objective is also a driving force behind this move. 7 million South Africans have no bank account. The private sector, and banks at the forefront through the *Financial Services Black Economic Empowerment Charter*, is developing numerous projects to get citizens registered in the economic transactions. In that regard, they back up the State effort to formalise people's identities. In June 2004, the Unemployment Insurance Fund launched a web-based process to officialise the employment of domestic workers. The campaign was led by the Department of Labour (DL) and more than 600 000 workers got registered, who thus could claim their unemployment insurance (SAPA, 2004c). In a more comprehensive manner, aiming at formalising its knowledge about the working population of South Africa, the DL is computerising its 125 *labour centres* and 529 *satellite offices* and installing an ERP-like system called *Lesedi* that ought to accelerate the resolution of work-related conflicts and the response time to these. The installation is a core piece of the Public-Private Partnership that ties the Department and Siemens Business Services. The State also wants to introduce 20 mobiles offices units targeting Limpopo, Eastern Cape, KwaZulu-Natal and Northern Cape provinces.

All these initiatives recall that the constitution of data sets is closely "associated with the construction of the State, its unification, its administration" (Desrosières, 1993, p.16; Desrosières, 1997). South Africa provides us with an example of the British tradition of "political arithmetic" that aims at registering, referencing, codifying people and their activities. This enterprise reveals the tight connection of "knowledge instruments" with "discipline and constraint instruments" (Lascoumes & Le Galès, 2004, p.27). The best example thereof is the Apartheid State apparatus.

THE APARTHEID REGIME: THE "MANIA OF MEASUREMENT" FOR TOTALITARIAN POWER

The Statistical Frenzy at the Core of Apartheid State-Building

For totalitarian purposes, the Apartheid regime has generated an incredible statistical apparatus, answering to the huge data appetite of the segregationist administration.

The use of ICTs was aiming at three objectives: developing a comprehensive *knowledge* of the population; reinforcing the *surveillance* capacity of the State; consequently, ensuring its ability to *control* people. The construction of the Apartheid regime did not absolutely follow a grand plan but was rather a collection of discrete administrative and political acts (Bonner & al., 1993; Fauvelle-Aymar, 2006; Posel, 2000).

In this scenario, the "power of number" was crucial for the nascent administration to build "totalising modes of racialised knowledge" (Posel, 2000, p.116). The obsession of numbering, counting and classifying was remarkable within the Department of Native Affairs (DNA): the formalisation and categorisation of people was aiming at normalising them because the "feasibility of apartheid came to rest on the pervasive presence of the State in every facet of life" (Evans, 1997, p.1). To put it shortly, this Department provided the DNA to the Apartheid State! The government and particularly the DNA were thus characterised by "an explosion of 'scientific' empirical investigation into the lives of Africans[1] in the urban and rural areas" (Evans, 1997, p.11).

The better the State knows the population and its members and activities, the easier the control thereof can be. The "mania of measurement" is supporting the conquest of omniscience by the Apartheid state which aims at developing "'modern' modes of political rationality" (Posel, 2000, p.116). The quest for numbers is fuelled by an administrative scientism and a racialist technocracy. The State becomes a super-calculator, a computer in its original meaning i.e. compiling data to process them for specific purposes.

However, this mania did begin before the start of the apartheid regime with the *Census Act* (1910) which enabled the first national census to take place in 1911; then, the *Statistics Act* (1914) established the Office of Census and Statistics. In 1937, the *Native Laws Amendment Act* forced the local authorities to conduct a bi-annual census of the Black people. These actions pushed for a "scientific statecraft" (Posel, 2000, p.122) which was all the more necessary for the apartheid regime that is was deeply based on a geographical approach of reality: the influx control policy, that was inaugurated in 1923 with the *Natives (Urban Areas) Act*, was to realise the following equation: *n* African people authorized to work in urban areas = *n* working African people = *n* African people lodging in townships. This transcription of reality into State mathematics enabled the administration to calculate the "surplus" of Black population to be rejected from "White spots" (Posel, 2000, p.122). Though the Apartheid regime does not inaugurate the birth of statistics, it does implement it with a remarkable systematisation. For François-Xavier Fauvelle-Aymar (2006, p.33), the Apartheid is a "delirious ideal of organising the world and of indexing the human diversity". It is

an engineering of the population that is crossed by a delirious ideal, that of rationalising diversity, that of summing it up to a series of social objects that are differentiated, homogeneous, singular, consisting as many as separate races, ethnies or tribes, in an unchangeable hierarchy. (Fauvelle-Aymar, 2006, p.102)

This "political arithmetic" (Sadie, 1950) is powerful because the "Apartheid statecraft represented a hankering for 'total order'" (Posel, 2000, p.127). It is fully implemented with the *Population Registration Act* (1950) and the National Population Register. To depict this "statistical frenzy", Posel (2000, p.134) and Breckenridge (2002a) use the term "hubris", qualifying an administration that is obsessed with its measurement tools.

In the 1960s, the cognitive system of the administration is getting more and more complex because the State is determined to settle a new order of "social engineering". The extension and the sophistication of the measurement system foster the creation of the Department of Planning in 1965. It complements the Department of Bantu Administration and Development whose statistical routine was involving

monthly counts of the numbers of work-seekers' permits issued on national basis; the number of service contracts registered (showing the area of origin of each of the registered workers); the number of registered vacancies in urban areas and in bantustans and on farms; annual counts of the numbers of 'Bantu' seeking assistance from special 'aid centres' which assisted with the work of the labour bureaux; the compilation of an 'occupation register'... in respect of each national unit... containing certain particulars of the Bantu concerned'; the creation of a separate data base of 'existing and anticipated employment opportunities in the homelands for Bantu with advanced qualifications'; annual counts of the number of workers recruited by various recruiting organisations and the sites of recruitment; annual counts of the numbers of curfew proclamations issued; the number of 'new Bantu residential regulations' promulgated as well as amendments to the existing regulations; annual counts of the number of bodies designated as urban local authorities and the number of 'promulgations and redefinitions of Bantu residential areas'; annual counts of the numbers of permits issued in terms of the Group Areas Act for 'recreation and health services', such as cinema and hospitals; numbers of applications to 'conduct' church services for Bantu in white residential areas'; amounts of money accruing in respect of the Bantu Services Levy Act; annual income from the sale of 'Bantu Beer' by local authorities and employers; numbers of inspections undertaken of a whole range of different types of sites on which particular types of developments were being considered (each type of site being enumerated separately); 'numbers of cases in which comments... from inspectors were furnished on group areas planning'; numbers of townships planned and developed, per ethnic group; numbers of sub-economic houses constructed; numbers of families 'removed' from urban townships, on an ethnic basis; numbers of families removed from 'black spots' and white rural areas, per ethnic group; separate counts of numbers of 'Bantu traders, industrialists and professionals' resettled; numbers of children, placed in various welfare institutions; annual counts of numbers of reference books issued to males and females; numbers of duplicate reference books issued to males and to females; numbers of identification documents issued to 'foreign Bantu'; comparisons of 'numbers of Bantu whose identity numbers were known with numbers of existing records of fingerprints'; numbers of fingerprints 'classified and searched to determine whether the fingerprints of the persons concerned are not already on record'; numbers of sets of fingerprints 'added to existing record'; numbers of births, marriages and deaths registered each year; annual counts of the numbers of 'Bantu males' on the National Population Register; annual counts of the numbers of 'Bantu females' on the Population Register; numbers of inquiries into 'the tax particulars and movements of Bantu' on the Population Register; numbers of beneficiaries of social pensions, maintenance grant beneficiaries and pneumoconiosis grant beneficiaries; along with records of monies spent on salaries, capital equipment, land and other projects. (Posel, 2000, pp. 132-133)[2]

In that regard, the Apartheid apparatus is an "encyclopedian State". But the requirements that this statistical mania involves are far too demanding for the capacities of the administration. This has motivated the mechanisation of the system in order to satisfy the data appetite of the State (Breckenridge, 2002a; Breckenridge, 2002b). From the early 1970's, ICTs – at this time, computers – are considered as useful auxiliaries of the oppressive system (Chokshi & al., 1995). In that sense, the most rational instruments of obsessional statistical arithmetic are serving a folly made of violence, racism and authoritarianism. For Dubow (1995, p.248), "when discussing apartheid, it is important not to draw false distinctions between rationality and irrationality, sanity and madness: there was method in the madness of apartheid, just as there was madness in its method".

The Use of "Control Electronics" to Boost the State Knowledge

Formalising reality is an obsession of the Apartheid regime. This formalisation is aiming at strengthening the efficiency of the totalitarian power structure. It is heavily based on the use of ICTs, depicted at this time as "control electronics". At this time, the South African State has to rely on the intervention of foreign firms to get appropriate ICT products and services. The IT sector is the most advanced of the Apartheid business (First & al., 1972, p.106). In 1977, only the United States and Great Britain are spending more money than South Africa on the acquisition of ICT equipment (KSG, 2004).

In the 1970s, the country is totally dependent on foreign imports for ICTs. In 1980, 70% of computers are of American origin (KSG, 2004). Some U.S. firms involved in this business are Unisys, Hewlett Packard, IBM, Westinghouse (Slob, 1990). Knight (1986) lists 14 U.S. companies as main IT providers for the State. For example, in 1986, the State is the first buyer of computers, absorbing 42% of the total IT sales (KSG, 2004). The United Nations military embargo that was established in 1963 did not prevent the State from acquiring IT equipments for authoritarian purposes.

The prominent role that was played by telecom and ICT infrastructures is shown by the prosecution of firms like IBM, ICL and Fujitsu in the United States. Because they turned down the invitation to appear before the Truth and Reconciliation Commission (TRC), the South African non-governmental organisation Khulumani Support Group (KSG) ledged a complaint against these firms in the State of New York in 2002 but lost the case. Indeed, ICL equips the police, local authorities and the defence administration with hundreds of computers. It particularly provides IT equipment to the Bantu Reference Bureau in 1967 and to the Department of Plural Affairs (DPA) that is using a computer network to coordinate its 14 Bantu Administration Boards and store fingerprints and personal data of 16 million South Africans in 1978. The DPA digitalises the National Population Register (NPR) that is aiming at "managing" 25 million South Africans. An impressive database is constructed upon several criteria: name, sex, age, home address, identity picture, civil status, driving license, work address, fingerprints… The government can thus rationalise the influx control and its surveillance on political opponents. On that specific issue, the National Intelligence Service (NIS, succeeding to the Bureau of State Security in 1970) is using an advanced ICT infrastructure to document the activities of anti-Apartheid movements (Slob, 1990, p.26). The South African police is using Series 2900, 190s and 2960s ICL computers, notably in the implementation of the *pass laws* system. In the mid-1960s, ICL automatizes the system that is aiming at running the flows of Black people across the country. IBM, on its side, implements a system targeting at non-Black people (Crush, 1992). IBM started its activities in South Africa in 1952 when it provides the first supercalculator that is used in the ICT-based management of population databases. The Department of Interior is using IBM computers (notably 370/158 Series) to store the *Book of Life* files of 7 million people classified as "non-Black" (Coloured, Asians, Whites). The Department of the Prime Minister, Department of Statistics and Department of Prisons are also using IBM products, as well as the South African firm Infoplan that was the main provider of the national defence force. In the same time, Phillips is promoting its *Access Control System* that stores the identity and movements of people entering and leaving a public place. AEG-Daimler-Benz is proposing the same product.

This digital automatisation must improve the efficiency of the national identification system. This system is the basis of the formalisation process: to normalise the population under a totalitarian order, the State needs to know all people's personal data. In the 1950s and 1960s, the national identification system is based on three main pillars that are heavily reliant on ICTs. The first component is the Reference Book (*Bewysboeks*) nicknamed Dompas (the dumb pass), which is replaced in 1986 by the current green bare-coded I.D. book. According to Breckenridge (2002a, p.17), the bureau in charge of these passes (*Bewysburo*) has quickly been used by the regime as a key piece of the surveillance process. The second pillar is the national fingerprints registry. The last pillar is the attribution of an identity number (*persoonsnommer*) to each individual

In order to implement the cognitive and repressive process, the creation of fingerprint databases has been accelerated by the South African authorities. The first fingerprint was collected in 1900 by Sir Edward Henry in the Natal colony in order to establish the first fingerprint bureau in Pietermaritzburg. This administrative arrangement was concretising the "panoptic fantasy" of the government (Breckenridge, 2002a, p.2). In 1972, the Department of Home Affairs computerises the NPR. In 1979, the *National Research Institute for Mathematical Sciences* improves the system. Until 1986, all personal data and fingerprints of the Black population are stored by the Plural Affairs Department. In 1987, the Department of Home Affairs uses the Model 370/158 mainframe IBM computers to store the NPR and holds 18,9 million fingerprints (Slob, 1990, p.24-25).

This historical process shows that the ICT-based formalisation has served the totalitarian normalisation led by the Apartheid regime. The results that were obtained have an important impact on the nature of the post-Apartheid Welfare State.

E-GOVERNMENT IN DEMOCRACY: THE USE OF ICTS TO ENHANCE THE WELFARE STATE – AT WHAT PRICE?

The post-Apartheid South Africa is deeply ambivalent: since 1994, the country has experienced both radical changes and continuities with the former regime (Guillaume & al., 2004). As far as e-government is concerned and especially its formalisation and normalisation component, the ambivalence is clear: the sophisticated infrastructure deployed by the apartheid government is an opportunity for the Welfare State to be more efficient in the provision of public services but it can also be a burden for the population if this infrastructure is used on a non-democratic way.

The Formalisation of the South African Population

The formalisation of the population remains a priority for the government, especially in its push for the development of a Welfare State. We will focus on one specific project that is being implemented for that purpose: the Home Affairs National Identification System (HANIS).

This project is the major building block of the DHA renewal strategy to rationalise the e-Welfare State. The DHA IT "turn-around strategy" could reach a total amount of 2 billion Rands and will be implemented from January to December 2008. It is implemented by a public-private DHA "Turnaround Action Team". One project is the "track and trace" system for IDs. It must allow the DHA – and the public – to track each ID book transaction from application to collection by means of a website, SMS and a call centre. The system should start being implemented in February 2008 (Engelbrecht, 2007b).

In this context, HANIS aims at providing a centralised database detailing the profile and activities of the South Africa population. In 2003, the then DHA Director-General, Barry Gilder, describes the system in these terms:

as I understand the dream, it was that any client of the Department would be able to go into a Home Affairs office and have their fingerprints, photograph, signature and application taken on the spot electronically, checked immediately against the system and entered in real time onto the system. (Gilder, 2003)

HANIS was conceived by the DHA in 1993 and approved by the Presidency in January 1996 for an initial total amount of 930 million Rands[3]. In February 1999, the tender was awarded to the Marpless consortium to implement the first steps of the system. The consortium is combining the Japanese conglomerate Marubeni Inc. and the South African firm Plessey in a joint venture. In January 2000, the project was officially launched.

By officialising the existence of people in the eye of the State, HANIS is expected to strengthen the fight against crime and terrorism, to reduce fraud (notably in the payment of pensions and social grants[4]) and to improve the performances of the South African Welfare State. To fulfil these objectives, HANIS is made of three components. The first one is the Automated Fingerprint Identification System (AFIS) which is a fingerprint recognition technology developed by the firm NEC. About 40 million fingerprints must be digitalised: it is the back record conversion. New fingerprints (for instance, those of new born or people getting their first identity card) are directly stored in the HANIS database. In September 2007, almost all of the fingerprints that were stored at the DHA had been migrated to a digital format: the AFIS contained about 30 million sets of fingerprints, the target being 31 million out of the country's 45 million citizens. The remainder are either children younger than 16, who are not fingerprinted, or the estimated 6% of adults who do not have an identity document and whom the department is trying to reach (Engelbrecht, 2007c).

To secure the data, the DHA has spent 207 million Rands in 2005-2007 creating the HANIS disaster recovery system at a secure location which backs up all our records every fifteen minutes on a continual basis (Mapisa-Nqakula, 2007).

The second component of HANIS is the South African Multi-Application Identification Card (SAMID): a chip-based "smart card" must gather all personal data and facilitate the provision of everyday public services. This Secure Electronic National Identity Card (SENID) will combine fingerprints, an identity picture, an identity number and an electronic chip gathering the following information: unemployment insurance number, health data (blood group, allergies, the list of the 10 last medical treatments and prescriptions), social transfers (family and housing benefits), number and amount of pensions, driving license and car registration numbers, tax information (Brümmer, 2002a). In Foucault's terms (1995, p.163-165), this card will be a tool of the "political anatomy of the details" that characterizes the formalisation project of the State. The cost of the SAMID component is estimated to be around 2,5 billion Rands (Brümmer, 2002b). It is not yet decided whether the card will be free of charge. The first cards will be distributed to retired people in order for pensions to be paid more rapidly and waiting queues to be shorter (Department of Home Affairs, 2001, p.48). The other reason why the focus is put on retired people is the amount of fraud in the payment of pensions that would approximate one billion rand (Buthelezi, 1997).

The last component of the HANIS project is the integration of HANIS in the National Population Register (NPR) which generates birth and death certificates. The firm Unisys is expected to integrate the different elements of the system (identity image capture, identity checks in the database, deployment of the telecommunication infrastructures). Beyond this, all DHA databases must be integrated in the HANIS system: the Electronic Document Management System (EDMS[5]), the Movement Control System (MCS), the Visa System, the Refugee Database and the Illegal Foreigners Database (Gigaba, 2004).

As in the movie *Matrix*, one talks of HANIS *reloaded* because the database must now integrate all individuals that are interacting with the DHA: citizens, foreign residents, foreign visitors (tourists for instance), refugees[6], documented and undocumented immigrants, the total number of people concerned reaching about 300 million individuals. A priority will be given to the equipment of the 57 official points of entry into the country and particularly to the Lindela retention and repatriation centre that is dealing with undocumented people (Lambinon, 2003). According the government, the formalisation of flows in and out of the country must rationalise the management of the migrant population through a better interconnection of all entry points, in order to avoid a situation whereby a migrant candidate can jump from one entry point to the other in order to get a go-ahead. For Barry Gilder (2003), this situation is not totally without consequence on the levels of criminality.

Because the HANIS is not totally operational yet, it is too early to evaluate its achievements or shortages but one can already says that it is the first priority of the DHA in its formalisation-normalisation enterprise since it represents 22% of the IT expenditure of the DHA for the 2007/2008 budget period. However, prospects for an operational HANIS are still blur: the supplier contract for HANIS expired in June 2006 and by June 2007 no action had been taken whether to reopen it for tender or to extend it with the current supplier (Vecchiatto, 2007a). In July 2007, a contract was signed for equipment to be delivered in March 2008. Also, the HANIS maintenance contract with the Marpless consortium has been extended by a further three years. The introduction of the first smart cards could start in April 2008 (Engelbrecht, 2007c; Senne & Engelbrecht, 2007) and the government intends to introduce "e-passports" (containing a chip and, probably, a RFID) in April 2008 (Engelbrecht, 2007c).

The HANIS project is part of a bigger move illustrated by the fact that, in October 2007, the South African firm GijimaAst-led consortium won a 2 billion rand tender to implement an integrated citizen-centric documentation system called "Who am I online (I am I said)" which gives a "single view of the citizen" and visitors. The tender process started in March 2006. The consortium is expected to integrate birth, death and marriage certificates and identity documents with the existing AFIS (Engelbrecht, 2007d).

The Risk of a Non-Democratic Normalisation

The fact that the Welfare State needs to know as much as possible about the population as a whole for efficient policy making and about individuals for appropriate public service delivery is not surprising. However, more contentious is the way this formalisation of reality is being implemented: what data for whom? In other words, because the State is a huge machinery, a specific matrix must be designed so that all data can not be accessed by all parts of the administrative apparatus.

This is the heart of the ambivalence of e-

Welfare State: on the one hand, citizens can get the most of the formalisation of their existence by getting more visible to the State and thus having access to appropriate public service; on the other, they must accept to be transparent to the administration and get their lives scrutinized by the Leviathan. To put it Bourdieu's terms (1993, p.221-223), the "left hand" of the State can serve the population only if the "right hand" of the State knows and controls it.

Who knows what? That is the critical question in the formalisation and normalisation of reality. Two issues are at stake: the publicisation of private life and the commercialisation of personal data. They are both core pieces of the rise of the e-Welfare State that can be depicted as a collection of *Little Sisters* rather than a single *Big Brother.*

The construction of a seamless administrative apparatus implies that most of (if not all) parts of the administration can communicate with each other and notably exchange information about individuals that they have to deal with. The intimate life of everyone must be accessed and "networked" in the State databases. The protection of personal data can not be but relative: social life (i.e., to live within a society and get the most of it) and the Welfare State make absolute private life impossible (Davies, 1997; Lyon, 2002; Lyon, 2003). The individual can not be invisible; otherwise, he/she would lose any sociality. Moreover, Lyon (2002) recalls that the everyday private life is positively affected by State control because it is the condition for citizens to enjoy the benefits of a Welfare State.

Nevertheless, the publicisation of private life that results from the formalisation / normalisation dynamics is damaging when it's implemented unbeknown to the citizens. It is especially the case when citizens know that the State collects data about them but don't know where these data go to in the end: which administration? Judiciary? Social? Etc... What connections are being established between the different databases? The risk is that of an "administrative panopticism—the urgent desire to complete and centralise the state's knowledge of its citizens". Breckenridge even sees here the "informational legacy of Apartheid" (Breckenridge, 2002b, p.3 & 13; Brümmer, 2002a).Van Tonder (2003) and the NGO Bridges.org (2002) point at the lack of any suitable institutional arrangement which could protect citizens against the misuse of their personal data for political or other purposes. Indeed, there might be a necessity of an independent agency that could set limits to the formalisation/surveillance process. This agency could implement the dispositions that are contained in the *Promotion of Access to Information Act* (PAIA, n°2, 2000) and the *Electronic Communications and Transactions Act* (n°25, 2002).

In 2002, the South African Law Reform Commission started drafting a comprehensive national Data Privacy Act (De Kock, 2005). In October 2005, the South African Law Reform Commission published a Discussion Paper (n°109) on "Privacy and Data Protection" that invites the government to push for the introduction of legislation regulating the collection, storage and processing of personal information by the public and private sector. To that end, the Discussion Paper contains a draft Protection of Personal Information Bill that must improve the South African data protection system which is based on the Bill of Rights and developed by the PAIA Act.

One improvement must deal with the "compartmentalisation" of data within the administrative apparatus: in a networked Welfare State, who can access what? In the United States, Loïc Wacquant (2004, p.49) diagnoses the rising of such a State through the implementation of various instantaneous identification and surveillance instruments, what he calls an "infrastructural power" – the capacity for the State to "penetrate" within the population. The regulation of data access is particularly important with the crime situation in South Africa: ICTs are powerful tools in the fight against crime but they can foster the undue mixing up of civilian and criminal data. The

DHA and the South African Police Service (SAPS) databases are built upon the same architecture, in order to facilitate their joint exploitation. In 1997, the then Minister of Home Affairs, Mangosuthu Buthelezi, was declaring that the government was considering the integration of both databases (Buthelezi, 1997). In 2004, some high-rank police officials were also considering this option. The Court Process Project is sometimes using both systems so that the Departments of Justice, of Correctional Services and of Social Development can cross their information for court decisions to be taken quicker. This might illustrate what Wacquant (2004, p.19) calls the "criminalisation of social insecurity".

Beyond the data access within the Welfare State administration, another issue is crucial: the commercialisation of data outside the administration. Thus, the HANIS card is expected to have a financial functionality dedicated to banks, insurance and medical aid companies. The DHA intends to sale the HANIS-related data to the private sector, particularly for data mining initiatives. The fight against fraud is the main argument that must justify this commercialisation (Buthelezi, 2002). For Van Tonder (2003), this privatisation is a misuse of biometrics. In June 2004, according to the South African press, the Independent Electoral Commission (IEC) had used the national election roll to provide data to the South African Post Office (SAPO) and to the firm Intimate Data that was contracted by the SAPO in February 2004 to update, clean up and maintain its database of postal addresses. The IEC and the SAPO refuted this allegation (IEC, 2004).

The development of the e-Welfare State can not go without that of Little Sisters: this ambivalence of e-government is well illustrated in the South Africa case. We know that knowledge is power. In the e-Welfare State, information, data and knowledge are the building blocks on which both social services and surveillance systems can operate. Both need to deal with a social reality that is formalised and, *in fine*, normalised. In the e-Welfare State, standardised knowledge is power. And this standardisation is based on ICTs. The collection of *Little Sisters* (Armatte, 2002; Castells, 20000; Lyon, 2003) is essential to the existence of the social Leviathan : ICTs are improving the "reason of State" for knowledge production that is aiming at both social and control purposes. Citizens are compelled to "produce truth" about their lives (Foucault, 1997, p.22-23). In South Africa, the legacy that was left by the Apartheid regime is an opportunity for the Welfare State to make the most of e-government; but it is also a burden that must fit into the set of democratic requirements.

REFERENCES

Armatte, E. (2002). Informatique et libertés: de big brother à little sisters ? *Terminal, n°88*, 11-21.

Bonner, P., Delius, P., & Posel, D. (1993). The shaping of apartheid: Contradiction, continuity and popular struggle. In P. Bonner, P. Delius, & D. Posel (Eds.), *Apartheid's genesis 1935-1962* (pp. 1-41). Johannesburg: Wits University Press.

Bourdieu, P. (1993). La démission de l'Etat. In P. Bourdieu (Eds.), *La misère du monde* (pp. 219-228). Paris: Seuil.

Breckenridge, K. (2002a, May). *From hubris to chaos: The making of the bewysburo and the end of documentary government.* Paper presented at the Wits Interdisciplinary Research Seminar, WISER, University of the Witwatersrand, Johannesburg.

Breckenridge, K. (2002b, October). *Biometric government in the new South Africa.* Paper presented at the The State We Are In Seminar Series, WISER, University of the Witwatersrand, Johannesburg.

Brdiges.org (2002). *Government efficiency vs.*

citizens' rights: The debate about electronic public records comes to developing countries. Cape Town: Bridges.org.

Brümmer, S. (2002a). From dompas to smart card. Johannesburg: *Mail & Guardian*, 22 February.

Brümmer, S. (2002b). Buthelezi and Masetlha at it again. Johannesburg: *Mail & Guardian*, 22 March.

Buthelezi, M. (1997). *Department of Home Affairs budget vote 1997/1998.* Cape Town: National Assembly – Republic of South Africa, 17 April.

Buthelezi, M. (2002). *Parliamentary media briefing.* Cape Town, GCIS – Republic of South Africa, 11 February.

Buthelezi, M. (2003). *Home Affairs budget speech.* Cape Town, National Assembly – Republic of South Africa, 19 May.

Castells, M. (2000). *The information age: Economy, society and culture, vol.1 The rise of the network society*, Oxford: Blackwell.

Center for public service innovation. (2003). *Government unplugged – mobile and wireless technologies in the public service*, Pretoria: Center for public service innovation.

Cheneau-Loquay, A. (2005). Comment les nouvelles technologies de l'information et de la communication sont-elles compatibles avec l'économie informelle en Afrique. *Annuaire Français de Relations Internationales*, *5*, 345-375.

Chokshi, M., Carter, C., Gupta, D., Martin, T., & Robert, A. (1995). *Computers and the apartheid regime in South Africa.* Retrieved October 28, 2003, from http://www-cs-students.stanford.edu/~cale/cs201/

Crush, J. (1992). Power and surveillance on the South African goldmines. *Journal of Southern African Studies*, *18*(4), 825-844.

Davies, S. G. (1997). Re-engineering the right to privacy : How privacy has been transformed from a right to a commodity. In P. E. Agre & M. Rotenberg (Eds.), *Technology and privacy: The new landscape* (pp. 143-165). Cambridge MA: MIT Press.

De Kock, E. (2005). Data protection in South Africa. *De Rebus*, December.

Department of Home Affairs (2001). *Strategic Plan 2002/2003 to 2004/2005*, Document n°2/6/8/P, Pretoria : Department of Home Affairs.

Desrosières, A. (1993). *La politique des grands nombres: Histoire de la raison statistique.* Paris: La Découverte.

Desrosières, A. (1997). Du singulier au général: L'argument statistique entre la science et l'Etat. In B. Conein & L. Thévenot (Eds.), *Cognition et information en société* (pp. 267-282). Paris: Ecole des Hautes Etudes en Sciences Sociales.

Dubow, S. (1995). *Scientific racism in modern South Africa.* Cambridge: Cambridge University Press.

Engelbrecht, L. (2007a). Five in court for marriage fraud. Johannesburg: *ITWeb*, 10 December.

Engelbrecht, L. (2007b). Blank cheque for Home Affairs information technology. Johannesburg: *ITWeb*, 5 September.

Engelbrecht, L. (2007c). HANIS gets Rands 30m 'refresh. Johannesburg: *ITWeb*, 6 September.

Engelbrecht, L. (2007d). Getting Home Affairs cup-ready. Johannesburg: *ITWeb*, 5 November.

Evans, I. (1997). *Bureaucracy and race: Native administration in South Africa.* Berkeley, CA: University of California Press.

Sadie, J. L. (1950). The political arithmetic of the South African population. *Journal of Racial Affairs*, *1*(4), 3-8.

Fauvelle-Aymar, F. X. (2006). *Histoire de l'Afrique*

du Sud. Paris: Seuil.

First, R., Steele, J., & Gurney, C. (1973). *The South African connection: Western investment in apartheid*. Harmondsworth: Penguin.

Foucault, M. (1995). *Surveiller et punir: Naissance de la prison*, Paris: Gallimard.

Foucault, M. (1997). Cours du 14 janvier 1976. In F. Ewald, A. Fontana & M. Bertani Ed.), *Il faut défendre la société* (pp. 21-36). Paris: Gallimard-Seuil.

Gigaba, M. (2004). Home Affairs department budget vote 2004/2005. Cape Town: National Assembly – Republic of South Africa, 11 June.

Gilder, B. (2003). Media Briefing. Johannesburg: Department of Home Affairs – Republic of South Africa, 5 November.

Gilliom, J. (2001). *Overseers of the poor: Surveillance, resistance and the limits of privacy*. Chicago, IL: University of Chicago Press.

Guilaume, M. (1978). *Eloge du désordre*. Paris: Gallimard.

Guillaume, P., Péjout, N., & Wa Kabwe Segatti, A. (2004). *L'Afrique du Sud dix ans après: Transition accomplie?*. Johannesburg – Paris: Institut Français d'Afrique du Sud – Karthala.

Hart, K., (1973). Informal income opportunities and urban employment in Ghana. *The Journal of Modern African Studies*, *11*(1), 61-89.

International Labour Organisation. (1972). *Incomes and equality: A strategy for increasing productive employment in Kenya*, Geneva: International Labour Organisation.

Independent Electoral Commission. (2004). Independent Electoral Commission on reports about giving out details from voters' roll. Pretoria: Independent Electoral Commission, 15 June.

Khulumani Support Group. (2004). Complaint. South District Court, New York State, 15 December.

Knight, R. (1986). *US computers in South Africa*. Retrieved October 26, 2003, from http://richardknight.homestead.com/files/uscomputers.htm

Lambinon, I. (2003). *Briefing at the Home Affairs portoflio committee*. Cape Town, National Assembly – Republic of South Africa, 18 March.

Lascoumes, P., & Le Gales, P. (Ed.). (2004). *Gouverner par les instruments*. Paris: Presses de Sciences Po.

Lautier, B. (2004). *L'économie informelle dans le tiers monde*. Paris: La Découverte.

Lyon, D. (2002). Everyday surveillance : Personal data and social classifications. *Information, Communication & Society*, *5*(2), 242-257.

Lyon, D. (2003). Introduction. In D. Lyon (Ed.), *Surveillance as social sorting: Privacy, risk and digital discrimination* (pp. 1-9). London: Routledge.

Mapisa-Nqakula, N. (2004). *Home Affairs department budget vote 2004/2005*. Cape Town: National Assembly – Republic of South Africa, 11 June.

Mapisa-Nqakula, N. (2007). DHA budget speech for budget vote 2007. Cape Town: National Assembly – Republic of South Africa, 7 June.

Martin, D.-C. (1998). Le poids du nom: Culture populaire et constructions identitaires chez les 'métis' du Cap. *Critique Internationale*, n°1, 73-100.

Minnaar, C.-L. (2004). Putting maps to work. Johannesburg: *ITWeb*, 12 March.

Péjout, N. (2004). Big brother en Afrique du Sud? Gouvernement électronique et contrôle panoptique sous et après l'apartheid. In P. Guillaume, N. Péjout & A. Wa Kabwe Segatti (Ed.), *L'Afrique du Sud dix ans après: Transition accomplie?* (pp. 79-103). Johannesburg – Paris: Institut Français d'Afrique du Sud – Karthala.

Péjout, N. (2007). *Contrôle et contestation. Sociologie des politiques et modes d'appropriation des technologies de l'information et de la communication (TIC) en Afrique du Sud post-apartheid.* Unpublished doctoral dissertation, Ecole des Hautes Etudes en Sciences Sociales, Paris.

Posel, D. (2000). A mania for measurement: Statistics and statecraft in the transition to apartheid. In S. Dubow (Ed.), *Science and society in Southern Africa* (pp. 116-142). Manchester: Manchester University Press.

Senne, D., & Engelbrecht, L. (2007). Home Affairs admits ID inefficiencies. Johannesburg: *ITWeb*, 30 August.

Slob, G. (1990). *Computerizing apartheid: Export of computer hardware to South Africa.* Amesterdam: Holland Committee on Southern Africa.

South Africa Press Agency, 2004. More than half a million domestics registered with unemployment insurance fund (UIF). Pretoria: South Africa Press Agency, 2 June.

Statistics South Africa. (2002). *South African Statistics.* Pretoria: Statistics South Africa.

Van Der Berg, R. J. (2004). First ID for !Xhu and !Khwe communities, *Bua News*, Pretoria: Government Communication and Information System – Republic of South Africa, 9 June.

Van Tonder, K. (2003). Biometric identifiers and the right to privacy. *De Rebus*, 28 (8), August, Retrieved October 3, 2004, from http://www.derebus.org.za/archives/2003Aug/articles/Biometric.htm

Vecchiatto, P. (2007a). Languishing HANIS needs attention. Johannesburg: *ITWeb*, 7 June.

Vecchiatto, P. (2007b). SA trials smart ID cards. Johannesburg: *ITWeb*, 2 November.

Wacquant, L. J. D. (2004). *Punir les pauvres: Le nouveau gouvernement de l'insécurité sociale*, Marseille: Agone.

Weber, M. (1978). *Economy and society: An outline of interpretative sociology*, Berkeley: University of California Press.

ENDNOTES

1. For the Apartheid regime, the term "African" designates the Black population (Constant-Martin, 1998, p.75).
2. This list is compiled by Posel (2000) from : Republic of South Africa. (1971). *Report of the Department of Bantu Administration and Development for the Period 1 January 1971 to 31 December 1971.* Pretoria: Department of Bantu Administration and Development, RP 41/73.
3. 1 rand = 0.098 euro (15/01/2008).
4. The annual fraud would amount to about 2 billion Rands (DHA, 2001, p.47). Government lost more than 1 billion Rands to social grant fraud in 2006. The private sector is also suffering from ID-related fraud: for instance, in 2006, insurance lost in excess of 3 billion Rands due to their inability to verify information (Senne & Engelbrecht, 2007). In the United States, Gilliom (2001) analyses how ICTs are fuelling a "welfare surveillance": the Client Registry and Information System – Enhanced (CRIS-E) refines the meticulous control operated by the welfare administration upon beneficiary families.
5. Under its Electronic Data Management System, the DHA has scanned a total of 57 million records including birth, marriage and death records (Mapisa-Nqakula, 2007).
6. The Department of Home Affairs was planning to roll out a refugee online verification system in January 2008 (Vecchiatto, 2007b).

This work was previously published in Electronic Constitution: Social, Cultural, and Political Implications, edited by F. Amoretti, pp. 99-114, copyright 2009 by Information Science Reference (an imprint of IGI Global).

Compilation of References

2007 Asia - Telecoms, Mobile and Broadband in India: http://www.researchandmarkets.com/reportinfo.asp?report_id = 481013

Abdulahi, A. (2007, March). Spatial and vertical price transmission in food stapples market chains in eastern and southern Africa: What is the evidence? A paper presented at the FAO *Trade and Markets Division Workshop on Staple Food Trade and Market Policy Options for Promoting Development in Eastern and Southern Africa*, Rome.

About Zambia. (2007). *OpenSource Zambia*. Retrieved July 21, 2007, from www.opensource.org.zm/

About, E. -Government. (2005). *World Bank*. Retrieved April 27, 2007, from http://go.worldbank.org/6WT3UPVG80

ADB (2003). *Toward E-Development in Asia and the Pacific: A Strategic Approach to Information and Communication Technology*, www.adb.org, accessed 27 May 2007.

Adegite, V. (2006, Sept. 15). *E-Agriculture—Defining a new paradigm for agricultural development*. Retrieved Dec. 5, 2008 http://www.e-agriculture.org/.

Adeniyi, S. (2008, Aug. 6). WB consultant expresses fear over food crisis despite Nigeria's external reserves. *Nigerian Tribune*. Retrieved Nov. 9, 2008 from http://www.tribune.com.ng/.

Adeniyi, S. (2008, May 21). 'Without food security, Nigeria can't talk of national security.' *Nigerian Tribune*. [Online]. Retrieved June 1, 2008 from http://www.tribune.com.ng/

Adesisa, O. (2001). Governance in Africa: The role for information and communication technologies. *Economic Research Papers*, *65*, 5–30.

Adjei, E. (2004). Retention of medical records in Ghanaian teaching hospitals: Some international perspectives. *African Journal of Library Archives & Information Science*, *14*(1), 37–52.

Adjei, E., & Ayernor, E. T. (2005). Automated medical record tracking system for the Ridge hospital, Ghana Part 1: Systems development and design. *African Journal of Library Archives and Information Science*, *15*(1), 1–14.

Afrol News. (2008, June 7). New media policy being designed in Nigeria. *afrol News*. Retrieved Dec. 5, 2008 http://www.afrol.com/.

AgLeader Technologies (2008). http://www.agleader.com/products.php?Product=pfadvantage. (Verified January 23, 2008)

Aker, J. C. (2008). *Does digital divide or provide? The impact of cell phones on grain markets in Niger*. Job Market Paper, University of California, Berkeley, CA.

Akther, M. S., Onishi, T., & Kidokoro, T. (2007). E-Government in a developing: citizen-centric approach for success. *International Journal of Electronic Governance*, *1*(1). doi:10.1504/IJEG.2007.014342

Akula, V.B. (2000). *Putting Technology to Work for Poverty Alleviation: A Draft Proposal for $151,030 to Develop Smart Cards for Microfinance*, Swayam Krishi Sangam, Hyderabad.

Alford, T., Chasteen, T., & DeRosear, K. (2008). *It's the educonomy, stupid!: Redifining America's workforce sysetm: A practitioner's guide*. Kingston, Tennessee: Worldwide Interactive Network, Inc.

Alhassan, A. (2005). Market valorization in broadcasting policy in Ghana: Abandoning the quest for media democratization. *Media Culture & Society, 27*(2), 211–228. doi:10.1177/0163443705050469

Alhassan, A. (2007). Broken Promises in Ghana's Telecom sector. *Media Development, 3*, 45.

Allen, I.G. (2002). *Can we eradicate poverty while tolerating alienation?* Retrieved June 29, 2006, from the Christian Mission for the United Nations Community, http://www.Christianmission-un.com

Alozie, E. C. (2005). *Cultural reflections and the role of advertising in the socio-economic and national development of Nigeria*. Lewiston, NY: The Edwin Mellen Press.

Aluko-Olokun, I. (1997). Institutional building for effective capacity development. In ITF (Ed.). *Realizing National Vision by the Year 2010: The Human Resource Development Imperatives* (pp. 69 – 76). Jos, Nigeria: Industrial Training Fund (ITF).

American Library Association. (2000). *American Library Association Presidential Committee on Information Literacy*. Retrieved June 29, 2006, from http://www.ala.org/acrl/nili/ilit1st.html

Amoretti, F. (2006). *The digital revolution and Europe's constitutional process. E-democracy between ideological ad institutional practices*. Paper presented at the VII Congresso Espanol De Ciencia Politica Y De La Administration, Grupo De Trabajo 9 Communicacion Politica.

Amoretti, F. (2007). International Organizations ICTs policies: E-Democracy and E-government for political development. *Review of Policy Research, 24*(4), 331–344. doi:10.1111/j.1541-1338.2007.00286.x

Anderson, A., Virmarlund, V., & Timpka, T. (2002). Management demands on information and communication technology in process-oriented healthcare organizations: The importance of understanding managers' expectations during early phases of systems design. *Journal of Management in Medicine, 16*(2/3), 159–169. doi:10.1108/02689230210434907

Anderson, G. L., Herr, K., Nihlen, A. S., & Noffke, S. E. (2007). *Studying Your Own School: An Educator's Guide to Practitioner Action Research*. Thousand Oaks, CA: Corwin Press.

Anderson, J. D., Ward, C. E., Koontz, S. R., Peel, D. S., & Trapp, J. N. (1998). Experimental simulation of public information impacts on price discovery and marketing efficiency in the fed cattle market. *Journal of Agricultural and Resource Economics, 23*, 62–278.

Anheier, H. K. (2007). Bringing civility back in – Reflections on global civil society. *Development Dialogue*, (49): 41–49.

Aregu, R., Bagaya, M., & Nerbonne, J. (2008). An ICT-Based Digital Content Information Access Framework in Developing Countries: the Case of Agricultural Informatics Access and Management in Uganda. In *Proceedings of IST-Africa Conference*, Windhoek, Namibia.

Armatte, E. (2002). Informatique et libertés: de big brother à little sisters ? *Terminal, n°88*, 11-21.

Aroyo, T. (2006). Promoting the use of information and communication technologies (ICTs) in Nigeria's agricultural extension service. *Moor Journal of Agricultural Research, 7*(1&2), 100–106.

Artkinson, D. R. (2003). *Network Government for the Digital Age*. Washington, DC: Progressive Police Institute.

Ashraf, N., Gine, X., & Karlan, D. (2005). *Growing export oriented crops in Kenya: Evaluation of the DrumNet services*. Unpublished.

Asia-Pacific Journal on Information, Communication and Space Technology: Reviews and Updates. First Issue – 2006, Economic and Social Commission for Asia and the Pacific

Atkins, P., & Bowler, I. (2001). *Food in Society: Economy, Culture and Geography*. London, UK: Hodder Arnold.

Avgerou, C., & Walsham, G. (Eds.). (2000). *Information technology in context: Implementing systems in the developing world*. Brookfield, VT: Ashgate.

Ayers. S. (2005). *ICT, poverty and the global economy: Challenges and opportunities for developing countries*. Information for Development (InfoDev). Retrieved Nov. 4, 2008 http://www.infodev.org/en/.

Ayogu, M., & Ogbu, O. (2002). *Globalization and technology: Africa's participation and perspectives*. Nairobi, Kenya: African Technology Policy Studies Network (ATPS).

Babatunde, R., & Oyatoye, E. (2005, Oct. 11–13). *Food security and marketing problems in Nigeria: The case study of marketing maize in Kwara State*. Presentation at Deutscher Tropentag, Hohenheim. Retrieved Nov. 25, 2008 www.tropentag.de/2005/abstracts/.

Background notes—Zambia. (2007). United States State Department. Retrieved July 29, 2007, from www.state.gov/r/pa/ei/bgn/2359.htm

Badu, E. E. (2004). Academic library development in Ghana: Top managers' perspectives. *African Journal of Archives and Information Science*, *14*(2), 93–107.

Bage, L. (2005). *Statement delivered on the launch of MDG Report, 18 January, 2005*. Retrieved February 1, 2009, from http://www.ifad.org/events/mdg/ifad.htm

Bagetoft, P., & Olesen, H. B. (2004). *The Design of Production Contracts*. Denmark: Copenhagen Business School Press.

Bakos, Y. (1991). A strategies analysis of EM. *MIS Quarterly*, *15*(4), 295–310. doi:10.2307/249641

Bakos, Y. (1998). The emerging role of electronic marketplaces on the Internet. *Communications of the ACM*, *41*(8), 35–42. doi:10.1145/280324.280330

Baldwin, L. P., Clarke, M., Eldabi, T., & Jones, R. W. (2002). Telemedicine and its role in improving communication in healthcare. *Logistics Information Management*, *15*(4), 309–319. doi:10.1108/09576050210436147

Baramati Initiatives (2001). *SKS-Smart Cards: Case Study*. www.baramatiinitiatives. com/cases/case10. htm.

Barrett, C. (2008). Smallholder market participation: Concepts and evidence from eastern and southern Africa. *Food Policy*, *34*, 299–317. doi:10.1016/j.foodpol.2007.10.005

Basu, S. (2004). E-government and developing countries: An overview. *International Review of Law Computers & Technology*, *18*(1), 109–132. doi:10.1080/1360086041 0001674779

Batchelor, S., & Evangelista, S. Hearn, S., Peirce, M., Sugden, S. & Webb, M. (2003). *ICT for development contributing to the millennium development goals: Lessons learned from seventeen InfoDev Projects*. Washington, DC: World Bank.

Batte, M. T. (2005). Changing computer use in agriculture: Evidence from Ohio. *Computers and Electronics in Agriculture*, *47*, 1–13. doi:10.1016/j.compag.2004.08.002

Baym, N. K. (1998). The emergence of On-Line Community. In S. Jones (ED), *Cybersociety 2.0: Revisiting computer-mediated communication and community* (pp. 35-68). Thousand Oaks, California: SAGE Publications, Inc.

Bell, C., & Newby, H. (1972). *Community Studies: An Intriductionto the Sociology of the Local Community*. New York: Praeger Publishers.

Bender, T. (1978). *Community and Social Change in America*. New Brunswick, NJ: Rutgers University Press.

Bertolini, R. (2003). *Making Information and Communication Technologies Work for Food Security in Africa*. Washington, DC: International Food Policy Research Institute.

Bertolini, R., Dawson Sakyi, O., Anyimadu, A., & Asem, P. (2001). *Telecommunication Use in Ghana: Research from the Southern Volta Region*. Working Paper, University of Ghana- Center for Development Research, Bonn University.

Besley, T. (1998). How do financial failures justify interventions in rural credit markets? In C.K. Eicher & J.M. Staatz (Eds). *International Agricultural Development,* (3rd Ed.). (pp. 370-389).

Best, M., & Maclay, C.M. (2002). *Community Internet Access in Rural Areas: Solving the economic sustainability puzzle.* The Global Information Technology Report 2001-2002: Readiness for the networked world, ed., Geoffrey Kirkman (Oxford University Press, 2002). Available online at http://www.cid.harvard.edu /cr/gitrr_030202 .html (May 20, 2007)

Bhakoo, S. (2008). Even scientists recommend his farming methods. *The Tribune, May 14, 128*(133), 5.

Bharti Airtel set up joint venture with IFFCO to provide rural mobile phone services, (n.d.). Retrieved February 1, 2009, from http://www.domain_b.com/companies/companies_b/bharti tele_ventures/20080502_joint.

Bhatnagar, S. (2000). Empowering Dairy Farmers through a Dairy Information & Services Kiosk. Washington, DC: World Bank.

Bhatnagar, S. (2002). E-government: Lessons from implementation in developing countries. *Regional Development Dialogue, 24,* 164–174.

Bhatnagar, S. (2004). *E-government: from vision to implementation – a practical guide with case studies.* New Delhi: Sage Publications (pp. 17, 53, 54, 73).

Bhatnagar, S., & Schware, R. (Eds.) (2000). *Information and Communication Technology in Rural Development. Case Studies from India.* Washington, DC: World Bank.

Bhatnagar, S., & Vyas, N. (2001). Gyandoot: Community-Owned Rural Internet Kiosks. Washington, DC: World Bank.

Bhattacharya, A. (2008). India's food grain yield half of China's; scope for at least 50% increase in production. *Times of India, Wednesday, April 2, 7.*

Bingham, L. B., Nabatchi, T., & O'Leary, R. (2005). The new governance: Practices and processes for stakeholder and citizen participation in the work of the government. *Public Administration Review, 65*(5), 547–558. doi:10.1111/j.1540-6210.2005.00482.x

Binns, T., Kyei, P., Nel, E., & Porter, G. (2005)... *Africa Insight, 35*(4), 21–31.

Blanche, M. T., & Durrheim, K. (Eds.). (1999). *Research in practice: Applied methods for the social sciences.* Cape Town: University of Cape Town Press.

BMi-TechKnowledge Group. (2002). *BMi-T Approach to eReadiness,* Johannesburg: BMi-T.

Bockermann, A., Meyer, B., Omannc, I., & Spangenberg, J. H. (2005). Modeling sustainability: Comparing an econometric (PANTA RHEI) and a systems dynamics model (SuE). *Journal of Policy Modeling, 27*(2), 189-210.

Boehlje, M. (2000). Critical dimensions of structural change: policy issues in the changing structure of the food system. In *Proceedings of the American Agricultural Economics Association Preconference Workshop,* Tampa, FL, July 29, 2000. Retrieved from http://www.agbioforum.org

Bond, P. (2007). Linking below, across and against – World social forum weaknesses, global governance gaps and the global justice movement's strategic dilemmas. *Development Dialogue,* (49): 81–95.

Bonner, P., Delius, P., & Posel, D. (1993). The shaping of apartheid: Contradiction, continuity and popular struggle. In P. Bonner, P. Delius, & D. Posel (Eds.), *Apartheid's genesis 1935-1962* (pp. 1-41). Johannesburg: Wits University Press.

Bourdieu, P. (1993). La démission de l'Etat. In P. Bourdieu (Eds.), *La misère du monde* (pp. 219-228). Paris: Seuil.

Braga, F. & Baker, G. A. (2008). Parma Agrofood Research Management Knowledge Network: PARMA KN. *International Food and Agribusiness Management Review, 11*(3).

Brah, K. (2001). Ghana goes for IT lead. *African Business, July/August, 25.*

Breckenridge, K. (2002a, May). *From hubris to chaos: The making of the bewysburo and the end of documentary government.* Paper presented at the Wits Interdisciplinary Research Seminar, WISER, University of the Witwatersrand, Johannesburg.

Breckenridge, K. (2002b, October). *Biometric government in the new South Africa.* Paper presented at the The State We Are In Seminar Series, WISER, University of the Witwatersrand, Johannesburg.

Bridges.org (2002). *Government efficiency vs. citizens' rights: The debate about electronic public records comes to developing countries.* Cape Town: Bridges.org.

Brint, S. (2001). Gemeinschaft Revisited: A Critique and Reconstruction of the Community Concept. *Sociological Theory, 19*(1), 1–23. doi:10.1111/0735-2751.00125

Brümmer, S. (2002). Buthelezi and Masetlha at it again. Johannesburg: *Mail & Guardian*, 22 March.

Brümmer, S. (2002). From dompas to smart card. Johannesburg: *Mail & Guardian*, 22 February.

Bruwer, J. (2003). South African wine routes some perspectives on the wine tourism industry's structural dimensions and wine tourism product. *Tourism Management, 24*(4), 23–435. doi:10.1016/S0261-5177(02)00105-X

Bryceson, K. P. (2006). *'E' Issues in Agribusiness: The 'What', 'Why' and 'How.'* Gateshead, UK: Athenaeum Press.

Business Day Nigeria/Africa News Network (2007, Nov. 22). Why Nigeria imports $3 billion of food annually. *Business Day Nigeria/ Africa News Network.* Retrieved November 24, 2008 from http://africanagriculture.blogspot.com/

Buthelezi, M. (1997). *Department of Home Affairs budget vote 1997/1998.* Cape Town: National Assembly – Republic of South Africa, 17 April.

Buthelezi, M. (2002). *Parliamentary media briefing.* Cape Town, GCIS – Republic of South Africa, 11 February.

Buthelezi, M. (2003). *Home Affairs budget speech.* Cape Town, National Assembly – Republic of South Africa, 19 May.

Bynum, A. B., Cranford, C. O., Irwin, C. A., & Banken, J. A. (2006). Effect of telemedicine on patient's diagnosis and treatment. *Journal of Telemedicine and Telecare, 12*(1), 39. doi:10.1258/135763306775321407

Canadian International Development Agency. (2007). *Knowledge-Sharing Plan.* Retrieved February 19, 2008 from http://www.acdi-cida.gc.ca/CIDAWEB/acdicida.nsf/En/EMA-218122154-PR4

Cape Gateway. (2003). *Strategic plan for the Department of Agriculture, Western Cape.* Retrieved May 3, 2006, http://www.capegateway.gov.za/text/2003/strategic_plan_2003b.pdf

Carr-Hill, R., & Leach, F. (1995). *Education and training for the informal Sector.* London: Overseas Development Administration.

Castells, M. (2000). *The information age: Economy, society and culture, vol. 1 The rise of the network society,* Oxford: Blackwell.

Castells, M. (Ed.). (2004). *The network society: A cross-cultural perspective.* Northampton, MA: Edward Elgar.

Catts, R., & Lau, J. (2008). *Towards Information Literacy Indicators.* Retrieved December 20, 2008 from http://unesdoc.unesco.org/images/0015/001587/158723e.pdf

Cecchini, S. (2001). *Back to Office Report: Information and Communications Technology for Poverty Reduction in Rural India, Mimeo.* Washington, DC: World Bank.

Cecchini, S. (2002). *Back to Office Report: Evaluation of Gyandoot and Bhoomi and International Conferences on ICT for Development, Mimeo.* Washington, DC: World Bank.

Cecchini, S., & Raina, M. (2002, April). Warana: *The Case of an Indian Rural Community Adopting Information and Communications Technology in Information Technology in Developing Countries, 12*(1). www.iimahd.ernet.in/egov/ifip/apr2002/apr2002.htm.

Cecchini, S., & Scott, C. (2003). Can information and communication technology applications contribute to poverty reduction? Lessons from rural India. *Information*

Technology for Development, 10, 73–84. doi:10.1002/itdj.1590100203

Cecchini, S., & Shah, T. (2002). Information and Communications Technology as a Tool for Empowerment in Empowerment and Poverty Reduction: A Sourcebook. Washington, DC: World Bank. www.worldbank.org/poverty/empowerment/

Center for public service innovation. (2003). *Government unplugged – mobile and wireless technologies in the public service*, Pretoria: Center for public service innovation.

Centre of E-governance. (2003, June). IIM Ahmedabad, Evaluation Studies by the Centre for E-Governance. Indian Institute of Management, Ahmedabad (CEG-IIMA). *Information Technology in Developing Countries, 13*(1). Available online at http://www.iimahd.ernet.in/egov/ifip/jun2003/ article6.htm

Chambers, W., Hopkins, J., Nelson, K., Perry, J., & Pryor, S. Stenberg, & P., Worth, T. (2001). *E-Commerce in United States Agriculture*. ERS White Paper. http://www.ers.usda.gov.

Chan, O. J. (2005). Enterprise Information Systems Strategy and Planning. *The Journal of American Academy of Business, 2*(3).

Chandrasekhar, C. P., & Ghosh, J. (2001). Information and communication technologies and health in low income countries: the potential and the constraints. *Bulletin of the World Health Organization, 79*(9), 850–855.

Chen, D. (2006). *A Knowledge Economy Perspective on Malaysia: A comparative diagnostic.* Washington, DC: World Bank. Retrieved March 25, 2009 from http://info.worldbank.org/etools/docs/library/232719/D1_Malaysia_Benchmark.pdf

Chen, M. A., Jhabrala, R., & Lund, F. (2002). *Supporting workers in the informal economy: A policy framework* [Working Paper on the Informal Sector No. 2]. Geneva: Employment Sector, International Labour Office.

Cheneau-Loquay, A. (2005). Comment les nouvelles technologies de l'information et de la communication sont-elles compatibles avec l'économie informelle en Afrique. *Annuaire Français de Relations Internationales, 5,* 345-375.

Chigona, W., Beukes, J., Vally, D., & Tanner, M. (2009). Can Mobile Internet Help Alleviate Social Exclusion in Developing Countries? *The Electronic Journal of Information Systems in Developing Countries, 36,* 1–16.

Chokshi, M., Carter, C., Gupta, D., Martin, T., & Robert, A. (1995). *Computers and the apartheid regime in South Africa.* Retrieved October 28, 2003, from http://www-cs-students.stanford.edu/~cale/cs201/

Christenson, J. A. (1984). Gemeinschaft and gesellschaft: Testing the spatial and communal hypothesis. *Social Forces, 63,* 160–168. doi:10.2307/2578863

Ciborra, C., & Navarra, D. (2005). Good governance, development theory, and aid policy: Risks and challenges of e-government in Jordan. *Information Technology for Development, 11*(2), 141–159. doi:10.1002/itdj.20008

Closson, R. B., Mavima, P., & Siabi-Mensah, K. (2002). The shifting development paradigm from state centeredness to decentralization: What are the implications for adult education? *Convergence, 35*(1), 28–42.

Coates, N., & Rojas, M. (2006). China and e-government. In D. Remenyi (Ed.), *Proceedings of the 2nd International Conference on E-Government* (pp. 34-41). Pittsburgh: University of Pittsburgh Press.

Corcoran, K. (2008, March 24). Food prices soaring worldwide. *Associated Press.* Retrieved December 6, 2008 from http://abcnews.go.com/International/

Coulter, J., Millns, J., Tallontire, A. & Stringfellow, R. (1999). Marrying farmer co-operation and contract farming for agricultural service provision in liberalizing economies in Sub-Saharan Africa. *ODI Natural Resources Perspectives, 48.*

CRISP Group (2003, July). National Informatics Centre, Department of IT, Ministry of Communications & IT, Government of India, Information Needs Assessment for Rural Communities- An Indian Case Study. Available online at http://ruralinformatics.nic.in/files/4_12_0_229.pdf

Critical Success Factors for Rural ICT Projects in India: A study of n-Logue kiosk projects at Pabal and Baramati. Submitted in partial fulfillment of the requirements for the degree of Master of Management By Vivek Dhawan, IIT Bombay.

Crush, J. (1992). Power and surveillance on the South African goldmines. *Journal of Southern African Studies, 18*(4), 825-844.

CTA, (2002). *Gender and agriculture in the information society: A special report of a CTA meeting. Wageningen,* the Netherlands, September 11-13.

D'Costa, A. (Ed.). (2006). *The new economy in development: ICT challenges and opportunities.* Basingtoke, England: Palgrave Macmillan.

Dada, D. (2006). The failure of e-government in developing countries: A literature review. *Electronic Journal on Information Systems in Developing Countries, 26*(7), 1–10.

Dada, J. (2007, Jan. 25–27). *Nigeria.* Presented at the inauguration of the Telecentre Network of Nigeria. Retrieved November 25, 2007, http://www.giswatch.org/

Daly, J. A. (2003). *Information and communications technology and the eradication of hunger.* Washington, DC: Development Gateway.

Daly, P. (1993). Navstar GPS and GLONASS - global satellite navigation systems. *Electronics & Communication Engineering Journal, 5*(6), 349-357.

Davies, S. G. (1997). Re-engineering the right to privacy : How privacy has been transformed from a right to a commodity. In P. E. Agre & M. Rotenberg (Eds.), *Technology and privacy: The new landscape* (pp. 143-165). Cambridge MA: MIT Press.

Davision, M. R., Wagner, C., & Ma, C. K. L. (2005). From government to e-government: a transition model. *Information Technology & People, 18*(3), 280–299. doi:10.1108/09593840510615888

Dawes, S. S. (2002). *The Future of E-Government.* Retrieved July 3, 2008, from www.vinnova.se/upload/EPiStorePDF/vr-06-11.pdf

de Janvry, A., Fafchamps, M., & Sadoulet, E. (1991). Peasant household behavior with missing markets: some paradoxes explained. *The Economic Journal, 101,* 1400–1417. doi:10.2307/2234892

De Kock, E. (2005). Data protection in South Africa. *De Rebus*, December.

de Silva, H. (2008). Scoping Study: ICT and rural livelihoods in South Asia. Draft report to IDRC.

Dean, G. J. (1994). *Designing instruction for adult learners.* Malabar, Australia: Krieger Publishing Company.

Department of Home Affairs (2001). *Strategic Plan 2002/2003 to 2004/2005,* Document n°2/6/8/P, Pretoria : Department of Home Affairs.

Dept of Agriculture & Cooperation, Ministry of Agriculture, Govt. of India (n.d.). Retrieved from http://agricoop.nic.in/policyincentives/BRIEF%20ON%20AGRISNET.htm Dept of Agriculture, Ministry of Agriculture, Govt. of India, (n.d.). Retrieved January 3, 2009, from http://agricoop.nic.in/Annualreport0607/INFORMATION%20TECHNOLOGY.pdf/

Desrosières, A. (1993). *La politique des grands nombres: Histoire de la raison statistique.* Paris: La Découverte.

Desrosières, A. (1997). Du singulier au général: L'argument statistique entre la science et l'Etat. In B. Conein & L. Thévenot (Eds.), *Cognition et information en société* (pp. 267-282). Paris: Ecole des Hautes Etudes en Sciences Sociales.

Dhaliwal, S. (2006). Helping farmers to change their fortune. *The Tribune, October 12, 2006, 126*(283)

DiMaggio, P., Hargittai, E., Neuman, W. R., & Robinson, J. P. (2001). Social Implications of the Internet. *Annual Review of Sociology, 27,* 307–336. doi:10.1146/annurev.soc.27.1.307

Doczi, M. (2005, December 13). *What "access" and "e-literacy" mean.* Retrieved December 20, 2008 from Information and Communication Technologyies and Social Economic Inclusion http://www.med.govt.nz/templates/MultipageDocumentPage_9666.aspx#P27_7816

Dorward, A., Poole, N., Morrison, J., Kydd, J., & Uray, I. (2003). Markets, institutions and technology: missing links in livelihoods analysis. *Development Policy Review*, *21*(3), 319–332. doi:10.1111/1467-7679.00213

Dose, H. (2007). Securing household income among smallscale in Kakamega district: possibilities and limitations of diversification. *GIGA Working Paper No. 41*. Hamburg. Germany.

Dou, W., & Chou, D. (2002). A structural analysis of business-to-business digital markets. *Industrial Marketing Management*, *31*, 165–176. doi:10.1016/S0019-8501(01)00177-8

Drucker, P.F. (2001). *The essential Drucker*. Harvard Business School.

Dubow, S. (1995). *Scientific racism in modern South Africa*. Cambridge: Cambridge University Press.

Dul, J., & Hak, T. (2007). *Case Study Methodology in Business Research*. Oxford, UK: Butterworth-Heinemann.

Dupont, J. K., Willers, J. L., Seal, M. R., & Hood, K. B. (1999). Precision pesticide applications using remote sensing. *Proc. 17-th Biennial Workshop on Color Photography and Videography in Resource Assessment, Reno, NV.*

Dzvimbo, K. P. (n.d.). *The international migration of skilled human capital from developing countries*. Retrieved November 1, 2008, from The World Bank http://siteresources.worldbank.org/INTAFRREGTOPTEIA/Resources/peter_dzvimbo.pdf

E-agriculture. (2008). *ICT-based Agri-hub in Mokopane to aid subsistence farmers*. Retreived February 25, 2009 from http://www.e-agriculture.org

Ebrahim, Z., & Irani, Z. (2005). E-government adoption: Architecture and barriers. *Business Process Management Journal*, *11*(5), 589–611. doi:10.1108/14637150510619902

e-choupal. Retrieved January 1, 2009, from http://www.echoupal.com/frontcontroller.ech

Economic and Social Commission for Asia and the Pacific. (2008). Information and communications technology for food security and sustainable agriculture in the knowledge economy. In *Proceedings of Word summit on the information society*, November 19-21, 2008.

Economic Commission for Africa. (2008). *Africa Information Society Initiative: A Decade's Perspectives*. Addis Ababa, Ethiopia: Author.

Economic Times (2008, February 29). Rs 644 crore for agriculture insurance scheme. Retrieved December 29, 2008, from http://economictime.indiatimes.com/articles show/2826829.cms

Edwin, L. (2003, November). Project Leader, OECD E-Government Project. Challenges for E-Government Development. *5th Global Forum on Reinventing Government*, Mexico City.

eJournalUSA, (2007, September). U.S. food aid: Reducing world hunger. Retrieved Nov. 24, 2007, from http://usinfo.state.gov/journals/

El Obeid, S., Mendlsohn, J., Lajars, M., Forster, N., & Brule, G. (2001). *Health in Namibia: Progress and Challenges*. Windhoek, Namibia: Raison Research and Information Services of Namibia.

Elsenburg, (2007). Department of agriculture: Western Cape. Retreived June 28, 2007 from http://www.elsenburg.com/economics/ statistics/start.htm

Elshorbagy, A., Jutla, A., & Barbour, L. (2005). System dynamics approach to assess the sustainability of reclamation of disturbed watersheds. *Canadian Journal of Civil Engineering*, *3*(1), 144-158.

Engelbrecht, L. (2007). Blank cheque for Home Affairs information technology. Johannesburg: *ITWeb*, 5 September.

Engelbrecht, L. (2007). Five in court for marriage fraud. Johannesburg: *ITWeb*, 10 December.

Engelbrecht, L. (2007). Getting Home Affairs cup-ready. Johannesburg: *ITWeb*, 5 November.

Engelbrecht, L. (2007). HANIS gets Rands 30m 'refresh. Johannesburg: *ITWeb*, 6 September.

Entsua-Mensah, C. (2005). Revitalizing the indigenous agricultural marketing system in Ghana through the e-commerce project: A performance appraisal. *IAALD Quarterly Bulletin*, *50*(3), 141–147.

Equipment%20Sept%202006%20AFTracker.pdf. (Verified January 23, 2008)

Eribo, F. (2001). *In search of greatness: Russia communications with Africa and the world.* Westport, CT: Ablex.

Ericon (2007). *Malaysia and the Knowledge Economy.* Retrieved March 25, 2009 from http://blog.beekens.info/index.php/2007/12/malaysia-and-the-knowledge-economy/

Ernst, D., & Lundvall, B. (1997). *Information technology in the learning economy: Challenges for developing countries.* DRUID, Copenhagen Business School, Department of Industrial Economics and Strategy/Aalborg University, Department of Business Studies in its series DRUID Working Papers with number 97–12. Retrieved Nov. 4, 2008, http://www.druid.dk/

Ertmer, P.A., & Addison, P, L., Molly, R, E. & Woods, D. (1999). Examining teacher's' beliefs about the role of technology in the elementary classroom. [Fall.]. *Journal of Research on Computing in Education*, *32*(1), 54–72.

Esselaar, S., Gillwald, A., & Stork, C. (2006). South African Telecommunications Sector Performance Review 2006. *LINK Centre Public Policy Research Paper No. 8.* Johannesburg: LINK.

Esterhuizen, D., & Van Rooyen, C. J. (2006). An enquiry into factors impacting on the competitiveness of the South African wine industry. *Agrekon*, *45*(4), 467–485.

European Commission. (2003*). Linking-up Europe: the importance of interoperability for e-Government services* [European Commission working document]. Retrieved July 20, 2007 from http://europa.eu.int/ISPO/ida/

Evans, D., & Yen, D. C. (2005). E-government: An analysis for implementation: Framework for understanding cultural and social impact. *Government Information Quarterly*, *22*, 354–373. doi:10.1016/j.giq.2005.05.007

Evans, D., & Yen, D. C. (2006). E-government: Evolving relationship of citizens and government, domestic and international development. *Government Information Quarterly*, *23*, 207–235. doi:10.1016/j.giq.2005.11.004

Evans, I. (1997). *Bureaucracy and race: Native administration in South Africa.* Berkeley, CA: University of California Press.

Ezednma, C., & Chukuezi, C. (1999). A comparative analysis of urban agro enterprises in Lagos and Port Harcourt, Nigeria. *Environment and Urbanization*, *11*(2), 135–144. doi:10.1177/095624789901100212

Fafchamps, M. (1996). The Enforcement of Commercial Contracts in Ghana. *World Development*, *24*, 427–448. doi:10.1016/0305-750X(95)00143-Z

Fafchamps, M. (2004). *Market institutions in sub-Saharan Africa.* Cambridge, MA: MIT Press.

Fafchamps, M., & Gabre Madhin, E. (2006). Agricultural markets in Benin and Malawi. *African Journal of Agricultural and Resource Economics*, *1*(1), 67–94.

Fafchamps, M., & Hill, R. V. (2005). Selling at the farm gate or traveling to the market. *American Journal of Agricultural Economics*, *87*(3), 717–734. doi:10.1111/j.1467-8276.2005.00758.x

Fall, A. S. (1998). Migrants' long distance relationships and social networks in Dakar. *Environment and Urbanization*, *10*(1), 135–145. doi:10.1177/095624789801000104

Fanon, F. (1963). *The wretched of the earth.* New York: Grove Press.

Fauvelle-Aymar, F. X. (2006). *Histoire de l'Afrique du Sud.* Paris: Seuil.

FCC. (1985). http://www.access.gpo.gov/nara/cfr/waisidx_01/47cfr15_01.html (Verified January 23, 2008)

Ferris, S., & Engoru, P. &. Kanganzi, K. (2006, October). *Making market information services work better for the poor in Uganda.* A paper presented at Research Workshop on Collective Action and Market Access for Smallholder Farmers, Cali, Columbia. *ICT to substantially reduce agricultural costs to farmers* (2008). *Financial Times, 2*(40), March 2.

First, R., Steele, J., & Gurney, C. (1973). *The South African connection: Western investment in apartheid.* Harmondsworth: Penguin.

Flockemann, H. M. E. (1972). South Africa. In A. H. Gold (eds.), *Wines and spirits of the world.* Surrey, UK: Virtue & Company Limited.

Forrester, J.W. (Ed.). (1961). *Industrial dynamics.* Cambridge, MA: MIT Press.

Foucault, M. (1995). *Surveiller et punir: Naissance de la prison*, Paris: Gallimard.

Foucault, M. (1997). Cours du 14 janvier 1976. In F. Ewald, A. Fontana & M. Bertani Ed.), *Il faut défendre la société* (pp. 21-36). Paris: Gallimard-Seuil.

Fountain, J. E. (2001). *Building the virtual state: information technology and institutional change,* Washington D.C: Brookings Institution Press.

Galloway, L., & Mochrie, R. (2005). The use of ICT in rural firms: A policy-orientated literature review. *Info, 7*(3), 33–46. doi:10.1108/14636690510596784

Gans, H. J. (2005). Urbanism and suburbanism as ways of life: A reevaluation of definitions. In J. Lin & C. Mele (Eds.). *The Urban Sociology Reader* (pp. 42 - 49). New York: Routledge.

Garsten, C., & Jacobsson, K. (2007). Corporate globalization, civil society and post-political regulation – Whither democracy? *Development Dialogue*, (49): 143–157.

Gartner estimate rates India as the fourth largest IT spender. *The Economic Times*, 13 July 2003. http://www.infochangeindia.org/archives1.jsp?secno=9&monthname=September&year=2003&detail=T

Gates, B. (1999). *Business @ the speed of thought: Using a digital nervous system.* Warner Books.

Gaur, R. K. (2003). Rethinking the Indian digital divide; present state of digitization in Indian management libraries. In T. A. V. Murthy, (Ed.) *Mapping technology on libraries and people* (p. 108). Ahmedabad, India: Inflibnet.

Gebremichael, M. D., & Jackson, J. W. (2006). Bridging the gap in Sub-Saharan Africa: A holistic look at information poverty and the region's digital divide. *Government Information Quarterly, 23*, 267–280. doi:10.1016/j.giq.2006.02.011

Geddes, G. (2006). Old dogs and new tricks: Teaching computer skills to adults. *eLIT, 5* (4). Retrieved December 29, 2008, from http://www.ics.heacademy.ac.uk/italics/vol5iss4/geddes.pdf

Geir, J. (2002). *Wireless Lans, Implementing High Performance IEEE802.11 Networks.* Indianoplis, IN: SAMS Publishing.

Geospatial Systems, Inc. (2008). http://www.geospatialsystems.com/imaging-products/ (Verified February, 6, 2008)

Gerster, R., & Zimmermann, S. (2003). *ICTs for poverty reduction: Lessons for donors.* Retrieved June 29, 2006, from http://www.commint.com/strategicthinking/st2003/thinking-187.html

Gichoya, D. (2005). Factors Affecting the Successful Implementation of ICT Projects in Government. *The Electronic . Journal of E-Government, 3*(4), 175–184.

Gigaba, M. (2004). Home Affairs department budget vote 2004/2005. Cape Town: National Assembly – Republic of South Africa, 11 June.

Gilder, B. (2003). Media Briefing. Johannesburg: Department of Home Affairs – Republic of South Africa, 5 November.

Gil-Garcia, J. R., & Luna-Reyes, L. F. (2003). Towards a definition of electronic government: A comparative review. In A. Mendez-Vilas, et al. (eds.), *Techno-legal aspects of information society and new economy: An overview.* Extremadura, Spain: Formatex Information Society Series.

Gil-Garcia, J. R., & Pardo, T. (2005). E-government success factors: Mapping practical tools to theoretical foundations. *Government Information Quarterly*, 187–216. doi:10.1016/j.giq.2005.02.001

Gilliom, J. (2001). *Overseers of the poor: Surveillance, resistance and the limits of privacy*. Chicago, IL: University of Chicago Press.

Gillman, H. (2003). *Fighting rural poverty—The role of ICTs*. Paper presented at IFAD side event at the WSIS-Geneva. Retrieved June 29, 2006, from http://www.ifad.org/events/wsis/synthesis/index.htm

Goering, L. (2006). *Mauritius set to become first 'Cyber Island.'* Government Technology. Retrieved March 25, 2009 from http://www.govtech.com/gt/articles/98096

Government of Kenya. (2002). *Kakamega District Development Plan 2002-2008*. Nairobi, Kenya.

Granovetter, M. (1983). The strength of weak ties: A network theory revisited. *Sociological Theory*, *1*, 201–233. doi:10.2307/202051

Grieger, M. (2003). Electronic marketplaces: A literature review and a call for supply chain management research. *European Journal of Operational Research*, *144*, 280–294. doi:10.1016/S0377-2217(02)00394-6

Guardian.co.uk (2008, May 10). Food: A perfect storm | Leader: The rocketing price of food | Comment is free | *The Guardian*. Retrieved Dec. 5, 2008 from http://www.guardian.co.uk/commentisfree

Guilaume, M. (1978). *Eloge du désordre*. Paris: Gallimard.

Guillaume, P., Péjout, N., & Wa Kabwe Segatti, A. (2004). *L'Afrique du Sud dix ans après: Transition accomplie?*. Johannesburg – Paris: Institut Français d'Afrique du Sud – Karthala.

Gumm, D. (2008, May 8). Nigeria can feed her citizens, become world food exporter—Govt. *Vanguard*. Retrieved Dec. 8, 2008 from http://www.vanguardngr.com/

Gunawardene, N. (n.d.). *ICT for poverty reduction: Think big, act boldly*. Retrieved June 29, 2006, from http://www.teriin.org/terragreen/issue71/essay.html

GUS (Polish Central Statistical Office). (2007). Information Society. Research results of 2004-2006. Retrieved from http://www.stat.gov.pl

GUS (Polish Central Statistical Office). (2008) *Characteristics of agricultural holdings in 2007*. Retrieved from http://www.stat.gov.pl

GUS (Polish Central Statistical Office). (2008). *The usage of information technologies in households in 2007*. Retrieved from http://www.stat.gov.pl

Haan, C. H. (2002). *Training for work in the informal sector: New evidence from eastern and southern Africa*. Retrieved March 20th 2003 from www.itcilo.it/english/bureau/turin/whatisnew/flyers/Training%20for%20work.pdf

Hakikazi Catalyst. (2001). *Tanzania without poverty—A plain language guide to Tanzania's poverty reduction strategy paper*. Retrieved June 29, 2006, from http://www.hakikazi.org/eng/

Hakikazi Catalyst. (2003). *Millennium development goals—No more broken promises*. Retrieved June 29, 2006, from http://www.srds.co.uk/mdg/nmbp-draft-04.pdf

Hamelink, C. (1997). *New information and communication technologies, social development and cultural changes*. United Nations Research Institute for Social Change. [Online]. Retrieved Nov. 4, 2008, from http://www.unrisd.org/

Haricharan, S. (2005). *Knowledge Management in the South African Public Sector*. Retrieved July 5, 2008 from http://www.ksp.org.za/holonl03.htm

Harris, R. W. (2004). *Information and communication technologies for poverty alleviation*. Kuala Lumpur, India: UNDP-APDIP.

Harris, R.W. (2004). Information and Communication Technology for Poverty Alleviation. *E-primers for the Information, Economy, Society and Polity*, Kuala Lumpur (UNDP-APDIP), (pp. 21, 35).

Hart, K., (1973). Informal income opportunities and urban employment in Ghana. *The Journal of Modern African Studies*, *11*(1), 61-89.

Haruna, P. F. (1999). *An empirical analysis of motivation and leadership among career public administrators: The case of Ghana*. Unpublished doctoral dissertation, University of Akron, Ohio.

Heeks, R. (1999). *Information and Communication Technologies, Poverty and Development, Development Informatics Working Paper Series, Paper No. 5*. www.idpm.man.ac.uk/idpm/di wp5.htm

Heeks, R. (2001). *Understanding e-governance for development*. Manchester, UK: Institute for Development Policy and Management. Retrieved July 15, 2007, from http://idpm.man.ac.uk/ publications/wp/igov/igov_wp11.pdf

Heeks, R. (2002). eGovernment in Africa: Promise and Practice. *iGovernment Working Paper Series*, Paper 13.

Heeks, R. (2002). Information systems and developing countries: Failure, success, and local improvisations. *The Information Society*, *18*(2), 101–112. doi:10.1080/01972240290075039

Heeks, R. (2003). *e-Government Special – Does it Exist in Africa and what can it do?* Retrieved September 30, 2007 from http://www.balancingact-africa.com/news/back/balancing-act93.html#headline

Heeks, R. (2003). Most e-government-for-development projects fail: How can risks be reduced? *iGovernment Working Paper Series*, No. 14. Retrieved July 21, 2007, from http://unpan1.un.org/ intradoc/groups/public/documents/CAFRAD/UNPAN011226.pdf

Heeks, R. (2006). Benchmarking e-government: Improving the national and international measurement, evaluation and comparison of e-government. *iGovernment Working Paper Series*, No. 18. Retrieved July 21, 2007, from www.sed.manchester.ac.uk/idpm/research/publications/wp/igovernment/documents/ iGWkPpr18.pdf

Hesselmark, O., & Miller, J. (2002). A country ICT survey for Namibia. Retrieved from http://www.cyberzoo.co.za/download/milless/newdocs/Namrep%20v5.pdf

Hinson, R. E. (2005). Internet adoption among Ghana's SME nontraditional exporters. *Africa Insight*, *35*(1), 20–27.

Hjelm, N. M. (2005). Benefits and drawbacks of telemedicine. *Journal of Telemedicine and Telecare*, *11*(2), 60–70. doi:10.1258/1357633053499886

HLL Shakti changing lives in rural India. (n.d.). Retrieved December 1, 2008, from http://www.hllshakti.com/sbcms/templ.asp?pid=46802246

Hoffman, D., & Novak, T. (1996). Marketing in hypermedia computer-mediated environments: conceptual foundation. *Journal of Marketing*, *60*(July).

Hu, W. (2007, May 4). Seeing no progress, some schools drop laptops. *New York Times*. Retrieved May 4, 2007, from www.nytimes.com/2007/05/04/education/04laptop.html?ex=1180670400&en=fae7a1a9906 70 e83&ei=5070

Hueth, B. (1999). Incentive instruments in fruits and vegetables: Input control monitoring, measuring and price risk. *Review of Agricultural Economics*, *2*, 374–398. doi:10.2307/1349886

Huggins, R., & Izushi, H. (2002). The Digital Divide and ICT Learning in Rural Communities: Examples of Good Practice Service Delivery. [May.]. *Local Economy*, *17*(2), 111–122. doi:10.1080/02690940210129870

I. C. T. Update (2002). *Agricultural market information services, Issue 9*. Wageningen, The Netherlands: CTA.

I. C. T. Update (2007). *Financial services, Issue 36*. Wageningen, The Netherlands: CTA.

i4d (2008, April). Anambra government to train youths in agriculture. Retrieved Dec. 6, 2008, http://www.i4donline.net/

i4d (2008, February). Nigeria's agrovision. [Online]. Retrieved Dec. 5, 2008, http://www.i4donline.net/.

ICTPORTAL. (2009). *Information and communication technology penetration in South Africa*. Retrieved March 27, 2009 from http://www.ictportal.org.za/news.php

Idacha, F. (1980). *Agricultural research policy in Nigeria*. Research Report 17. Washington, DC: International Food Policy Research Institute.

Ifinedo, P. (2005). Measuring Africa's e-readiness in the global networked economy: A nine-country analysis. *International Journal of Education and Development using ICT, 1*(1), 53-71.

IICD. (2004). *Telemedicine hampered by infrastructure and awareness—says Tanzanian health expert during seminar by SWOPNet*. Retrieved June 29, 2006, from International Institute for Communication and Development (IICD), http://www.iicd.org/articles/iicd-news.2004-09-02.7493425067

Independent Electoral Commission. (2004). Independent Electoral Commission on reports about giving out details from voters' roll. Pretoria: Independent Electoral Commission, 15 June.

Information for Development, (2005). Agriculture: Computerised system of irrigation in South Africa. *I4D, 3*(8), 23-26.

Information Technologies Group of Harvard University's Centre for International Development. (2002). *Readiness for the Networked World: A Guide for Developing Countries.* Cambridge, MA: Centre for International Development, Harvard University.

International Finance Corporation/World Bank. (2008). *Information & Communications Technology in Africa.* Retrieved Nov. 30, 2008, from http://www.ifc.org/ifcext/africa.nsf/

International Labour Organisation. (1972). *Incomes and equality: A strategy for increasing productive employment in Kenya*, Geneva: International Labour Organisation.

International Labour Organization. (2008). *Global employment trends: January 2008.* Geneva: International Labour Organization.

International Monetary Fund. (2005, December 23). *IMF to extend 100 percent debt relief to Zambia under the multilateral debt relief initiative.* Press Release. Retrieved June 20, 2007, from http://imf.org/ external/np/sec/pr/2005/pr05306.htm

International Telecommunication Union (ITU). (2007). *Telecommunication/ ICT markets and trends in Africa.* Geneva: ITU.

International Telecommunication Union, Global Survey on Rural Communications. (2004, June). *ITU News, No. 5.* www.itu.int/itunews, Accessed 28 May 2007.

Intime, Inc. (2007). http://www.gointime.com/ (Verified January 23, 2008)

IRIN (2002, July 4). Nigeria: IRIN Focus on shift towards offshore oil production. Retrieved Sept. 14, 2008 from http://www.irinnews.org and http://www.africaaction.org/

Isaacs, S. (2007). *Survey of ICT and Education in Africa: Zimbabwe Country Report.*

i-watch a wake up call for India. (n.d.). Retrieved February 1, 2009, from http://www.wkeupcall.org/employment/improving business.php.

Jacobs, S.J. & Herselman, M.E. (2005). An ICT-hub model for rural communities. *International Journal of Education and Development using ICT, 1*(3), 57-93.

Jaeger, P. T., & Thompson, K. M. (2003). E-government around the world: Lessons, challenges, and future directions. *Government Information Quarterly, 20*(4), 389–394. doi:10.1016/j.giq.2003.08.001

Jallow, B. (2004). Community radio for empowerment and impact. *The Journal of Development Communication, 15*(2), 56–68.

James, A. (2008, April 24). World food crisis hits home. *Seattleepi.com.* Retrieved December 6, 2008 from http://seattlepi.nwsource.com/business/

James, J. Pro-Poor. (2000). *Modes of Technical Integration into the Global Economy in Development and Change, 31*, 765–783.

James, T. (2001). *An information policy handbook for Southern Africa.* Ottawa, Canada: International Development Research Centre.

Jayne, T. S., & Jones, S. (1997). Food marketing and pricing policy in Eastern and Southern Africa: A survey. *World Development*, *25*(9), 1505–1527. doi:10.1016/S0305-750X(97)00049-1

Joint, N. (2003, June 11). eLiteracy and social exclusion: A global perspective. In *eLit2003:2 International Conference on Information and IT Literacy*. Retrieved December 20, 2008, from http://www.strathprints.strath.ac.uk/2357/1/strathprints002357.ppt

Juma, C. (2005). The way to wealth. *New Scientist*, *185*(21), 15–21.

Kaaya, J. (2004). Implementing e-Government Services in East Africa: Assessing Status Through Content Analysis of Government Websites. *The Electronic . Journal of E-Government*, *1*(2).

Kalusopa, T. (2006). The challenges of utilizing information communication technologies (ICTs) for the small scale farmers in Zambia. *Library Hi Tech*, *23*(3), 414–424. doi:10.1108/07378830510621810

Kanpur-Luknow media lab @ IIT Kanpur media lab Asia (2003). Retrieved November 12, 2008, from http://www.iitk.ac.in/MLAsia/infothela.htm

Kao, J. (1996). *Jamming: The art and discipline of corporate creativity*. Harper Business Publishers.

Kapange, B. (2006). ICTs in agricultural development: The case of Tanzania. In *Proceedings of the IST-Africa 2006 Conference*, Pretoria, May 3-5, 2006.

Kapange, B. W. (2008). Using ICTs for Managing Network Genebank: The Case of SADC Plant Genetic Resources Centre. In *Proceedings of IST-Africa Conference*, Windhoek, Namibia.

Kasumbalesa, F. (2005). The untapped potential of e-governance. *Times of Zambia*. Retrieved April 24, 2007, from www.times.co.zm/news/viewnews.cgi?category=8&id=11309

Kaur, G. (2007). *From APMCs to electronic markets*. Retrieved from http://www.indiatogether.org/2007/jul/agr-spotex.htm

Kaur, S. & Mathiyalagan, N. (2007, December 28-30). Adoption of e-governance: Social dimensions of e-government: Poverty perspective. In *Proceedings of the 5th International Conference on E-Governance*, Hyderabad, India.

Keller, S. (1998). The American dream of community: An unfinished agenda. *Sociological Forum*, *3*, 167–183. doi:10.1007/BF01115289

Keniston, K. (2000). *The Four Digital Divides*. Available online at http://web.mit.edu/~kken/Public/ PDF/Intro_Sage_1_.pdf (May 11, 2007)

Keniston, K. (2002). *Grassroots ICT Projects in India: Some Preliminary Hypotheses ASCI Journal of Management, 31*(1&2). Available online at http://web.mit.edu/~kken/Public/PDF/ASCI _Journal_Intro __ASCI_version_. pdf (May 11, 2007)

Keniston, K. (2007). *IT for Masses: Hope or Hype*? Available online at http://web.mit.edu /~kken/Public/PDF/EPW_paper.pdf (May 11, 2007)

Kenny, C., Navas-Sabater, J., & Quiang, C. (2001). *Information and Communication Technologies and Poverty in World Bank Poverty Reduction Strategies Sourcebook*. www.worldbank.org/poverty/ strategies/ict/ict.htm

Key, N., & Runsten, D. (1999). Contract farming, smallholders, and rural development in Latin America: The organization of agroprocessing firms and scale of outgrower production. *World Development*, *27*, 381–401. doi:10.1016/S0305-750X(98)00144-2

Khajuria, R. K. (2008). Strawberry growers laugh their way to bank. *The Tribune*, April 17,2008.

Khaketla, M. (2004, October). *A national ICT policy for Lesotho: Discussion draft*. Retreived April 26, 2006 from http://www.lesotho.gov.ls/articles/ 2004/Lesotho_ICT_Policy_Draft.pdf

Khulumani Support Group. (2004). Complaint. South District Court, New York State, 15 December.

Kindness, H., & Gordon, A. (2001). *Agricultural marketing in developing countries: the role of NGOs and CBOs*. Policy Series No. 13. Social and Economic Development

Department, Natural Resources Institute, University of Greenwich.

Kirsten, J. (in press). The case for interdisciplinary research and training in agricultural economics in Southern Africa. *Nature Sciences Sociétés*.

Kljajić, M, Legna, C., Škraba, A., & Peternel, J. (2003). Simulation model of the Canary Islands for public decision support: Preliminary results. In *Proc. of the 20th International Conference of the System Dynamics Society*. Albany, NY: The System Dynamics Society.

Kljajić, M., Legna, C., & Škraba, A. (2002). System dynamics model development of the Canary Islands for supporting strategic public decisions. In *Proc. of the 20th International Conference of the System Dynamics Society* (p. 16). Palermo, Italy: The System Dynamics Society.

Knight, R. (1986). *US computers in South Africa*. Retrieved October 26, 2003, from http://richardknight.homestead.com/files/uscomputers.htm

Kössler, R., & Melber, H. (2007, November). International civil society and the challenge of global solidarity. *Development Dialogue*, (49): 29–39.

Kraut, R., Lundmark, V., Patterson, M., Kielser, S., Mukopadhay, T., & Scherlis, W. (1998). Internet paradox: A social technology that produces social involveemnt and psychological well-being? *The American Psychologist*, *53*(9), 1017–1031. doi:10.1037/0003-066X.53.9.1017

Krishna, S., & Walsham, G. (2005). Implementing public information systems in developing countries: Learning from a success story. *Information Technology for Development*, *11*(2), 123–140. doi:10.1002/itdj.20007

Kumar, B. A. (2007). Technology transfer paramount. *The New Indian Express*, Chennai, January 04.

Kumar, V., Mukerji, B., Butt, I., & Persaud, A. (2007). Factors for Successful e-Government Adoption: a Conceptual Framework. *The Electronic . Journal of E-Government*, *5*(1), 63–76.

Kwapong, O.A.T. (2007). Problems of policy formulation and implementation: The case of ICT use in rural women's empowerment in Ghana. *International Journal of Education and Development using ICT*, *3*(2), 1-21.

Kydd, J., & Dorward, A. (2004). Implications of market and coordination failures for rural development in least developed countries. *Journal of International Development*, *16*, 951–970. doi:10.1002/jid.1157

Lall, S., Navaratti, G. B., Teitel, S., & Wiggnaraja, G. (1994). *Ghana under structural adjustment*. New York: St Martin's Press.

Lambinon, I. (2003). *Briefing at the Home Affairs portoflio committee*. Cape Town, National Assembly – Republic of South Africa, 18 March.

Lanvin, B. (2002). *The e-government handbook for developing countries*. Retrieved December 20, 2008, from Center for Democracy & Technology, http://www.infodev.org

Larson, D. F. (2006). Measuring the efficiency of food markets: The case of Nicaragua. In A. Sarris & D. Hallam (Eds), *Agricultural Commodity Markets and Trade: New Approaches to Analyzing Market Structure and Stability*. Rome: FAO.

Lascoumes, P., & Le Gales, P. (Ed.). (2004). *Gouverner par les instruments*. Paris: Presses de Sciences Po.

Lautier, B. (2004). *L'économie informelle dans le tiers monde*. Paris: La Découverte.

Layne, K., & Lee, J. (2001). Developing fully functional E-government: A four stage model. *Government Information Quarterly*, *18*(2), 122–136. doi:10.1016/S0740-624X(01)00066-1

Lee, E., Lee, J., & Schumann, D. W. (2002). The influence of communication source and mode on consumer adoption of technological innovations. *The Journal of Consumer Affairs*, *36*(1), 1–27.

Leroux, N., Wortman, M., & Mathias, E. (2001). Dominant factors impacting the development of business-to-business (B2B) e-commerce in agriculture. *International Food and Agribusiness Management Review*, *4*, 205–218. doi:10.1016/S1096-7508(01)00075-1

Lessiter, F. (2006). Auto-Steer's New Frontier – Precise implement steering. *Farm Eguipment.* http://www.gps-farm.com/_resources/uploads/files/Farm%20

Liimatainen, M. (2003. *Training and skills acquisition in the informal sector: A literature review*. Geneva: International Labour Office.

Lio, M., & Liu, M. C. (2006). ICT and agricultural productivity: evidence from cross-country data. *The Journal of the International Association of Agricultural Economics, 34*(3), 221–228.

Lollar, X. (2006). Assessing China's e-government: Information, service, transparency and citizen outreach of government websites. *Journal of Contemporary China, 15*(46), 31–41. doi:10.1080/10670560500331682

Lucking-Reiley, D., & Spulber, D. F. (2001). Business to Business Electronic Commerce. *The Journal of Economic Perspectives, 15*(1), 55–68.

Lyon, D. (2002). Everyday surveillance : Personal data and social classifications. *Information, Communication & Society, 5*(2), 242-257.

Lyon, D. (2003). Introduction. In D. Lyon (Ed.), *Surveillance as social sorting: Privacy, risk and digital discrimination* (pp. 1-9). London: Routledge.

Majcen, M.H., & Jurcan, S. (Eds.). (2006). *Action organic farming development plan in Slovenia to year 2015 (ANEK).* Ljubljana: Government of the Republic of Slovenia. ISBN 961-6299-73-5.

Malakata, M. (2005). Zambia's leap into e-banking. *iConnectAfrica: ICTs at Work*, 2(2). Retrieved April 24, 2007, from www.uneca.org/aisi/iconnectafrica/v2n2.htm#5

Malaysia kEconomy, (2000). *Transcending The Divide*. Speech by the Prime Minister the Hon Dato Seri Dr Mahathir bin Mohamad at the Second World Knowledge Conference at the Ballroom, Mandarin Oriental, KLCC, Kuala Lumpur, Wednesday 8 March 2000, 5:00 PM.

Malaysia kEconomy, (2006). *Knowledge-based masterplan (Malaysia).* Retrieve March 25, 2009 from http://unpan1.un.org/intradoc/groups/public/documents/apcity/unpan013973.pdf

Malaysia kEconomy, (2007). *Developing Malaysia into a knowledge-based economy.* Retrieved March 25, 2009 from http://www.epu.gov.my/new%20folder/development%20plan/opp3/cont_chap5.pdf

Malaysia kEconomy, (2007). *Malaysia and the Knowledge Economy: Building a World-Class Higher Education System. Human Development Sector Reports*. Washington DC: World Bank. Retrieved March 25, 2009, from http://publications.worldbank.org/online

Malaysia, Multimedia Super Corridor (2009). Retrieved March 25, 2009 from http://en.wikipedia.org/wiki/multimedia_super_corridor

Mambwe, K. (2005). Boosting local authorities: E-governance in Zambia. *IConnectOnline*. Retrieved June 29, 2007, from www.ftpiicd.org/iconnect/ICT4D_Governance/EN_Governance_ZM.pdf

MANAGE. (n.d.). *Making inroads into rural India through mobile VSAT videoconferencing van.* Retrieved November 17, 2008, from http://www.manage.gov.in/managelib/NewEvents/mobile%20VSAT.htm

Mandela, N. (2005, February 3). *Make poverty history* (Speech during The Global Campaign for Action Against Poverty). London's Trafalgar Square.

Mangesi, K. (2007). *ICT in education in Ghana, Survey of ICT and Education in Africa: Ghana Country Report*. Retrieved from www.Infodev.org/ict4edu-Africa

Mani, S. (2001). *Globalization, markets for technology and the relevance of innovation policies in developing economies* (ATPS Special Paper No. 2). Nairobi, Kenya: ATPS.

Mapisa-Nqakula, N. (2004). *Home Affairs department budget vote 2004/2005.* Cape Town: National Assembly – Republic of South Africa, 11 June.

Mapisa-Nqakula, N. (2007). DHA budget speech for budget vote 2007. Cape Town: National Assembly – Republic of South Africa, 7 June.

Marche, S., & McNiven, J. D. (2003). E-Government and E-Governance: The Future isn't what it used to be.

Canadian Journal of Administrative Sciences, *20*(1), 74–86.

Margetts, H., & Dunleavy, P. (2002). Cultural Barriers to EGovernment, published by the National Audit Office, April 4, 2002 (HC 704-III) in conjunction with the Value for Money report "Better Public Services Through EGovernment."

Marker, P., McNamara, K., & Wallace, L. (2002). *The significance of Information and Communication Technologies for Reducing Poverty*. London: DFID.

Markoff, J. (2006, January 30). Microsoft would put poor online by cellphone. *New York Times*. Retrieved May 23, 2007, from www.nytimes.com/2006/01/30/technology/30gates.html?ex=1178683200 &en=f7fded26ce9f0dc9&ei=507

Markoff, J. (2006, November 30). For $150, third-world laptop stirs big debate. *New York Times*. Retrieved May 23, 2007, from www.nytimes.com/2006/11/30/technology/30laptop.html?ei=5070&en=040c21fc7ec4bec4&ex=1178683200

Marotta, D., Mark, M., Bloom, A., & Thorn, K. (n.d.). *Human capital and university-industry linkages' role in forstering firm innovation: An empirical study of Chile and Colombia.* Retrieved November 1, 2008, from The World Bank http://www-wds.worldbank.org/external/default/WDSContentServer/IW3P/IB/2007/12/13/000158349_20071213130640/Rendered/PDF/wps4443.pdf

Martin, A. (2005). DigEuLit - a European framework for digital literacy: A progress report. *Journal of eLiteracy, 5*(4), 130 – 136. Retrieved December 29, 2008, from http://www.jelit.org/65/01/JeLit_Paper_31.pdf.

Martin, D.-C. (1998). Le poids du nom: Culture populaire et constructions identitaires chez les 'métis' du Cap. *Critique Internationale*, n°1, 73-100.

Martin, L. (2006). Enabling eliteracy: Providing non-technical support for online learners. *eLIT, 5* (4), 97 – 108. Retrieved December 20, 2008, from http://www.ics.heacademy.ac.uk/italics/vol5iss4/martin.pdf.

Matei, S., & Ball-Rokeach, S. J. (2001). Real and virutal social ties: connections in the Everyday lives of seven ethnic neighborhoods. *The American Behavioral Scientist, 45*(3), 550–564.

Matthee, A. (2004). *Traceability in the wine industry. Wynboer: A technical guide for wine producers*. Paarl, South Africa: Wineland.

Maumbe, B. M., & Owei, V. T. (2007). E-government Policy Development in South Africa: Current Status, Distributional Issues, and Future Prospects. In *Proceedings of the IST-Africa Conference*, 9th -11th May, Maputo, Mozambique.

Maumbe, B. M., Owei, V., & Alexander, H. (2008). Questioning the Pace and Pathway of e-Government Development in Africa: Case of South Africa's Cape Gateway Project. Government Information Quarterly, 25.

Maumbe, B. M., Taylor, W. J., Erwin, G., & Wesso, H. (2006). E-Value Creation for Government Web Portal in South Africa. In A.Tutnall (Ed.), *Encyclopedia of Portal Technologies and Applications*, Hershey, PA: IGI Global.

Mayur, R., & Daviss, B. (1998). The technology of hope: Tools to empower the world's poorest peoples. *The Futurist, 32*(7), 46–51.

McDonald, S., & Punt, C. (2001). *The Western Cape of South Africa: Export opportunities, productivity growth and agriculture*. A paper prepared for the 4th Global economic analysis conference, Global trade analysis project. Purdue University, West Lafayette, Indiana. June 2001.

McKenzie, D., & Sasin, M. J. (2007). *Migration, remittances, poverty, and human capital: Conceptual and empirical challenges*, World Bank Working Paper No. 4272. Retrieved on October 15, 2008 from http://www.worldbank.org/migration/wps4272.pdf.

McKinion, J. M., Turner, S. B., Willers, J. L., Read, J. J., Jenkins, J. N., & McDade, J. (2004). Wireless technology and satellite internet access for high-speed whole farm connectivity in precision agriculture. *Agricultural Systems, 81*(2004), 201-212.

McKinion, J. M., Willers, J. L., & Jenkins, J. N. (2007). Wide area wireless network (WAWN) for supporting Precision Agriculture. *Proceedings of the 2007 Beltwide Cotton Production Research Conferences, New Orleans, LA*. Unpaginated CDROM.

McNamara, K. S. (2003). Information and communication technologies, poverty and development: Learning from experience. *A background paper for the InfoDev Annual Symposium*. Geneva, Switzerland. December 9-10, 2003.

Meeira, S. N., Jhamtani, A. & Rao, D. U. M. (2004). *Information and communication technology in agricultural development: A comparative analysis of 3 projects from India. AgREN*, Network paper 135.

Mele, C. (1999). Cyberspace and disadvantaged communities: the internet as a tool for collective action. In M. Smith & P. Kollock (eds.), *Communities in Cyberspace*. New York: Routledge.

Menda, A. (2005). *ICT experts probe methods in Tanzania to train medics outside hospitals*. Retrieved June 29, 2006, from International Institute for Communication and Development (IICD), http://www.iicd.org/articles/iicdnews.2005-07-15.2614286290

Menda, A. (2005). *Stakeholders build strategy to integrate ICT in the Tanzanian secondary school*. Retrieved June 29, 2006, from International Institute for Communication and Development (IICD), http://www.iicd.org/articles/iicdnews.2005-07-15.3965031729

Mensah, J. V. (2005). Problems of district medium-term development plan implementation in Ghana. *International Development Planning Review*, *27*(2), 245–270.

Milestones in India's Internet Journey (2007). http://www.amitranjan.com/category/5/

Minasyan, G. (2006, Oct. 13). *Environment and ICT: "enemies or friends."* Retrieved November 4, 2008, http://www.athgo.org/downloads/position_papers/Minasyan_Gohar

Ministry of Information and Communication Technology. (2008). *Information technology Policy for the republic of Namibia 2008*. Windhoek, Namibia: Author.

Minnaar, C.-L. (2004). Putting maps to work. Johannesburg: *ITWeb*, 12 March.

Minot, N. (1999). *Effects of transaction costs on supply response and marketed surplus: Simulations using non-separable household models*. IFPRI Discussion Paper No. 36. Washington DC.

Mir, R. M. (2008). http://www.cs.wustl.edu/~jain/cis788-97/ftp/satellite_data.pdf. (Verified January 23, 2008)

Miranda, M. (2007). Global civil society and democracy – A difficult but unavoidable task: Visions from the south. *Development Dialogue*, (49): 97–107.

Mission 2007 India http://www.mission2007.org/

Moemeka, A. (1994). Development communication: a historical and conceptual overview. In A. Moemeka (ed.), *Communicating for development: a new pan-disciplinary perspective*, (pp. 3–12). Albany, NY: State University of New York Press.

Mohammed, I. (2008, May 5). Road map to attaining food security in Nigeria. *Nigerian Tribune*. Retrieved Nov. 24, 2008 from http://www.tribune.com.ng/.

Moloi, M. J. (2007). Commercialisation of emerging vegetable producers in the Western Cape province of South Africa. *Proceedings of the 17Th International Food and Agribusiness Management Association (IAMA) Conference*, Parma, Italy. 23-26 June 2007. Moodley, S. (20050. The promise of e-development? A critical assessment of the state ICT poverty reduction discourse in South Africa. *Perspectives on Global Development and Technology*, *4*(1), 1–26.

Moser, C., Barrett, C., & Minten, B. (2005, July). *Missed opportunities and missing markets: Spatio-temporal arbitrage of rice in Madagascar*. Selected Paper presented at the American Agricultural Economics Association Annual Meeting, Providence, Rhode Island, July 24-27, 2005.

Munyua, H. (2000). Application of ICT in Africa's agricultural sector: A gender perspective. In Rathgeber, E. M., & Adera, E. O (eds). *Gender and the information Revolution in Africa*. Ottawa: IDRC.

Munyua, H. (2007). *ICTs and smallscale agriculture in Africa: ascoping study.* Unpublished Report 1, Submitted to International Development Research Center.

Mupuchi, S. (2003, August 23). Government should consider e-governance. *Post of Zambia.* Retrieved April 2, 2007, from http://unpan1.un.org/intradoc/groups/public/documents/un/unpan014273.htm

Mushinge, G. (2005). ICTs help transform lives of rural women in Zambia. *ConnectAfrica: ICTs at Work, 2*(2). Retrieved July 24, 2007, from www.uneca.org/aisi/iconnectafrica/v2n2.htm#5

Mytelka, L. K., & Tesfachew, T. (1999). *The role of policy in promoting enterprise learning during early industrialization: Lessons for African countries.* Geneva: United Nations Conference on Trade and Development (UNCTAD).

Namibia. MOHSS (2004b). *The technical efficiency of District Hospitals in Namibia.* Windhoek: Directorate: Policy, Planning and Human Resources Development.

National Computer Board. (1997). *IT2000 – A Vision of an Intelligent Island.* Singapore: NCB.

National Computer Board. (1997). *Towards an Intelligent Island: NCB 10th Anniversary 1981-1991.* Singapore: NCB.

National Computer Board. (1997). *Transforming Singapore into an Intelligent Island.* Singapore: NCB.

National Telecommunication Policy (2000, September). Federal Republic ofNigeria: National Telecommunication Policy, 2000. Retrieved November 25, 2008 from http://www.bpeng.org/NR/rdonlyres/

Ndou, V. D. (2004). E-government for developing countries: Opportunities and challenges. *The Electronic Journal of Information Systems in Developing Countries, 18*(1), 1–24.

News, B. B. C. (2002, Jan. 16). Nigeria's economy dominated by oil. Retrieved Sept. 14, 2008 from http://news.bbc.co.uk/ and http://www.africaaction.org/

News, B. B. C. (2007, July 12). Police stop Nigeria hunger march. Retrieved November 9, 2008 from http://news.bbc.co.uk/2/hi/africa/

News, B. B. C. (2008, Nov. 13). Nigerian satellite fails in space. Retrieved Dec. 6, 2006 from http://newsvote.bbc.co.uk/

Nie, N., & Erbring, L. (2000). *Internet and Society (Preliminary report).* Palo Alto, CA: Stanford Institute for the Qualitative Study of Society.

Nigeria's Agriculture Sector (2007, May 21). Opportunities in Nigeria's agriculture sector. Retrieved November 24, 2008 from http://agriculturepro.blogspot.com/2007/05/nigerias-agriculture-sector.html/.

Nigerian Embassy Vienna (n.d.). Yar'adua seven point agenda to transform Nigerian. Retrieved December 2, 2008 from http://www.nigeriaembassyvienna.com/YarAduasSevenPointAgenda/

Nigerian Tribune (2008, April 29). Food crisis in Nigeria. Retrieved December 8, 2008 http://www.tribune.com.ng/

Nikam, K., Ganesh, A. C., & Tamizhchelvan, M. (2004). The changing face of India. Part 1: bridging the digital divide. *Library Review, 53*(4), 213–219. doi:10.1108/00242530410531839

North, D. C. (1990). *Institutions, Institutional Change and Economic Performance.* Cambridge, MA: Cambridge University Press.

Obasanjo, O. (2004, March 8). *Transparency, accountability and good governance through eGovernment.* President Speech. Retrieved from September 14, 2008 http://www.negstglobal.com/sitefiles/Retrieved.

Obayelu, A. E. & Ogunlade, I. (2006). Analysis of the uses of information and communication technology for gender empowerment and sustainable poverty alleviation in Nigeria. *International Journal of Education and Development using Information and Communication Technology, 2*(3), 45-69.

Obeng, K. W. (2003). Ghana pursues justice and development through computer training. *Choices (New York, N.Y.), 12*(4), 20–21.

Obi, (2008, October 29). A panacea to global food crisis. *The Tide Online*. Retrieved November 9, 2008 from http://www.thetidenews.com/

OECD (2005). *The contribution of ICTs to Achieving the Millennium Development Goals (MDGs)*. Forthcoming. Good Practice Paper on ICTs for Economic Growth and Poverty Reduction. http://www.oecd.org/dataoecd/15/38/34662239.pdf, accessed 23 May 2007.

Ofori-Dankw, J., & Julian, S. D. (2004). Conceptualizing social science paradoxes using diversity and similarity curves model: Illustrations from the work/play and theory novelty/continuity paradoxes. *Human Relations*, *57*(11), 1449–1477. doi:10.1177/0018726704049417

Ofori-Dankwa, J., & Julian, S. (2002). Toward diversity and similarity curves: Implications for theory, research and practice. *Human Relations*, *55*, 199–224. doi:10.1177/0018726702055002183

Ogen, O. (2007). The agricultural sector and Nigeria's development: Comparative perspectives from the Brazilian agro-industrial economy, 1960–1995. *Nebula 4.1*(March). Retrieved November 25, 2008 from http://www.nobleworld.biz/

Ogwo, B. A. (2004). *Technological developments in the maintenance operations of imported used automobiles and their policy implications for the automobile industry in Nigeria*. Nairobi, Kenya: African Technology Policy Studies Network (ATPS). Rodney, W. (1973). *How Europe Underdeveloped Africa*. London: Bogle-L'Ouverture Publications.

Ogwo, B. A. (2007). Informal sector technical skills development experiences in the maintenance of modern automobiles in Nigeria. *Network for Policy Research Review and Advice on Education and Training (NORRAG) News*, (pp. 25 – 27).

Ogwo, B. A., Oranu, G. C., & Oranu, R. N. (2008). Application of ICT-based open learning principles for market/mechanic village schools in south eastern Nigeria. In *IST-Africa 2008 Conference proceedings held in Windhoek, Namibia*. http://www.ist-africa.org/Conference2008/default.asp?page=my-page

Ohrtman, F., & Roeder, K. (2003). *Wi-Fi Handbook: Bulding 802.11b Wireless Networks*. New York, NY: McGrawHill.

Okello, J. J. (2005). *Compliance with international food safety standards: The case of green bean production in Kenyan family farms*. Unpublished Dissertation, Michigan State University, East Lansing, MI.

Okot-Uma, R. W. O. (2000). *Electronic Governance – Reinventing Good Governance*. London: Commonwealth Secretariat. Retrieved March 25, 2009 from http://web-world.unesco.org/publications/it/EGov/wordbank%20**okot-uma**.pdf

Okot-Uma, R. W. O. (2003). *Capacity Building in ICT and Development, Background Paper commissioned for the Microsoft Government Leaders' Forum*, Johannesburg, 21-23 September 2003, Okot-Uma, R. W. O., (2005). *The Roadmap to eGovernance implementation: Selected Perspectives*. Retrieved March 25, 2009 from http://citeseerx.ist.psu/viewdoc/download?doi=10.1.1.84.1656&type=pdf; http://www.rileyis.com/publications/research_papers/rogers2.html

Okot-Uma, R. W. O. (2003). *Electronic governance masterplan for the government of the Republic of Mauritius*. London: Commonwealth Secretariat.

Okuneye, P. (2002). Rising cost of food prices and food insecurity in Nigeria and its implication for poverty reduction. *CBN Economic & Financial Review 39*(4). Retrieved November 25, 2008 from http://www.cenbank.org/out/Publications/

Omamo, S. W. (1998). Farm-to-market transaction costs and specialization in smallscale agriculture: explorations with non-separable household model. *The Journal of Development Studies*, *35*, 152–163. doi:10.1080/00220389808422568

Omamo, S. W. (1998). Transport costs and smallholder cropping choices: an application to Siaya district, Kenya. *American Journal of Agricultural Economics*, *80*, 116–123. doi:10.2307/3180274

One laptop per child. (2007). *One Laptop Per Child*. Retrieved May 23, 2007, from http://laptop.org/en

Ortmann, G. F. (2000). Use of information technology in South African agriculture. *Agrekon, 39*(1), 26–35.

Ouma, S. (2007, November). Civil society and the nation-state - The case of Kenya. *Development Dialogue*, (49): 109–118.

Over, E. (2001). *Social justice in world cinema and theatre*. London: Ablex Publishing.

Owusu, G. (2005). The role of district capitals in regional development. *International Development Planning Review, 27*(1), 59–89.

Oxfam (n.d.). World food crisis. Retrieved Dec. 6, 2008 from htttp://www.oxfam.org.uk/Oxfam/

Parajuli, J. (2007). A content analysis of selected government web sites: A case study of Nepal. *Electronic Journal of e-Government, 5*(1), 87. Retrieved September 9, 2007, from www.ejeg.com/volume-5/vol5-iss1/v5-i1-art9.htm

Parajuli, J. (2007). A Content Analysis of Selected Government Web Sites: A Case Study of Nepal. *The Electronic . Journal of E-Government, 5*(1).

Patterson, C. (n.d.). Development communication. The Museum of Broadcast Communication (MBC). [Online]. Retrieved Nov. 24, 2008 http://www.museum.tv/archives/etv/D/

Pažek, K., Rozman, Č., Borec, A., Turk, J., Majkovič, D., Bavec, M., & Bavec, F. (2006). The Use of multi criteria models for decision support on organic farms. *Biological agriculture & horticulture, 24*(1), 73-89.

Pažek, K., Rozman, Č., Turk, J., Bavec, M., & Pavlovič, M. (2005). Ein Simulationsmodell für Investitionsanalyse der Nahrungsmittelverarbeitung auf ökologischen Betrieben in Slowenien. *Bodenkultur 56*(2), 121-131.

Péjout, N. (2004). Big brother en Afrique du Sud? Gouvernement électronique et contrôle panoptique sous et après l'apartheid. In P. Guillaume, N. Péjout & A. Wa Kabwe Segatti (Ed.), *L'Afrique du Sud dix ans après: Transition accomplie?* (pp. 79-103). Johannesburg – Paris: Institut Français d'Afrique du Sud – Karthala.

Péjout, N. (2007). *Contrôle et contestation. Sociologie des politiques et modes d'appropriation des technologies de l'information et de la communication (TIC) en Afrique du Sud post-apartheid.* Unpublished doctoral dissertation, Ecole des Hautes Etudes en Sciences Sociales, Paris.

Pendakur, M., & Harris, R. (Eds.). (2002). Citizenship and participation in the information age. Aurora, Ontario: Garamond Press.

People's Daily Online. (2008, Nov. 7). World Bank to adopt Nigeria's food security policy. English: People's Daily Online. [Online]. Available at http://english.peopledaily.com.cn/ Retrieved Dec. 7, 2008.

Pimbert, M. P., & Tom, W. (n.d.). *Prajateerpu; a citizen jury- scenario on food and farming futures for Andhra Pradesh.* Retrieved January 13, 2009, from http://www.farming solutions.org/pdf.db/EPWPrajateerpu.pdf.

Porter, M. E. (n.d.). *Workforce Development in the global economy.* Retrieved December 30, 2008, from http://www.gwit.us/global.asp#main.

Portes, A. (1998). Social Capital: Its Origins and Applications in Modern Sociology. *Annual Review of Sociology, 24*, 1–24. doi:10.1146/annurev.soc.24.1.1

Portes, A., & Landolt, P. (1996). The downside of social capital. *The American Prospect Online.*

Portes, A., & Sensenbrenner, J. (1993). Embeddedness and Immigration: Notes on the Social Determinants of Economic Action. *American Journal of Sociology, 98*(6), 1320–1350. doi:10.1086/230191

Posel, D. (2000). A mania for measurement: Statistics and statecraft in the transition to apartheid. In S. Dubow (Ed.), *Science and society in Southern Africa* (pp. 116-142). Manchester: Manchester University Press.

Poulton, C., Dorward, A. & Kydd, J. (1998). The revival of smallholder cash crop in Africa: public and private roles in the provision of finance. *Journal of international International Development, 10*(1), 85-103.

Poulton, C., Kydd, J., & Doward, A. (2006). Overcoming market constraints on pro-poor agricultural growth in sub-saharan Africa. *Development Policy Review*, *24*(3), 243–277. doi:10.1111/j.1467-7679.2006.00324.x

Proenza, F.J. (n.d.). e-ForALL: *A poverty reduction strategy for the information age*. Retrieved June 29, 2006, from http://communication.utexas.edu/college/digital divide symposium/papers/index.html

Prolinnova (n.d.). Retrieved January 13, 2009 from http:/www.oaklandinstitute.org/?g=node/view/159

Punch On the Web. (2008, Oct. 8). Nigeria, fifth world hunger prone nation—FIIRO DG. Punch on the Web. [Online]. Available at http://www.punchng.com/ Retrieved Dec. 11, 2008.

Putnam, R. D. (2000). *Bowling Alone*. New York: Simon & Schuster

Rao, N. H. (2006). A framework for implementing information and communication technologies in agricultural development in India. *Technological Forecasting and Social Change*, *74*, 491–518. doi:10.1016/j.techfore.2006.02.002

Rao, S. S. (2001). Integrated health care and telemedicine. *Work Study*, *50*(6), 222–229. doi:10.1108/EUM0000000006034

Rao, S. S. (2004). Role of ICTs in India's rural community information systems. *Info*, *6*(4), 261–269. doi:10.1108/14636690410555663

Rao, S., Truong, D., Senecal, S., & Le, T. (2007). How buyers' expected benefits, perceived risks, and e-business readiness influence their e-marketplace usage. *Industrial Marketing Management*, *36*, 1035–1045. doi:10.1016/j.indmarman.2006.08.001

Raven Industries. (2008). http://www.ravenprecision.com/Manuals/pdf/016-0171/147A.pdf. (Verified January 23, 2008)

Recommendation (2004*). Record of the Committee of Ministers to member states on electronic governance.* Adopted by the Committee of Ministers on December 15, 2004 at the 909th meeting of the Ministers' Deputies.

Rehber, E. (1998). *Vertical integration in agriculture and contract farming*. Working Paper No.46. Department of Agricultural Economics, Uludag University, Turkey.

Reid, N. P. (2001). *Broadband Fixed Wireless Networks*. New York, NY: McGrawHill/Osborne.

Rich, K. M., & Narrod, C. (2005). *Perspectives on supply chain management of high value agriculture: The role of public-private partnerships in promoting smallholder access*. Unpublished paper.

Richardson, D. (2005). How can agricultural extension best harness ICTs to improve rural livelihoods in developing countries? In E. Geld, & A. Offer, (eds), *ICT in agriculture: Perspectives of technological innovation.* Jerusalem: EFITA.

Rimando, L. (2008). *Electronic Market For Farmers*. Retrieved from http://www.b2bpricenow.com/pr/WhatIsB2B.htm

Robertson, R. (1992). *Globalization: Social theory and global culture*. London: Sage.

Rozman, Č., Pažek, K., Bavec, M., Bavec, F., Turk, J., & Majkovič, D. (2006). The Multi-criteria analysis of spelt food processing alternatives on small organic farms. *Journal of sustainable agriculture, 2*, 159-179.

Rozman, Č., Škraba, A., Kljajić, M., Pažek, K., Bavec, M., & Bavec, F. (in press). The system dynamics model for development of organic agriculture. In *D. M. Dubois* (Ed.), *Eighth international conference on computing anticipatory systems*, HEC-ULg. Liege, Belgium: CASYS'07.

Rozman, Č., Škraba, A., Kljajić, M., Pažek, K., Bavec, M., & Bavec, F. (2008). The Development of an organic agriculture model: A system dynamics approach to support decision-making processes. In *Symposium on Engineering and Management of IT-based organizational systems* (pp. 39-44). Baden-Baden, Germany: The International Institute for Advanced studies in System Research and Cybernetics.

Ruhode, E., Owei, V., & Maumbe, B. (2008). Arguing for the Enhancement of Public Service Efficiency and

Effectiveness Through e-Government: The Case of Zimbabwe.' In *Proceedings of IST-Africa 2008 Conference*, Windhoek, Namibia.

Sadie, J. L. (1950). The political arithmetic of the South African population. *Journal of Racial Affairs, 1*(4), 3-8.

SAITIS. (2002). *International Scan for ICT Usage and Diffusion*. Pretoria, South Africa: Miller Esselaar.

Sander, C., & Mainbo, S. M. (2003). *Migrant Labor Remittances in Africa: Reducing Obstacles to Developmental Contributions.* World Bank Working Paper No. 64. Retrieved on November 6, 2008 from http://www.worldbank.org/afr/wps/wp64.pdf Scholte, J. A. (2007). Global civil society – Opportunity or obstacle for democracy? *Development Dialogue*, (49), 15 – 27.

Sargeant, J. M. Medical education for rural areas: Opportunities and challenges for information and communications technologies. *Journal of Postgraduate Medicine, 51*(4), 301-307. Retrieved June 20, 2007 from http://www.jpgmonline.com/text.asp?2005/51/4/301/19244

Saysel, A.K., Barlas, Y., & Yenigum, O. (2002). Environmental sustainability in an agricultural development project: a system dynamics approach. *Journal of Environmental Management, 64*(3), 247-260.

Schelin, S. H. (2003). E-Government: An Overview. In G.D. Garson (ed) *Public Information Technology: Policy and Management Issues,* (pp. 120 - 137). Hershey, PA: Idea Group Publishing.

Schowengerdt, R. A. (1997). *Remote Sensing: Models and Methods for Image Processing.* New York, NY: Academic Press.

Scott, R. (2004). Investigating e-health policy-tools for the trade. *Journal of Telemedicine and Telecare*, (10): 246–248. doi:10.1258/1357633041424377

Scott, R. E., Ndumbe, P., & Wootton, R. (2005). An e-health needs assessment of medical residents in Cameroon. *Journal of Telemedicine and Telecare*, 11.

Seddon, P. B. (2005). Are ERP Systems a Source of Competitive Advantage? *Strategic Change,* John Wiley & Sons, Ltd.

Selamat, J., & Shamsudin, M. N. (2005). *Sustainability of the Malaysian agri-food sector: issues at hand and the way forward.* University of Putra, Malaysia.

Senne, D., & Engelbrecht, L. (2007). Home Affairs admits ID inefficiencies. Johannesburg: *ITWeb*, 30 August.

Shaffer, P. (2001). *New thinking on poverty dynamics, implications for policy.* Retrieved June 29, 2006, from http://www.un.org/esa/socdev/poverty/paper_shaffer.pdf

Sharma, M. (n.d.) *Information technology for poverty reduction* (Proposal to Asian Development Bank, Manila, Philippines). Retrieved June 29, 2006, from http://topics.developmentgateway.org/ict

Sharma, V. P. (2003). *Cyber extensions: connecting farmers in India-some experiences.* Retrieved January 13, 2009 from http://wwwgisdevepoment.net/pdf/i4d003.pdf.

Shepherd, A. W. (1997). *Market information services: Theory and practice*. FAO Agricultural Services Bulleting No. 125. Rome.

Shi, T., & Gill, R. (2005). Developing effective policies for the sustainable development of ecological agriculture in China: The case study of Jinshan County with a systems dynamics model. *Ecological Economics, 53*(2), 230-246.

Singer, P. A., Salamnca-Buentello, F., & Daar, A. (2005). Harnessing nanotechnology to improve global equity. *Issues in Science and Technology, 4*, 57–64.

Singh, A. M. (2004). Trends in South African Internet banking. *Aslib Proceedings, 56*(3), 187–196. doi:10.1108/00012530410539368

Singh, S. H.(2000, September 26). Ways and Means of Bridging the Gap between Developed and Developing Countries. High-Level Panel on Information Technology and Public Administration, United Nations, New York.

Singhal, S., & Domatob, J. (2004). The field of development communication: An appraisal (a conversation with Professor Everett M. Rogers). *The Journal of Development Communication, 15*(2), 51–55.

Sinyangwe, C. (2007). Zambia's trade surplus to reach $1.2 billion. *Times of Zambia*. Retrieved July 6, 2007, from www.zamnet.zm/newsys/news/viewnews.cgi?category=6&id=1183627066

Škraba, A., Kljajić, M., & Kljajić-Borštnar, M. (2007). The role of information feedback in the management group decision-making process applying system dynamics models. *Group decision and negotiation, 16*(1), 77-95.

Slob, G. (1990). *Computerizing apartheid: Export of computer hardware to South Africa*. Amesterdam: Holland Committee on Southern Africa.

Smith, L. E. D. (2004). Assessment of the contribution of irrigation to poverty reduction and sustainable livelihoods. *Water Resources Development, 20*(2), 243–257. doi:10.1080/0790062042000206084

Smith, M. K. (2002). Globalization and the incorporation of education. In *The encyclopedia of informal education*. Retrieved December 24, 2008, from http://www.infed.org/biblio/globalization.htm

Sociocultural Barriers to E-Government: An Analysis from a Poverty Perspective: María Inés Salamancal, FLACSO Chile, DIRSI (Regional Dialogue on the Information Society), funded by the IDRC.

Songan, P., Yeo, A. W., Hamid, K. A., & Jayapragas, G. (2006). Implementing Information and Communication Technologies for Rural Development: Lessons Learnt From the eBario Project. In *International Symposium on ICT for Rural Development*, Faculty of Computer Science and Information Technology, UNIMAS, Kota Samarahan, Sarawak, Malaysia.

Sood, A.D. (2003, May-June). The Kiosk Networks: Information nodes in the rural landscape, I(1). Available online at http://www.i4donline.net/issue/may03/aditya_full.htm

South Africa Press Agency, 2004. More than half a million domestics registered with unemployment insurance fund (UIF). Pretoria: South Africa Press Agency, 2 June.

South Africa. (2005). *South Africa yearbook 2005/2006*. Pretoria, South Africa: GCIS.

Spence, R. (2003). *ICTs, the Internet, development and poverty reduction—Background paper for discussion, research and collaboration*. Retrieved June 29, 2006, from http://www.mimap.org/

Spence, R. (2003). *Information and Communications Technologies (ICTs) for Poverty Reduction: When, Where and How*? OECD International Development Research Center (IDRC). http://netwo rk.idrc.ca/uploa ds/user-S/10618469203RS_ICT-Pov_18_July.pdf

Sraku-Lartey, M. (2003). The role of information in decision making in the forestry sector: Developing a computerized management information system (MIS) for forestry research activities in Ghana. *IAALD Quarterly Bulletin, 48*(1), 105–108.

Sraku-Lartey, M. (2006). Building capacity for sharing forestry information in Africa. *IAALD Quarterly Bulletin, 51*(3), 186–190.

Sraku-Lartey. (2006). Developing the professional skills of information managers in the forestry sector in Africa. *IAALD Quarterly Bulletin, 51*(2), 75–78.

Srinivasin, L. (2005). *HLL, ITC draw up two prolonged strategies to woo customers*. Retrieved February 1, 2009, from http:/www.ictportal.com/newsroom/press_apr_05.htm/p5.

Stake, R. E. (1995). *The Art of Case Study Research*. Thousand Oaks, CA: Sage Publications.

Statistics South Africa. (2002). *South African Statistics*. Pretoria: Statistics South Africa.

Statistics South Africa. 2009. *Abstract of agricultural statistics*. Pretoria, South Africa: DoA.

Sterman, J. (2000). Business dynamics: Systems thinking and modeling for a complex world. Boston, MA: McGraw-Hill.

Stienen, J., Bruinsma, W., & Neuman, F. (2007). *How ICT can make a difference in agricultural livelihoods. The commonwealth ministers book-2007*. Netherlands: International Institute for Communication and Development.

Stiglitz, J. E. (2002). *Globalization and its discontents.* New York: W. W. Norton & Company, Inc. The World Bank Africa Region (AFR), (September, 2007). *African diaspora initiative concept note.* The World Bank: Washington DC

Suryammurthy, R. (2005). ICAR revamp to benefit farmers. *The Tribune, September 27, 2.*

Sushil, M. (August 1, 2008). Farmers dump parmal for muchhal. *The Tribune, August 1, 2008, 7.*

Syed, M. (2007). Zambia's first wireless ISP offers high speed broadband solution. *Balancing Act News Update, 250.* Retrieved September 9, 2007, from www.balancingact-africa.com/news/back/ balancing-act_250.html

Tacoli, C. (1998). Rural-urban interactions: a guide to the literature. *Environment and Urbanization, 10*(1), 147–166. doi:10.1177/095624789801000105

Tadégnon, N. (2007, Sept. 3). E-agriculture for Togolese farmers. Highway Africa News Agency. Retrieved Dec. 6, 2008 from http://hana.ru.ac.za & http://www.africa.upenn.edu/afrfocus/

Tamukong, J. (2007). *Analysis of information and communication technology policies in Africa.* Retrieved November 25, 2008 from http://www.ernwaca.org/panaf/

Tapscott, D. (1995). Leadership Needed in Age of Networked Intelligence. *Boston Business Journal, 11*(24).

Taylor, J., & Burt, E. (2005). Voluntary organizations as e-democratic actors: political identity, legitimacy and accountability and the need for new research. *Policy and Politics, 33*(4), 601–616. doi:10.1332/030557305774329127

Taylor, R. (2008). *Mauritian cyber island dream.* Retrieved March 25, 2009 from http://www.news.bbc.co.uk/2/hi/programmes/click_online/7169467.stm

Tellis, W. (1997). Introduction to Case Study. *Qualitative Report, 3*(2).

Texas Department of Information Resources. (2006, August 28). *TexasOnline.com named best state e-government site in the nation.* Press Release. Retrieved July 4, 2007, from www.dir.state.tx.us/dir_over view/pressreleases/20060825txo/index.htm

Texas web site surpasses $1 billion in state revenue collection. (2004, April 27). *Government Technology.* Retrieved July 4, 2007, from www.govtech.com/gt/articles/90060

The Economist (2008, April 17). Food: The silent tsunami. Retrieved Dec. 6, 2008 http://www.economist.com/opinion/

The E-Government Divide: Mutual Effects of Digital Divide and E-Government. (2003, October 12-19). *World Congress on Engineering and Digital Divide*, Tunis.

The Guardian (2008, Dec. 11). Africa, Asia most affected by food shortages, says FAO chief. Retrieved December 13, 2008 from http://www.ngrguadian.com/

The Guardian Online (2001, August 13). Nigeria is world's 10th populous nation, says report. Retrieved August 15, 2001 from http://www.ngrguadian.com/.

The Guarduan, (3rd February, 2006). *Govt ready to adopt ICTs in health services.* By Gardian Reporter. Retrieved from http://www.ipp.co.tz/ipp/guardian/2006/02/03/59218.html

The International Records Management Trust. (2004). *Developing a module for Assesssing Electronic Records as a Component of E-Government Readiness Assessments*: London: Commonwealth Secretariat.

The ITU Telecom Africa. (2004, May 4–8). *Has African renaissance begun?* Presented at The ITU Telecom Africa, Cairo Egypt. [Online]. Retrieved November 25, 2008 from http://www.itu.int/AFRICA2004/

The New York Times (2008, April 10). The world food crisis. *The New York Times.* Retrieved November 28, 2008 http://www.nytimes.com/2008/04/10/opinion/

The World Bank (2004). Millennium Development Goals. Retrieved December 2, 2008 from http://web.worldbank.org.

The World Bank. (2007). *The Little Green Data Book 2007.* Washington DC: The International Bank for Reconstruction and Development/THE WORLD BANK.

The World Bank. (November, 2007). *Mobilizing the African Diaspora for Development.* Working document presented at The African Diaspora Open House held at the World Bank Headquarters in Washington DC on November 29, 2007.

Tlabela, K., Roodt, J., Paterson, A., & Weir-Smith, G. (2007). *Mapping ICT access in South Africa.* Cape Town, South Africa: HSRC Press.

Tollens, E. F. (2006). Market information systems in sub-Sahara Africa challenges and opportunities. *Poster* paper prepared for presentation at the International Association of Agricultural Economists Conference, Gold Coast, Australia August 12-18, 2006.

Tönnies, F. (1957) *2002. Community and Society.* Trans. and ed. by Charles P. Loomis. East Lansing, MI: The Michigan State University Press.

Town, J. S. (2004). *Modelling eLiteracy: Politics, truth and beauty.* Retrieved December 20, 2008 from http://www.sconul.ac.uk/groups/information_literacy/publications/papers/st_elit_2004.ppt%20

Tradenethome (n.d). *Market information on your mobile.* Retrieved Dec. 6, 2008 from www.tradenet.biz

Tripp, R. (2001). Agricultural technology policies for rural development. *Development Policy Review, 19*(4), 479–489. doi:10.1111/1467-7679.00146

Turban, E. (2006). *Electronic Commerce 2006: A Managerial Perspective.* Upper Saddle River, NJ: Pearson Prentice Hall.

UK. *e*Strategy, (2000). *The UK vision as described in an April 2000 eGovernment Strategic Framework, was of modernised, efficient government, focused on better services for citizens and businesses and more effective use of information resources, source.* Retrieved May 1, 2003 from http://www.eEnvoy.gov.uk/ukonline/progress/estrategy/contents.htm

Uldal, S.B., & Amerkhanov, J., Bye, Manankova, S., Mokeev, A., & Norum, J. (2004). A mobile telemedicine unit for emergency and screening purposes: experience from north-west Russia. *Journal of Telemedicine and Telecare, 10*(1), 11–15. doi:10.1258/135763304322764121

UN. Kabwe, Africa's most toxic city. (2006, November 10). *IRIN.* Retrieved July 8, 2007, from www.irinnews.org/reporttest.aspx?ReportId=61010

UNDP (2006). *Human Development Report.* India.

UNDP-APDIP (2004). *ICT for Development: A Sourcebook for Parliamentarians* (p. 39), New Delhi: Elsevier.

UNESCO. (n.d.). *United Nations literacy decade (2003 - 2012).* Retrieved December 29, 2008, from UNESCO Education Plans and Policies: http://portal.unesco.org/education/en/ev.php-URL_ID=53553&URL_DO=DO_TOPIC&URL_SECTION=201.html

UNESCO-UIS/OECD. (2005). *Education Trends in Perspective – Analysis of the World Education Indicators.* Retrieved October 6, 2008 from http://www.uis.unesco.org/TEMPLATE/pdf/wei/WEI2005.pdf

United Nations, E-Government Survey Report (2008). Retrieved June 10, 2008, from unpan1.un.org/intradoc/groups/public/documents/UN/UNPAN028607.pdf

United Republic of Tanzania (URT). (2000). *Poverty reduction strategy paper (PRSP).* Dar es Salaam: Government Printer. Retrieved June 29, 2006, from http://www.tanzania.go.tz

United Republic of Tanzania (URT). (n.d.). *IDT/MDG progress—The United Nations and the International/Millenium Declaration development goals (MDG)—on United Republic of Tanzania.* Retrieved June 29, 2006, from http://www.undp.org/mdg/Tanzania.pdf

URT. (1997). *National poverty eradication strategy.* Dar es Salaam: Government Printer.

URT. (2001). *Poverty monitoring master plan.* Dar es Salaam: Government Printer.

URT. (2002). *Small and medium enterprise development policy.* Dar es Salaam: Government Printers.

URT. (2003). *National information communication technologies policy.* Dar es Salaam: Government Printer.

URT. (2003). *The Cooperative Societies Act—No. 20 of 2003*. Dar es Salaam: Government Printers. Retrieved June 29, 2006, from http://www.parliament.go.tz

URT. (n.d.). *The Tanzania development vision 2025*. Retrieved June 29, 2006, from http://www.tanzania.go.tz

Uzoka, F. E., Shemi, A. P., & Seleka, G. G. (2007). Behavioural Influences on E-Commerce Adoption in a Developing Country Context. *The Electronic Journal of Information Systems in Developing Countries, 31*(4), 1–15.

Van Der Berg, R. J. (2004). First ID for !Xhu and !Khwe communities, *Bua News*, Pretoria: Government Communication and Information System – Republic of South Africa, 9 June.

Van Tonder, K. (2003). Biometric identifiers and the right to privacy. *De Rebus*, 28 (8), August, Retrieved October 3, 2004, from http://www.derebus.org.za/archives/2003Aug/articles/Biometric.htm

Vanguard (2008, Sept. 2). FG urges telecom firms to invest more in Nigeria with their profits. *Vanguard.* Retrieved Nov. 25, 2008, from http://www.vanguardngr.com

Vecchiatto, P. (2007). Languishing HANIS needs attention. Johannesburg: *ITWeb*, 7 June.

Vecchiatto, P. (2007). SA trials smart ID cards. Johannesburg: *ITWeb*, 2 November.

Vesely, M. (2003). New technology for an old continent. *African Business*, July, 20-21. Website: http://www.ict.gov.gh

Villano, M. (2006, November 13). Wireless technology to bind an African village. *New York Times*. Retrieved November 13, 2006, from www.nytimes.com/2006/11/13/us/13tech.html?ei=5070&en=93831ac1f7064429&ex=1178683200

Wacquant, L. J. D. (2004). *Punir les pauvres: Le nouveau gouvernement de l'insécurité sociale*, Marseille: Agone.

Walker, W. R. (1989). *Guidelines for Designing and Evaluating Surface Irrigation Systems*. FAO, 1989: ISBN 92-5-102879-6, http://www.fao.org/docrep/T0231E/t0231e01.htm#preface (Verified January 23, 2008)

Warschauer, M. (2003). Demystifying the digital divide. *Scientific American, 289*(2), 42–47.

Weber, M. (1978). *Economy and society: An outline of interpretative sociology*, Berkeley: University of California Press.

Weigel, G., & Waldburger, D. (Eds.). (2004). *ICT4D—Connecting people for a better world—Lessons, innovations and perspectives of information and communication technologies in development*. Swiss Agency for Development and Corporation (SDC) and the Global Knowledge Partnership (GKP).

Wellman, B. (2001). Physical place and cyberplace: The rise of personalized networking. *International Journal of Urban and Regional Research, 25*(2), 227–252. doi:10.1111/1468-2427.00309

Wellman, B., Salaff, J., Dimitrova, D., Garton, L., Gulia, M., & Haythornthwaite, C. (1996). Computer networks as social networks; Collaborative work, telework, and virtual community. *Annual Review of Sociology, 22*, 213–238. doi:10.1146/annurev.soc.22.1.213

Welsh, W. (2004). Texas-size opportunity: Hard-charging CTO Larry Olson aims to make the lone star state a national leader in IT innovation. *Washington Technology, 19*(18). Retrieved July 4, 2007, from www.washingtontechnology.com/print/19_18/25090-1.html?topic=cover-stories

Willers, J. L., Seal, M. R., & Luttrell, R. G. (1999). Remote Sensing, Line-intercept Sampling for Tarnished Plant Bugs (Heteroptera: Miridae) in Mid-south Cotton. *The Journal of Cotton Science, 3*, 160-170.

Williams, B.K., Sawyer, S.C., & Hutchinson, S.E. (1999). *Using information technology—A practical introduction to computers & communications* (3rd ed.). Irwin McGraw-Hill.

Wines, M. (2007, July 29). Toiling in the dark: Africa's power crisis. *New York Times*. Retrieved July 29, 2007, from www.nytimes.com/2007/07/29/world/africa/29power.html

Winrock, J. (2003). *Future directions in agriculture and Information and Communication Technologies (ICT) at USAID*. Arkansas: Winrock.

Wirth, L. (1938). Urbanism as a Way of Life. *American Journal of Sociology*, *44*, 1–24. doi:10.1086/217913

Wise, R., & Morrison, D. (2000). Beyond the Exchange: The Future of B-to-B. *Harvard Business Review*, (November – December): 86–96.

Woodburn, M. R., Ortmann, G. F., & Levin, J. B. (1994). Computer use and factors influencing computer adoption among commercial farmers in Natal Province, South Africa. *Computers and Electronics in Agriculture*, *11*, 183–194. doi:10.1016/0168-1699(94)90007-8

Woolcock, M. (1998). Social Capital and Economic Development: Toward a Theoretical Synthesis and Policy Framework. *Theory and Society*, *27*(2), 151–208. doi:10.1023/A:1006884930135

World Bank (2003). *ICTs and MDGs: A World Bank Perspective*. www.worldbank.org, accessed 26 May 2007.

World Bank Group (2004). *World Development Report 2004: Making Services Work for Poor People* (p. 7). Washington, D.C., World Bank.

World Bank. (2001). *World Development Report 2000/01: Attacking Poverty*. Oxford University Press. www.worldbank.org/poverty/wdrpoverty/.

World Bank. (2002). *Information and communication technologies: A World Bank group strategy*. Washington, DC: World Bank.

World Bank. (2002). *World Development Report 2002: Building Institutions for Markets*. New York: Oxford University Press.

World Bank. (2003). *ICT and MDGs: A World Bank group perspective*. Washington, DC: World Bank.

World Bank. (2004). *2004 world development indicators*. Retrieved June 29, 2006, from http://www.worldbank.org/data/wdi2004/pdfs/table2-5.pdf

World Development Report. (2008). *Agriculture for Development*. Washington, DC: The World Bank.

World Economic Forum. (2001/2002). *World Economic Forum Consultation on SADC eReadiness*. Geneva: World Economic Forum.

World factbook—Zambia. (2007). *CIA*. Retrieved June 22, 2007, from www.cia.gov/library/publications/ the-world-factbook/geos/za.html

Wright, B. (2004). Telecoms around the continent. *African Business*, May, 16-17.

Wuthnow, R. (1998). *Loose Connections: Joining Together in America's Fragmented Communities*. Cambridge, MA: Harvard University Press.

Yamuah, L. K. (2005). ICT in the African health sector; towards healthy nations with ICT wealth. *I4d magazine*. Retrieved June 20, 2006 from http//:www.i4donline.net/may05/africahealth.asp

Yildiz, M. (2007). E-government research: Reviewing the literature, limitations, and ways forward. *Government Information Quarterly*, *24*(3), 646–665. doi:10.1016/j.giq.2007.01.002

Yin, R. K. (2003). *Case Study Research Design and Methods* (Vol. 5, in Applied Social Research Methods Series).

Yonah, Z.O. (1999). Orienting engineers in exploiting applied engineering and information technology in Tanzania: Challenges, opportunities and practical solutions. In *Proceedings of the ERB Press Seminar on Engineers as a Resource for Sustainable National Development* (pp. 64-76). Arusha, Tanzania.

Zachary, P. G. (2003). A program for Africa's computer people. *Issues in Science and Technology*, (Spring): 79.

Zachary, P.G. (2002) Ghana's digital dilemma. *Technological Review, July/August*, 66-73.

Zama, S., & Weir-Smith, G. (2006). *ICT Penetration in South Africa*. Pretoria, South Africa: Universal Service Agency.

Zambia: Govt cautions about spending debt savings. (2006, March 21). *IRIN*. Retrieved June 1, 2007, from www.irinnews.org/PrintReport.aspx?ReportID=58494

Zambian ICT policy fails to address key issues. (2007, April 4). *SciDev.Net*. Retrieved June 15, 2007, from www.scidev.net/News/index.cfm?fuseaction=readNews&itemid=3537&language=1

Zelnick, N. (2000). Colonialism? Not again. *Internet World*, *6*(18), 15.

Zimbabwe e-readiness survey report. (2005). *Information and communication technologies project Zim/03/003*. Retrieved August 21, 2007 http://www.ict.org.zw/Zim%20E-Readiness%20Survay%20Report.pdf

Zolfo, M. & Lynen (2008).Telemedicine for HIV/AIDS care in Low Resource Settings: Proven Practices. In *Proceedings of the 2008 conference of the IST Africa, Windhoek*, 07-09 May 2008, Ireland, IIMC International Information Management Corporation, Ltd.

Zwass, V. (2003). Electronic Commerce and Organizational Innovation: Aspects and Opportunities. *International Journal of Electronic Commerce*, *3*, 7–37.

About the Contributors

Blessing Mukabeta Maumbe is a leading expert on e-agriculture and e-government policy development, global food supply chain management, and rural financial service delivery in Africa. Professor Maumbe earned his PhD in Agricultural Economics from Michigan State University, USA in 2001 with a specialization in Agribusiness Management. He has received competitive research grants from the W.K. Kellogg Foundation, Rockefeller Foundation and the Southern Africa Center for Cooperation in Agricultural Research/Germany Technical Agency for Development (SACCAR/GTZ) for his research field work in Southern Africa. His current research interests are in ICT applications in agribusiness, rural development and policy. He has served as a consultant in smallholder agriculture technology adoption, agribusiness supply chain management, emerging African banking sector, and the privatization of higher education. Professor Maumbe has co-authored several publications in various national and international journals and proceedings of reputable conferences. He has taught numerous courses in Agricultural Economics and Agribusiness Management at various Universities and colleges in South Africa, Zimbabwe, and the United States (i.e., Michigan and Kentucky) for more than 15 years. He was the founding Lecturer and Coordinator of Agribusiness studies at Africa University between 1993 and 2002. He is currently an Associate Professor in the College of Business and Technology at Eastern Kentucky University (EKU), where he is in charge of the Agribusiness Management program. Before joining EKU, he was an Associate Professor in the Faculty of Business Informatics at Cape Peninsula University of Technology in Cape Town, South Africa where his responsibilities involved the supervision of post-graduate students in e-government and ICT for socioeconomic development. Professor Maumbe has presented numerous papers and chaired a number of conference sessions in South Africa, Canada, Australia, Spain, and the United States. He is currently the ***Editor-in-Chief*** of the *International Journal of Information and Communication Technology for Research and Development in Africa (ICTRDA)* and is an Associate Editor of the *International Journal of Health Care Delivery Reform Initiatives*. Professor Maumbe's work has advanced international agriculture and policy development in Sub-Saharan Africa.

* * *

Mieczysław Adamowicz is a professor of agricultural economics and a member of the Polish Association of Agriculture and Agribusiness Economists (SERiA). He got master's degree in 1962, PhD in 1969, and habilitation in 1975. In 1989 he got a full professor of economics title. For many years professor Adamowicz served as a dean of Agricultural Economics Faculty and as the head of Agricultural Politics and Marketing Department of SGGW (Warsaw University of Life Sciences). His research interests focus on common agricultural policy of European Union, multifunctional rural development, agriculture of developing countries, and trade of agricultural products. He is an author of about 600 publications and directed many national and international research projects. In 2008 professor Adamowicz was conferred honoris causa degree of the University of Life Sciences in Lublin.

Emmanuel C. Alozie (Ph.D., University of Southern Mississippi) has worked professionally in public relations, advertising and journalism and has taught mass communication for two decades. His research interests are in development communication, international/cultural journalism, advertising and public relations. A former assistant editor with Democratic Communiqué, Alozie published Cultural reflections and the role of advertising in the socio-economic and national development of Nigeria (2005, Edwin Mellen Press), and co-edited, Toward the common good: Perspectives in international public relations (2004, Ally & Bacon). He has published and presented several refereed articles and has won numerous awards in research, teaching and professional activities.

Neal Coates is an associate professor of political science at Abilene Christian University. He holds a J.D. from the University of Kansas and a M.A. and Ph.D. in Political Science from the University of Connecticut. He has worked as an attorney for the Texas Department of Insurance and for the City of College Station, Texas. Coates has published and presented on a range of topics including China's e-government, the Bush Doctrine, and the United Nations Law of the Sea Convention. Since 1999, he has taught at ACU, and will teach at ACU's Oxford campus during the fall semester of 2009. For the last decade, Coates has volunteered many hours working on two medical missions in Zambia and has helped collect and load humanitarian goods sent to that country in numerous container shipments from Texas.

Dawn Hinton is an associate professor of Sociology at Saginaw Valley State University has her doctorate degree from Western Michigan University. Her current research interests are in on the intersecting issues of race and ethnicity and also on modern urban sociology

Lisa Nikolaus was raised in Guatemala City. She has a B.S. in Nursing from Abilene Christian University. While at ACU she participated in its Study Abroad Program and matriculated in the United Kingdom, and was selected as a LYNAY Scholar and a McNair Scholar. Since 2007 she has been a registered nurse at a major hospital in Abilene, Texas, specializing in pediatrics and maternity. Nikolaus is fluent in Spanish and is beginning her graduate work in Fall 2009 at King's College London, in Conflict, Security, and Development. She has served as a volunteer for the nonprofit International Rescue Committee by helping refugees assimilate to life in the United States, and has served as a delegate to the Texas State Democratic Convention.

Stella Nwizu is a Senior Lecturer, Department of Adult Education, University of Nigeria, Nsukka. She is also the Secretary, Centre for Educational Development and Skills Empowerment (CEEDSE), a non-governmental and non-profit organization focusing on Education and Training for Nigerians. Dr Stella Nwizu holds B.Sc. (hons), M.Ed, and PhD in Economics Education, Educational Technology and Extension and Distance Education respectively. She has 14 years experience as a lecturer in Nigeria. She has worked as Secretary, Publications Committee, Member, Teaching Practice Committee, Member, and Research Grant Committee. She has published over 26 papers in various learned journals and presented several workshop papers. Her research interests are in distance education, extension education, non-formal education, economics of education, women studies, programme design and development, e-Education, leadership training, financing adult education.

Edith Ofwona-Adera has over 15 years experience gained from working in government and international development organizations. She is currently a Senior Program Specialist with the International Development Research Centre (IDRC) of Canada. Edith has worked with IDRC for over 11 years in various senior capacities including as Program Manager for the ICT4D Program in Africa. Her specific focus is on Information and Communication Technologies for Development (ICT4D), where she has been pivotal in supporting the development of Uganda's and Kenya's telecommunication access strategies and implementation plans. She has managed various innovative community-based ICT projects in Africa in diverse fields such as health, governance, education, gender, agriculture, poverty reduction, and community-owned ICT networks. In the field of mobile communication for development, Edith has managed innovative projects in: mobile use for linking farmers to markets in Kenya; mobile application to enhance health care delivery (Rwanda and Uganda) and agricultural production and marketing (Uganda). Before joining IDRC, she served, for 4 years, in the Kenyan Ministry of Agriculture as a Senior Agricultural Officer responsible for planning, developing and evaluating donor-funded agricultural development projects and programs. Mrs Ofwona-Adera holds an M.Sc degree in Agricultural Economics and BSc in Agriculture from the University of Nairobi; a diploma in "ICT Policy, and Network Management & Development" from the United States Telecommunication Training Institute (USTTI); and a certificate in "Telecommunication Policy, Regulation and Management" from Witwatersrand University, South Africa. She has published several papers and recently co-authored a book on "Gender and the Information Revolution in Africa". She is also a recipient of several international fellowships and awards. Active in professional and philanthropic initiatives, Edith holds several leadership positions including Council member, Meru University College of Science and Technology - Kenya and board member, SiSi Kwa Sisi Inc (an international charity for Kenyan AIDs orphans with headquarters in the U.S.); committee member, Kenya ICT Initiative for the National Council for Science and Technology; a co-founder of the Kenya Information Society (KIS); a member of the International Telecommunication Union's Gender Working Group, and member of the International Association of Agricultural Economists (USA). In 2001-2002, Mrs. Ofwona-Adera represented IDRC on the global G8 Dotforce working group on "e-Content Development. Email: eadera@idrc.or.ke

Julius Juma Okello is Lecturer and Coordinator of Agribusiness Program in the Department of Agricultural Economics, University of Nairobi, Kenya. He earned his PhD in Agricultural Economics from Michigan State University in 2005 with specializations in international agricultural development, agricultural marketing and resource and environmental economics. His current research focuses on the role of collective action, public-private partnerships, and use of ICT in linking smallholder farmers to high value domestic and international markets. He has conducted several value chain studies and coordinated research surveys in several African countries in the last three years. His other research activities are on soil and water management and the management of beneficial and invasive animal species. His latest publications are "Compliance with international food safety standards in Kenya's green bean industry: a paired case study of small and large family farms." *Review of Agricultural Economics*, 29 (2007): 269-285; "Impact of International Food Safety Standards on Smallholders: Evidence from three cases in Africa. In E. McCulloh, P. Pingali and K. Stamoulis (Eds) *The Transformation of Agrifood Systems: Globalization, Supply Chains, and Smallholder Farmers*, FAO, Rome (2008, 416 pages); and

"Public-private partnerships in high value fruit and vegetable supply chain", *Food Policy* 34(2009): 8-15. Dr. Okello has also been research team leader in various projects funded by international organizations including the World Bank, Food and Agriculture Organization (FAO) and the Rockefeller Foundation.

Ruth Mutuli Okello is the Executive Director of Tito Musyoki Literacy Foundation and a part-time lecturer at the Department of Agricultural Economics, University of Nairobi. Ruth is also the founder and patron of the Matungulu Farmers Self Help Groups. She is also currently registered as a PhD student in Michigan State University where she has been pursuing a PhD degree Human Ecology. Ruth holds an M.Sc in Agricultural Extension Education from Michigan State University and a B.Sc. in Agriculture from the University of Nairobi. Ruth is actively involved rural community development programs where is currently coordinating efforts to empower rural communities by building their capacity to tackle spiritual, educational, and health problems. She currently runs a rural-based community library and a rural community health center. Ruth's teaches education extension planning; seminars in agricultural education and extension; and special projects in planning and analysis of extension projects to final year undergraduate students at the University of Nairobi. Her latest publication is "Do EU standards promote environmentally-friendly fresh vegetable production? Evidence from Kenyan green bean industry", *Environment, Development and Sustainability* (in press). Email: okelloru@msu.edu

Joseph Ofori-Dankwa is a professor of management at Saginaw Valley State University and had his doctorate degree from Michigan State University. His current research interests are on the concept of Diversimilarity and also in identifying determinants of firm performance in African economies.

Benjamin Ogwo is an Assistant Professor, Department of Vocational Teacher Preparation, State University of New York (SUNY). Recently he was an Adjunct Visiting Scholar, Workforce Development & Education Program, Department of Learning & Performance Systems, Pennsylvania State University. He holds a PhD degree in Industrial Technical Education. As a consultant Dr Ogwo has worked for various organizations including the Education Tax Fund (ETF), Nigeria, and Global Unification, Australia, UNESCO-HP project on Reversing Brain Drain to Brain Gain, World Bank STEPB Project in Nigeria. He is an Associate fellow, African Institute for Applied Economics (AIAE), Member, African Technology Policy Studies Network (ATPS), Nairobi, Editor, Nigerian Vocational Journal, Nigerian Journal of Teacher Education and Teaching and Journal of Technology and Education. He has published more than 40 papers in various learned journals and presented several international workshop papers in Nigeria, Kenya, Botswana, Mozambique, Namibia, South Africa, India and United States of America.

Vincent Eze Onweh is a Senior Lecturer, Department of Vocational Education, University of Uyo-AkwaIbom State Nigeria and has been a teacher educator for the past 15 years. He teaches courses in Industrial Technical and Vocational Education. Dr Onweh holds the City and Guilds of London: the certificate in Construction Craft Supplementary Studies in six components; B.Sc, M.Ed and Ph.D in Technical Education with emphasis in Building/ Woodwork Technology. He possesses versed industrial experiences in the Construction Industries at variously capacities: Technical Officer, General Building Supervisor and as a Project Manager. His research interests are in Instructional System Design, Information Technology Education, Training Program Development, Industrial Technical Education (Building and Woodwork Technology).

Vesper Owei received his BS in electrical and electronic engineering from Ohio State University, USA, and a Master's in electrical and electronic engineering and in operations research from the Georgia Institute of Technology, USA. He also holds a PhD from Georgia Tech. He has practiced as a project, design, and consulting engineer. His research work has spanned such areas as data management, data modeling, concept-based query languages, HCI, knowledge and expert systems, OLAP, information systems for disabled people, and IT for health care delivery. His current research interests include: ICT and society, mobile ICT, and developing countries, and HCIs and end user issues.

Ephias Ruhode is currently a full-time lecturer and doctoral research candidate at the Cape Peninsula University of Technology (CPUT) in South Africa. He was educated in Zimbabwe to Masters in Business Administration (MBA) level at the Zimbabwe Open University in 2004. He holds a Higher National Diploma (HND) in Computer Studies (1994) from the Harare Polytechnic in Zimbabwe. He also received further training in computer technology and computer software applications from Kyoto Computer Gakuin (KCG) in Japan and CMC in India. His research interest is in information systems in general and e-government applications in particular. He received research seed funding from CPUT in 2009 to carry out his research on e-government with the Government of Zimbabwe. Ephias has more than 12 years work experience in Information and Communication Technologies (ICTs). Prior to joining CPUT in 2007, Ephias worked in various organisations in Zimbabwe as Analyst Programmer, ICT Operations Manager, Software Development Manager and General Manager e-Commerce.

Meke I Shivute works as a Lecturer at Polytechnic of Namibia, School of Information Technology, in the department of Business Computing. Meke teaches a number of courses which including Management Information Systems, Project Management, and Web Development Fundamentals. She obtained her Masters degree in Information Technology from Cape Peninsula University of Technology in Cape Town, South Africa in 2007. Meke has co-authored articles in international journals; her recent publications have appeared in the *International Journal of Telemedicine and Telecare* and *International Journal of Healthcare Delivery Reform Initiatives*. Meke's current research interests include e-health, medical informatics, and ICT for health service delivery.

Joseph K. Ssewanyana holds a PhD in Business Administration (ICT) and an MBA. He is a Senior Lecturer at Makerere University Business School, and Faculty of Computing and IT - Makerere University. He is involved in supervision and examination of Master's and PhD students in management and computing disciplines, and has several publications in the area of ICT for development. He is a Business Information Technology Specialist – Consultant and specialist in data analysis, business process modeling, information systems and technology, system analysis and design, project management, strategic planning and management, and financial management. He has spent many years studying, designing, developing, implementing and maintaining various types and sizes of information systems for various organizations.

Dariusz Strzębicki is a senior lecturer of Warsaw University of Life Sciences (SGGW). He got master's degree in 2000 and doctor degree of economics sciences in 2005. He teaches marketing and e-business. His research interests focus on agricultural marketing, e-business, entrepreneurship, and microeconomics. Currently he is working on the research project on vertical coordination in the Polish agribusiness.

Rachael Tembo holds a Master of Technology in Business Information Systems (*Cum laude*) from Cape Peninsula University of Technology in Cape Town, South Africa and a BSc Agriculture economics, Honours (*Cum laude*) from University of Fort Hare, Alice, South Africa. Her research interests include e-agriculture, ICT for socio-economic development, monitoring and evaluation of development programmes and e-commerce. Rachael received the African Scholarship award from the National Research Foundation in South Africa.

Rogers W'O Okot-Uma holds a Doctorate in systems engineering from Delft University of Technology, The Netherlands. He is a Chartered Physicist and Member of the Institute of Physics, UK. He has previously lectured at Makerere University, Uganda, and worked at the Commonwealth Secretariat as Adviser in various areas of development including science and technology innovation, global environmental issues, and public sector informatics. He has to date directed the design, launch and implementation of a number of academic programmes and 'centres of excellence' in Botswana, Canada, India, Malta and Zimbabwe. He has undertaken several consultancy assignments on *e*Governance worldwide. He has authored/co-authored, edited/co-edited, and published/co-published over 15 books in diverse areas, including mathematics, environment, and *e*Governance. His research interests include *e*Collaboration in the international relations context and multi-agent Z specification. Rogers W'O Okot-Uma is currently consultant, writer and publisher with Studies Forum International (SFI) in the United Kingdom.

Index

E

F

G

H

I

K

L

M

N

O

P

R

S